Design Engineering

Design Engineering

Harry Cather
Richard Morris
Mathew Philip
Chris Rose

OXFORD AUCKLAND BOSTON JOHANNESBURG MELBOURNE NEW DELHI

Butterworth-Heinemann
Linacre House, Jordan Hill, Oxford OX2 8DP
225 Wildwood Avenue, Woburn, MA 01801-2041
A division of Reed Educational and Professional Publishing Ltd

A member of the Reed Elsevier plc group

First published 2001

British Library Cataloguing in Publication Data
A catalogue record for this book is available from the British Library

ISBN 0 7506 5211X

Composition by Genesis Typesetting, Laser Quay, Rochester, Kent
Printed and bound in Great Britain

Contents

Series Preface

'There is a time for all things: for shouting, for gentle speaking, for silence; for the washing of pots and the writing of books. Let now the pots go black, and set to work. It is hard to make a beginning, but it must be done' – Oliver Heaviside, *Electromagnetic Theory*, Vol 3 (1912), Ch 9, 'Waves from moving sources – Adagio. Andante. Allegro Moderato'.

Oliver Heaviside was one of the greatest engineers of all time, ranking alongside Faraday and Maxwell in his field. As can be seen from the above excerpt from a seminal work, he appreciated the need to communicate to a wider audience. He also offered the advice 'So be rigorous; that will cover a multitude of sins. And do not frown.' The series of books that this prefaces takes up Heaviside's challenge but in a world which is quite different to that being experienced just a century ago.

With the vast range of books already available covering many of the topics developed in this series, what is this series offering which is unique? I hope that the next few paragraphs help to answer that; certainly no one involved in this project would give up their time to bring these books to fruition if they had not thought that the series is both unique and valuable.

This motivation for this series of books was born out of the desire of the UK's Engineering Council to increase the number of incorporated engineers graduating from Higher Education establishments, and the Insitution of Incorporated Engineers' (IIE) aim to provide enhanced services to those delivering Incorporated Engineering courses and those studying on Incorporated Engineering Courses. However, what has emerged from the project should prove of great value to a very wide range of courses within the UK and internationally – from Foundation Degrees or Higher Nationals through to first year modules for traditional 'Chartered' degree courses. The reason why these books will appeal to such a wide audience is that they present the core subject areas for engineering studies in a lively, student-centred way, with key theory delivered in real world contexts, and a pedagogical structure that supports independent learning and classroom use.

Despite the apparent waxing of 'new' technologies and the waning of 'old' technologies, engineering is still fundamental to wealth creation. Sitting alongside these are the new business focused, information and communications dominated, technology organisations. Both facets have an equal importance in the health of a nation and the prospects of individuals. In preparing this series of books, we have tried to strike a balance between traditional engineering and developing technology.

The philosophy is to provide a series of complementary texts which can be tailored to the actual courses being run – allowing the flexibility for course designers to take into account 'local' issues, such as areas of particular staff expertise and interest, while being able to demonstrate the depth and breadth of course material referenced to a framework. The series is designed to cover material in the core texts which approximately corresponds to the first year of study with module texts focusing on individual topics to second and final year level. While the general structure of each of the texts is common, the styles are quite different, reflecting best practice in their areas. For example *Mechanical Engineering Systems* adopts a 'tell – show – do' approach, allowing students to work independently as well as in class, whereas *Business Skills for Engineers and Technologists* adopts a 'framework' approach, setting the context and boundaries and providing opportunities for discussion.

Another set of factors which we have taken into account in designing this series is the reduction in contact hours between staff and students, the evolving responsibilities of both parties and the way in which advances in technology are changing the way study can be, and is, undertaken. As a result, the lecturers' support material which accompanies these texts, is paramount to delivering maximum benefit to the student.

It is with these thoughts of Voltaire that I leave the reader to embark on the rigours of study:

'Work banishes those three great evils: boredom, vice and poverty.'

Alistair Duffy
Series Editor
De Montfort University, Leicester, UK

Further information on the IIE Textbook Series is available from
bhmarketing@repp.co.uk
www.bh.com/iie

Please send book proposals to:
rachel.hudson@repp.co.uk

Other titles currently available in the IIE Textbook Series

Mechanical Engineering Systems	0 7506 5213 6
Business Skills for Engineers and Technologists	0 7506 5210 1

1 The design process

Summary

The body of a fire extinguisher? A motorcycle cylinder head? A plastic milk bottle? A ball bearing? A drinks can? Everything that we use from a safety pin to a jet engine has been created by someone. The purpose of this book is to provide information that enables the creation of components and products, focusing particularly on metals and mechanically engineered parts. It sets out to achieve this in two ways. First, it illustrates how components and products are created. It therefore provides overviews, case studies and theoretical information to show how this is done. Secondly, it provides practical advice, techniques and exercises so that actual design skills can be improved. The book therefore provides a mix of tools and reference materials that is intended for beginners and experienced design professionals alike.

The purpose of this chapter in particular is to introduce the overall design process and to explore in detail the early stages of the process.

The overview is given in Section 1.1. It starts by exploring the depth to which design penetrates society. For example, a designer might be asked to create anything from a simple component like a battery clip inside a mobile phone, to a product like a mobile phone itself, to an entirely innovative way of communicating, such as a WAP-based phone system. It considers how the design of new components, the creation of new products and the strategy of innovation are all linked. The section also looks at the scope of design. Many people think that design is just concerned with shape and how well an article performs, but 'total design' involves considering a number of other factors such as manufacturing and marketing. Understanding both the depth and scope of design will help to place a perspective on how to design and why it is vital to people, companies and countries.

Section 1.2 explores the first stage of the design process. This involves establishing the initial concept for a design, defining exactly what that concept is and asking why it needs to be designed at all. There is, after all, little point in spending time and money designing something that is not needed (although this has been seen to be done in industry time and time again). Solutions include exploring the market need and using tools that capture market information.

Section 1.3 explores the second stage of the process, which is finding design solutions that arise in the form of schemes. The section also explores the issues behind how to make correct and timely decisions in choosing the correct schemes to develop. Section 1.3 hence serves as an introduction to the remaining chapters of the book that cover how to turn schemes into working designs. These will include looking at how to present, illustrate, develop, optimize, produce and manufacture a design.

Throughout the chapter, the terms components, products, innovation, designers and engineers will all be referred to. 'Product' and 'designer' will be used most often as they are deemed to be umbrella terms that capture all the inclusive elements of 'people who make things'.

Objectives

At the end of this chapter, you should:

- possess a greater understanding of the design process, what it involves and why it is important (Section 1.1);
- have available and know how to use a number of techniques to improve the process of generating new concepts (Section 1.2);
- have available and know how to use a number of techniques that will help to create and select schemes/designs (Section 1.3);
- understand how this chapter leads into the remaining chapters of the book.

1.1 Design and innovation

'Whatever made you successful in the past won't in the future.'

(Lew Platt, Hewlett Packard)

People who create components and products might be classified in a variety of ways: inventors, engineers, designers, engineering designers or innovators, for example. We might call all of these creators 'designers' of one form or another.

This section will explain why design is important to a nation and why the way a design looks and performs (its function and form) are key but not sole elements of design.

Designers are important to a country in two ways. First, they help existing businesses to be more successful. Secondly, they help to create new businesses. The better these two elements perform, the better an economy performs. New designs stimulate sales, stimulating growth, from which unemployment reduces, money circulates, supply chains are stimulated, balance of payments improve, and the quality of life becomes generally better for all.

Key point

In the UK, design-related activity employs 300 000 people (around 1.2% of employees). Manufacturing alone allocates 173 000 employees to the creation of new products.

A catalyst for new companies

A new design creates a chance for new sales and is hence an opportunity for the creation of a new business. Sometimes designers grasp this opportunity, while at other times it is left to other people. Designers who have been successful at either designing or exploiting their designs have often achieved respect or even notoriety in society.

Activity 1.1.1

Identify what the people in the list below are famous for designing:

John Kay
James Hargreaves
Richard Trevithick
George Stephenson
Isambard Kingdom Brunel
Thomas Edison
Guglielmo Marconi
Frank Whittle
James Dyson
Kenneth Grange
Trevor Bayliss
Renzo Piano
Marcel Breuer
Philippe Starck
Terence Conran

Assignment 1.1.1

Select six famous designers either from the list above or from any others that you know and admire.

Write down three attributes that you think each one has. An attribute could be the ability to draw, persistence, knowledge of people, organizational skills, humour, wit, perception, analytical and computational skills, understanding of materials, clear perception of problems, imagination, expressing thoughts, patience.

Are there any attributes common to your designers? Do you have any of these attributes? Which attributes do you need?

Create an action plan, with a timescale, as to how you will obtain these attributes? *Undertake your action plan.*

People who exploit business opportunities are known as entrepreneurs. They usually create a business which is a small or medium-sized enterprises (SMEs).

A small business is defined as having less than 100 people and a medium-sized business has less than 500. Anything less than 10 people is classed as 'teeny'.

> 99% of Europe's businesses are SMEs. This is around 3.6 million businesses in the UK. A further 186 000 new businesses were added to this in 1999, which is around one new business per 250 working age members of society.
>
> This growth rate of business start-ups is 50% higher than it was 20 years ago. Most of the new businesses selling products were new technology-based firms (NTBF).

SMEs may not receive the same publicity as large corporations but they are just as important to a nation's economy.

> Europe's SMEs generate 65% of European employment and turnover. In the UK between 1986 and 1993, small companies created 2.5 million jobs and accounted for 23% of Gross Domestic Product (GDP). The top 100 new start businesses in 1998 alone generated full-time employment for 16 400 and total sales of £1.4 billion.

A catalyst for growth

Existing businesses face a number of conflicting problems. Workers want more money, shareholders want more money, customers want better products at lower cost, sales may be falling, machinery is constantly wearing out, competition is increasing, and uncertainty is rife. Companies may manipulate and strategize, may introduce new systems and new schemes, but often they are only tinkering with one problem at the expense of others. An improved or new design can instantly create more sales, more money, more investment and can therefore offer a solution to all of these problems – and all in one go. Glaxo moved from being the world's 16th ranked pharmaceutical company to the world's second with the introduction of one new drug. Cadbury's, despite spending £5 million developing and £3 million advertising its new Fuse chocolate bar, had a payback period of just one week. It is easy to understand why the desire to design new components and create new products can become a driving force throughout existing companies. When this is planned and done regularly it forms a strategy of innovation and creates enterprising companies.

Intrepreneuring is the application of entrepreneur principles within an existing organization.

Nine out of 10 companies believe they need to be more creative. Companies like 3M, Intel, Microsoft, Motorola, ABB, Philips, Siemens, Nokia, Honda, Sony and Toyota are turning to 'innovation' as their fundamental corporate strategy. Some companies like Boots and Bulmers, for example, have appointed 'Directors of Innovation'. Not only do these companies benefit from new and creative products but they are also able to project strong brand images of success with consumers.

Conversely, kitchen utensil producer Prestige, despite having a product in 90% of UK homes, had a poor record of innovation. Its failure to produce any significant new products in a 17 year period prompted a collapse and takeover by a rival company.

Neither is innovation a passing management fad. As the world moves into the twenty-first century companies become ever more global and competition will continue to intensify. Products will become more and more similar, and easier and quicker to copy. The pressures on companies will become greater and the desire and ability to innovate will be the surest way to survive.

Key point

Forty per cent (and rising) of company sales are 'new' products. In IT 78% of income comes from products under two years old. UK manufacturing allocates around £10 billion annually to product creation.

Function and form

Knowing that design is important is one thing, but how do you go about it? (Figure 1.1.1). For example, how do you make the head of a hammer? It is essential to go about it in the right way as even simple products may require investments of millions of pounds, and considerably more if they are defective.

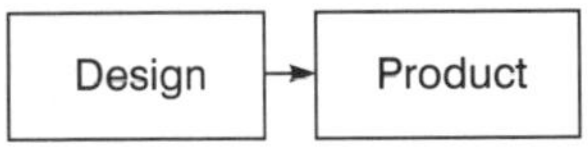

Design → Product

Figure 1.1.1 *The role of design*

Example 1.1.1

The head of a hammer must be shaped to perform the duties it is required for – and account for treatment it may not be intended for. It must be hard enough not to dent when striking and soft enough to absorb the blow without shattering. The head is first shaped by hot forging and then allowed to cool at room temperature. The grinding face and ball pein are then finished by grinding and the striking faces are puddled in molten lead at around 900°C. This is a process called austenization and changes the molecular structure of the iron to make it more ductile and more receptive to hard carbon atoms. Only the end parts of the hammer head that have

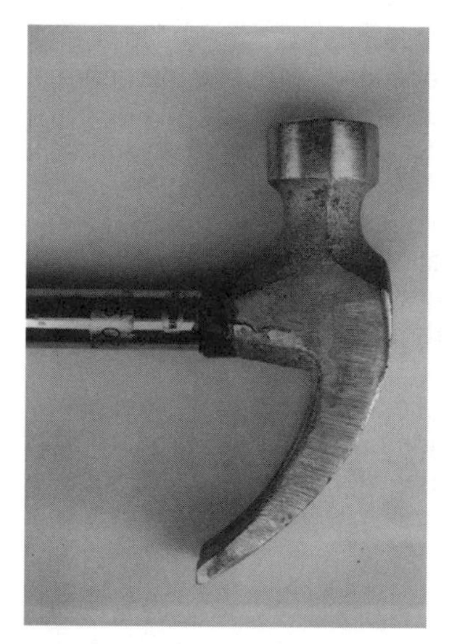

Figure 1.1.2 *Hammer head: poorly engineered design and manufacture has led to a fracture in the main steam of the head.*

been heated to above 723°C are consequently hardened when the hammer is quenched in cold water. The heads are then dried and immersed in a bath of molten salt at 450°C. This imparts toughness giving the head high strength and high resistance to cracking. Finally the heads are washed in cold water leaving a hammer head which has hard striking faces but which will not break when used.

Activity 1.1.2

The following chapters in this book will explain the processes and materials that are available to designers. See if you can guess now though how a fire extinguisher, motorcycle cylinder head, plastic milk bottle, ball bearing and a drinks can are made. Write down what additional information you might require to actually undertake the design.

A designer has a number of thoughts and questions when confronted with a new challenge.

Example 1.1.2

Typical design questions:

What properties should it have?
What size should it be?
What weight should it be?
How strong should it be?
Will temperature affect it?
Will humidity affect it?
Is reliability important?
Is ease of carrying an issue?
Is it controlled by legislation?
Are there any safety requirements?
Does it need power?
Will it be a pollutant?
Will it consume energy?
How complex to use can it be?
How long should it last?
What will break it?
Will it corrode or be corroded?
How will it break?
What happens if it breaks?
What should it be made of?
What should it look like?
Are ergonomics important?
Are aesthetics important?
Is there a fashion element?
Is there a certain style needed?

Activity 1.1.3

Select an item from around you that has been manufactured. See if you can answer, even by guessing, the solutions to the questions the designer of this item might have come up with.

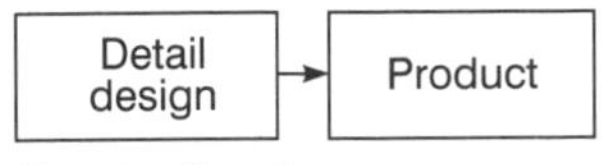

Figure 1.1.3 *Detailed design*

It's unlikely that a designer knows all of the answers to these questions and initially may not even be able to answer any of them. They can still, however, tackle the challenge to produce an engineered product because they know what questions to ask and how to find answers. The questions can be organized in a logical way which can then be repeated over and over again for any design challenge. This is hence a sequence which forms a process – the design process.

Notice that the questions asked in the example are mainly concerned with how the component or product looks and how it works. In other words they relate to 'function' and 'form'. These are considered as the detailed design (Figure 1.1.3). Detailed design is a major part of design. Answering the questions involved in detailed design is often achieved through research and development work.

If a company invests 1.4% of its sales in research and development it will lead to a 13% rise in profits in five years. A 1% investment in detailed design alone may raise a company's profits by 4%, particularly within the consumer goods industry. The US spends around $90 billion or 2.8% of its Gross Domestic Product (GDP) on R&D.

The UK spends around £12 billion or 1.8% of its GDP on design-related activity.

Case study

James Watt – part I, the function and form of the steam engine

James Watt was the chief inventor and designer behind the steam engine. He was born at Greenock, near Glasgow, on 19 January 1736. His great-grandfather was a farmer who was killed in religious wars. His grandfather was a teacher of mathematics who became a magistrate and kirk elder. It was his grandfather's pictures of Napier and Newton hanging in his father's house that James Watt grew up with. His father was a shipwright with a marine business. He was the first person to erect a crane in Greenock and prospered from the tobacco trade with Virginia.

Watt himself was a delicate child who suffered frequently from migraines. This had a profound influence on his character. He shrank from company and made irregular visits to school. At home, however, he was encouraged to read, study maths and work with tools. His desire was to be a mathematical instrument maker, working on the sort of navigational tools that Britain was using to expand its empire. Unfortunately, the loss at sea of his father's trading ship was a huge financial blow to the family and meant that he had to travel to Glasgow to finish work in order to study instrument making. Although he stayed with family relations, the Muirheads, Glasgow at this time was not a major port and Watt found it difficult to find work. He travelled on to London but because he was not a Londoner he was also not qualified to join a London guild of craftsmen and instead he worked as a craftsman for 10 hours a day in the shop of Mr John Morgan.

Despite poor working conditions he made great progress as a craftsman and was able to return to Glasgow. Despite his ability he was not able to find work as an instrument maker, reputedly again because he was not considered a craftsman of the town. Instead, on the advice of Professor Dick, he was engaged at Glasgow University to clean a large selection of apparatus the university had inherited from Alexander McFarlane, a wealthy merchant who had built a private observatory on his estate in Jamaica. Lack of finance or social standing mattered little in Glasgow's university and Watt's workshop became a lively place, attracting academics and students alike and including the many names that went on to be famous in their own right. One such person was Adam Smith, founder of political economy (despite this acquaintance, it is interesting to note that the word steam did not occur in Smith's work, *An Enquiry into the Nature and Causes of the Wealth of Nations*). During this period he jointly opened a small shop in Glasgow selling instruments he developed an interest in pottery, which later gave him a common interest with Josiah Wedgwood. He also married his cousin, Margaret Miller.

Towards the end of 1763, a Professor Anderson requested Watt to repair a small model Newcomen engine that belonged to his department. The engine would not work beyond four or five strokes and needed to be put into running order for demonstrations in lectures. Newcomen's engine had been invented in about 1702 for pumping water out of tin and coal mines. This industry was expanding rapidly and British production was about six times as large as that of the whole of continental Europe. Newcomen's engine was needed to remove water from the pits as they grew deeper and deeper and was the first effective machine using the energy of heat from coal. The engines consisted of a big cylinder and piston whose top was exposed to the atmosphere. Air under the piston inside the cylinder was blown away by steam and the steam was condensed by injecting cold water into the cylinder producing a vacuum under the piston. The pressure of the atmosphere on top of the piston pushed it down thereby operating the beam of a pump. The cycle was then repeated. This made it an atmospheric engine and not a steam engine. As the energy needed only inferior coal available in quantity at the pithead and almost valueless for any other purpose, its heavy coal consumption for the power produced was not felt to be a serious drawback. For this reason the engine developed little during the next 50 years.

Watt had made some early experiments at the university with steam which he did not pursue because he feared the danger of explosion. He was able to get the engine to work but did not remain satisfied. He wanted to know why it had not worked properly. He was not an engineer or scientist and had no experience with engines, but his approach is the characteristic method of modern engineering. He considered the details behind the design and began studying the function and form through a mixture of science and investigation. He first considered the relationship of size between the model and a full-sized engine, particularly looking at the relationship between surface area and volume.

Watt reasoned that ratio of surface area to volume was much larger for the small model cylinder than it was for the full-sized engine. In other words, more steam introduced into the smaller engineer would be needed to heat it up, which would then escape through, the greater proportion of cylinder wall. The smaller engine was therefore much more inefficient. When he studied the transference of heat, he also noted the cylinder in the model was made of brass whilst the full-size engine was made of cast iron. Heat can escape more easily through brass that it can through cast iron.

Watt tried to improve the vacuum inside the cylinder by throwing in more cold water but it cooled the cylinder so much that even more steam was needed on the next stroke. It was evident that there would be increased efficiency if the cylinder could be kept permanently hot and if a vacuum could be obtained without alternatively heating and cooling.

He then made an experimental investigation into the effects of temperature and pressure on the boiling point of water and plotted his results on a curve. He measured the volume of steam produced by boiling a given quantity of water and found that it was 1800 times as great. By making a special boiler he could measure the rate at which water was evaporated, and thus the rate at which steam was applied to the engine, and he was able to work out exactly what was happening inside the engine. He proved that in each stroke, much of the steam was used to heat the cylinder. Incredibly only a small amount of steam was necessary to raise the temperature of cold water. Water converted into steam can heat about six times its own weight of cold water up to boiling point. He was able to explain this in terms of the recently discovered latent heat which had been used to explain the boiling of water under reduced pressure.

Watt's work resulted in these profound discoveries yet he had no idea how they could be used to improve the engine design. It was a problem he struggled with for two years until a creative flash one day. 'It was in the green of Glasgow. I had gone to take a walk on a fine Sabbath afternoon. I entered the green by the gate of the foot of Charlotte Street – had passed the gate of the old washing house. I was thinking upon the engine . . . and gone as far as the Herds House when the idea came into my mind that as steam was an elastic body it would rush into a vacuum, and if a communication was made between the cylinder and an exhausted vessel, it would rush into it and might be there condensed without cooling the cylinder. I then saw that I must get quit of the condensed steam and injection water, if I used a jet as in a Newcomen engine.'

Watt's solution of condensing steam by taking it out of the cylinder and putting it into a separate condenser was the most important of all inventions. He swiftly constructed a new engine. He sealed the leaks around the piston and cylinder in the engine with a covering of water but this would not work if the cylinder was kept hot, so he decided to close the top of the cylinder and cover the piston with steam. It is the pressure of the steam and not that of the atmosphere that is used to force the piston down. He thus converted the Newcomen atmospheric engine into a steam engine, which was about four times more efficient.

Watt's design was deceptively simple but had been arrived at through the aid of research, practical and abstract experimentation, theoretical analysis, abstraction and new physical measurements that had never been conducted in such a manner. It kick started the continuous improvement of the steam engine, thereby bringing unlimited power within the prospect of mankind that fired an era of mechanical progress.

Watt's original engine is now in the Science Museum in London.

Maths in action

By comparing surface area with volume, repeat Watt's reasoning using maths to show why a smaller model engine is less efficient than a large engine.

Total design

Many people think that design is only concerned with the detailed aspects of function and form. Historically this is true. There was a time when function and form alone would sell products. In the eighteenth and nineteenth centuries new machines would make mass produced products available to people for the first time and new products would sell through novelty alone. Later, factories would produce the products and the marketing department would sell them just by making the public aware of availability by advertising.

Even in these circumstances, however, function and form would not guarantee that the design would be *commercially* successful.

Example 1.1.3

Consider the designers looked at in Activity 1.1.1:

John Kay invented the flying shuttle. This innovation allowed weavers to flick a weaving thread through a loom at much faster speeds and over greater distances. Overnight, the manufacturing capability of the British textile industry quadrupled. John Kay, however, died penniless in France.

Whilst weavers were able to use the flying shuttle to make cloth much faster, the suppliers of raw materials in the form of wool and cotton could not keep up. James Hargreaves invented the spinning jenny to solve this problem. He designed a rotating wheel which automatically spun and twisted yarn and because a number of threads could be attached to the wheel at the same time its output was far greater than spinning by hand on a traditional spinning wheel. The introduction of the flying shuttle and the spinning jenny gave a massive boost to the output of the textile industry. It kick started Britain's trading empire. Yet James Hargreaves made nothing from the spinning jenny itself.

The reciprocating shuttle and the rotating spinning jenny needed power and power needed engines. James Watt as we have already seen was ready to answer this call.

The large quantity of textile products being produced placed pressures on the incumbent cart and canal transport systems. Richard Trevithic advanced the cause for motive power by mounting high powered steam on wheels. Unfortunately he lacked the determination and commercial awareness to pursue it further. He too died penniless. It was left to George Stephenson to complete the task and develop the locomotive. He too, however, was a terrible businessman and it was left to the entrepreneurial George Hudson to capitalize on it and develop the railway itself.

New machinery, new engines and locomotives needed better materials. John Bessemer was a great Victorian engineer who answered this call. He discovered a method of mass producing steel and was the first person to make it available in large quantities. He was later bankrupted by cheap imports and quality problems.

Isambard Kingdom Brunel is perhaps recognized as the greatest engineer of all time in Britain, if not the world. His achievements included steam ships, screw propellers and underground tunnels. In spite of his successes, Brunel was never to be a rich man. The *Great Eastern* steam ship failed as a passenger vessel and has been described as the ship that was a laughing stock, a tragic heroine and ended her life as a floating circus.

Thomas Edison is recognized as the greatest inventor of all times. He generated over 1000 patents and his achievements included the light bulb, phonograph and valve. In fact Edison saw no potential in the light bulb, unlike a Professor Nernst in Germany who was able to sell his idea to AEG for £250 000. Edison also reinvested the majority of the profits from his inventions into a scheme to extract metal ore using magnetic induction. The scheme failed.

Guglielmo Marconi is credited with the creation of wireless communication. The company he established was beleaguered with scandals. Sir William Preece, communications expert for the government, reported to a House of Commons select committee that the Marconi company was the worst managed that he had ever had anything to do with.

Frank Whittle was a Royal Air Force pilot who knew that faster flight could be obtained at altitude where the air was thinner but that the engine itself would be less efficient in the thin air. The thin air needed to be compressed to make the engine work and this was done by a supercharging compressor. He conceived that a gas turbine could do this compression and then conceived that the engine exhaust could provide the power rather than the engine driving mechanical shafts. Whittle was never able to command the respect for the concept that it needed and as a result the protection of the design was lost and the exploitation potential severely damaged.

The growth of industrialization around the world and the availability of CNC (computer numerical control) machines that can make or copy designs has increased competition and reduced the chances of success through novelty and availability alone. The aims of selling designs is therefore not just on promotion but on ensuring that designs produce what the market wants (and not just what they are capable of producing on their machinery). Factories have had to become more flexible so that they can respond to changing customer demands.

To be commerically successful, the designer therefore needs to ask a range of additional questions that relate to business needs; such as the consumers' preferences and requirements, the capacity of organization, the assertiveness and control of the designers, and the opportunism, action orientation and risk taking attitude of the management.

Example 1.1.4

Typical business orientated design questions:

How will it be used?
Where will it be used?
Where else could it be used?
How well should it perform?
Who is it for?
Why do they want it?
Who will actually use it?
How many users will there be?
What is the motivation for its use?
Are there similar designs already?
Will a variety of designs be necessary?
How many parts to it should there be?
How accurate does it need to be?
How should it be made?
How many do we need?
What are the tooling costs?
Can it be rationalized?
Can it be standardized in the company?
Is it standardized within the industry?
Will we need suppliers?
Will we need subcontractors?
How will we evaluate it?
How can we protect it against design infringement?
When is the launch date?
Is packaging important?
Is presentation important?
Is it on display?
What is the shelf life?
How important is availability?
Is commissioning necessary?
Should it be maintainable?
How secure will it need to be?
Will it need servicing?
Will spare parts be necessary?
What will happen to it when it's finished?
How will it be implemented?
Do users need training?
Do users need operating manuals?
Is product support necessary?
Technical updates?
What about warranties?
What design budget is available?
Is the cash available?
What is the profit required?
What is the return on investment?
Is there a timescale requirement?

Activity 1.1.4

Using the same manufactured item you selected in Activity 1.1.3, try to answer the business-orientated design questions in Example 1.1.4.

This set of business-orientated questions acknowledge that design is a 'total process' that encompasses the entire life of a product, from conception through to abandonment. This is generally known as 'total design' (Pugh, 1990).

The entire range of questions needed to produce a successful design can be asked in a logical sequence called the total design process or the new product development (NPD) process.

Figure 1.1.4 *The total design process*

In the total design process, the first stage starts with exploring the concept behind the proposed new design. Where did that concept come from – a new idea, an invention, or a new market need? The first questions might not therefore be how do we design it, what should it do or what should it look like, but why should we design it at all? The second stage in the process is to find potential solutions to these concepts and these are called schemes. Scheme design is therefore concerned with finding solutions and being able to choose the best scheme. The next stage turns some of the schemes into detailed designs through experimentation, research and development (R&D). It is this stage that develops function and form. The final stages turn the design into reality through manufacturing and post-production stages.

Example 1.1.5

When confronted with the challenge to design a new hammer head, the first questions are not therefore what should it be made of, how should it be made or what should it look like? The first questions relate to the concept. What is the purpose of this hammer? Why is it needed? Perhaps the concept is to design an item that knocks a nail into a wall. Perhaps it is to produce something that helps to hang a picture on the wall. Concepts are therefore concerned with defining exactly what it is that needs to be designed and it is from this that the design process flows.

In moving to the second stage, a traditional hard mass hammer might be one obvious scheme. However, a compressed air gun might be another. A press tool might be a less obvious scheme. If the need was actually to hang a picture on the wall then maybe a glue-backed hook for picture hanging is needed. Unless the design process has been followed, money and resources might be spent creating a new hammer when in fact the challenge should have been to create a new type of glue.

Alternative view

A design process which is led by concepts generated by customer need can lead to claims that innovation is stifled. The argument is that designers only respond to customer needs and that customers have no knowledge of new technologies that might be of use to them. Strategists suggest that businesses should have a combination of market led design concepts and off-the-wall concepts with no apparent usage.

It is also suggested that as organizations seek to meet more and more individual customer demands, their product ranges become more modular and wider and factories organized to become flexible. It is possible to argue that this represents a move away from mass production back to crafted products.

The design process applies whether the design is a simple component or a complex product system. It also applies whether the work is undertaken by a single, lone designer or a large multinational organization. Larger organizations or more complex products, however, might include a whole range of additional activities. These can still be grouped into the main steps of the design process. For example, Figure 1.1.5 shows design tools and techniques organized within the basic design process.

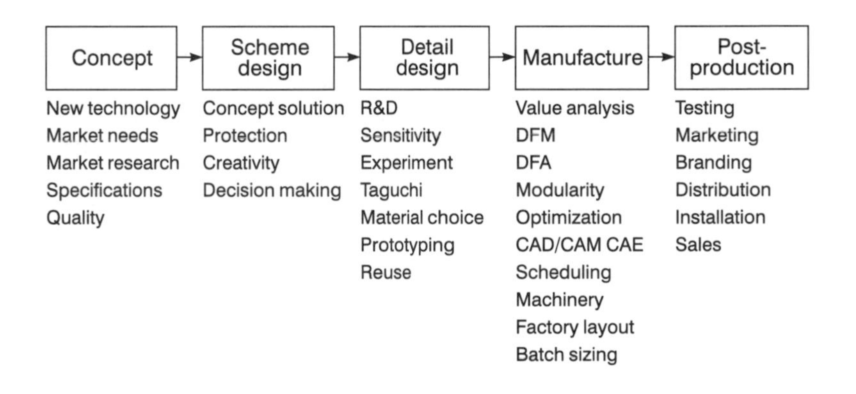

Figure 1.1.5 *Typical tools used in the design process*

James Watt – part II

Watt tried to construct a full-sized steam engine based on his ideas but he became beset with problems. Finding an adequate fit between the piston and cylinder would take him years to resolve. The larger scale of his experiments also required greater funding, and as his borrowing grew his anxiety grew and with this his migraines increased. He turned to financial support, first from Professor Joseph Black and then from Dr John Roebuck. Roebuck had invented the lead chamber process for making sulphuric acid and made his fortune by industrializing the process. He invested some of his fortune in the 'Carron' ironworks near the Firth of Forth that was to make superior canons called carronades for the British Navy. To do this the company needed coal from nearby deep coal pits. Watt installed an engine at the Carron works but his agony only increased as his roughly made engine performed poorly.

To increase his income Watt engaged in routine surveying and in 1767 he travelled to London to report on a survey for a proposed canal to the Carron Company. Acting on Roebuck's advice, he visited Dr William Small in Birmingham on his return journey. Small's scientific advice was poor but his psychological support was crucial and more importantly he was an adviser to the industrialist Matthew Boulton. Watt visited Birmingham again later that year and this time met Matthew Boulton himself. The two were well matched. Whereas Watt was despondent, timorous and retiring, Bolton was cheerful, confident and a brilliant man of affairs. Boulton's subdivision of handicrafts and labour in a large well-equipped, purpose-built factory laid the foundations for the methods of mass production and he was able to export goods throughout Europe. Boulton needed power for his factory and at the time it was perceived to be an engine which would pump water over a water wheel, providing consistency through all weathers. He was then clearly delighted with Watt's innovation.

He would liked to have acquired a share in the patent, but this was not possible. Watt had previously constructed an engine of sufficient capability that Roebuck himself acquired two-thirds of the rights in it and in doing so had relieved Watt's debts to Joseph Black. Roebuck merely offered Boulton the right to manufacture engines in three UK counties. Boulton replied that his motives for a share were a love of Watt's innovation and the love of money and he was not interested in being merely an engineer. He went on to explain that the engine would require finance, accurate workmanship and that 'an organisation and his idea had been to establish a special engine factory which would serve all the world with the engines of all sizes'. A period of intransigence followed with neither Boulton nor Roebuck willing to move. During this time Watt became poorer. Even worse, his wife died giving birth to a stillborn child.

The stalemate was broken in 1772 when a severe industrial depression hit. Both Boulton and Roebuck suffered but Roebuck was hit hardest and he became insolvent. Roebuck's creditors did not financially value the experimental engine. In this way, the engine became Watt's property, Boulton secured the rights and the engine was sent to Birmingham. In 1774, at the age of 38, Watt followed the engine and moved to Birmingham.

Based on confidence in the engine, Boulton settled Watt's debts. He also presented a Bill to the House of Commons for an extension of the original patent for a further 25 years. It was opposed as a monopoly, but Boulton's business brilliance saw the motion passed. They were able to use John Wilkinson's recent invention for boring canon as a perfect machine for making cylinders. Wilkinson was to make the cylinders for Watt's engines for the next 20 years and Wilkinson himself used Watt's engines for his blast furnaces.

Boulton also secured capable personnel who provided key support to Watt. William Murdock was one such assistant (who later left to found the modern gas industry). Watt and another assistant, John Southern, invented an instrument which measured the power engines produced and in so doing laid the foundation for the modern study of thermodynamics. He compared the English and French and measurements in researches on the composition of water and proposed a decimal system of weights and measures. He observed that a crank could be used to transfer the linear motion of the piston to rotary motion. He did not think it could be patented but when it was in 1780 by Pickard, invented five alternatives to the crank including the epicyclic gear. He developed a centrifugal governor for regulating the engine speed, which, along with Clerk Maxwell's theorising, later formed the modern science of cybernetics. He also invented a combination of rods so that the engines would be double acting, with pistons exerting a push as well as a pull.

During these times, industry was also picking up, mostly led by the Lancashire cotton industry. Priestley discovered oxygen. Scheele discovered chlorine. Berthollet put the two together and invented chlorine bleaching, and Charles Tennent made this practical in Britain by inventing bleaching powder – reducing the time for bleaching raw cloth from six months to a few hours. Mining, however, was still the main demand for engines. Boulton and Watt introduced a revolutionary system of charges based on the savings his engines made compared to Newcomen engines. Engines contained counters to monitor the water raised in another initiative that foreshadowed power measurement and developments of a century later.

Boulton and Watt were the perfect combination. Whilst Boulton attended the business, Watt was able to continue innovating. It is doubtful that Watt would have achieved his successes had it not been for Boulton's commercial skills.

Problems 1.1.1

(1) What is the first question a designer should ask when confronted with a new design challenge?
(2) What are the two ways in which design can benefit a country?
(3) Explain the difference between an attribute, a skill and a competence.
(4) Explain the difference between detailed design and total design.

1.2 Design concepts

> 'The best way to predict the future is to create it.'
> (Peter Drucker)

The first stage in the total design process is to clarify, understand and fully define the concept that is driving the design. This section will explore judging the value of a concept in terms of its market potential, and capturing that concept through market research, quality function deployment and a specification.

Needs and ideas

When considering concepts it is better to think big. Not just big but UNIVERSAL because it is only the utilization and organization of the phenomena of our universe that creates the boundaries for new design concepts.

Example 1.2.1

The universe was formed around four and a half billion years ago and now spreads out across 10 billion light years. The sun turns 600 million tonnes of hydrogen into 596 million tonnes of helium, from which four million tonnes of energy escape into space and some falls onto the earth creating life here. The earth itself has a 20 000 km diameter, spins and travels 100 000 km/h anticlockwise.

Key point

Fostering an innovative organisational culture and fostering serendipity are essential so that both scientists and designers can think of positive new uses for technology push discoveries instead of throwing them away.

As many as 18 people may have discovered penicillin before Alexander Fleming latched onto its potential properties. Even James Watt and Matthew Boulton failed to spot the significance of mounting engines onto wheels to create powered travel. Watt opposed the desire amongst his team to develop the steam driven locomotive and concentrate on driving machinery. He wrote to his partner Matthew Boulton stating 'I wish we have what could be brought to do as we do, to mind the business in hand and not to hunt shadows.'

When new discoveries are made and practical uses are found for them then this is known as 'technology push'. It is often scientists who discover the phenomena and engineers or designers who turn them into reality. At other times it is the demand from consumers for better products which results in new designs and this is known as 'market pull'.

Example 1.2.2

Scientists discovered that if a crystal with two opposing polished ends was stimulated by energy, photons inside the crystal it could be amplified, directed and focused into a strong beam of light. The discovery was titled Light Amplification by Stimulated Emission of Radiation or LASER. The original use of the laser was envisaged to be military. Laser guided weapons and metal cutting machinery might be regarded as technology push designs. Designers have, however, also utilized the technology for a host of other products. Guidance and navigation systems, measurement, bar coding, scanners and surgery could all be considered as market pull products.

Assignment 1.2.1

Select a recent scientific discovery. Magazines such as *Scientific American*, *Nature*, *Eureka* or inside academic journals and papers are a good place to look. See how many new applications you can think of for the discovery. Work one application into a fully designed new product.

Organizations are recommended to have a combination of market led designs and blue skies technology projects in their range of new developments. However, whether a concept stems from the 'technology push' or the 'market pull', there must be enough market demand to warrant developing the concept.

Activity 1.2.1

Imagine that you have formed the concept of a plastic pen that uses hard ink and wears down like a pencil. In theory there are six billion people in the world so to start with you have a potential market of six billion buyers. How many do you think you can actually sell? What factors do you need to know in order to be more confident in your prediction?

The number of factors that influence demand is very large but includes the type of product, where you are aiming your product, the number of people who are interested in the product generally and who are interested in your particular product, the degree of competition, likelihood of new competitors, the availability of your product, likelihood of repeat sales.

Market research is the term used to describe the collection of this type of information. Usually this is the job of the marketing department or experts, but sometimes the designer must undertake the process. They must certainly be aware of it. Market information is rarely readily available and must usually be gleaned from a number of sources.

Example 1.2.3

Suggested sources for finding market data

Surveys and questionnaires

Asking customers, shops and companies.

Trade directories
Kompass, Key British Enterprises, Kelly's Business Directory, D & B Europa, Engineers Buyers Guide, Dial Engineering, UK Directory of Manufacturing Businesses, Sells, Price Coopers Waterhouse European companies handbook, who owns whom, Directory of Directors, Directory of Business Biography, Business Leaders, International Directory of Companies History.

Service directories
Business Monitor, Yellow Pages, phone books.

Credit agencies
Fitch, Dun & Bradstreet, Standard & Poor, Moody's.

Statistics
Guide to Official Statistics, International Statistics Source, Sources of Unofficial Statistics, Official Business Statistics, Annual Abstract of Statistics, Monthly Digest of Statistics, social trends, regional trends, new earnings survey, IDS focus, Family Expenditure Survey, economic trends, financial statistics, national income and expenditure, Eurostat: Basic Statistics of the European Community, European Marketing, International Marketing, Pocket Book, Whitakers Almanac, Statesman's Yearbook, Europa World yearbook, Government Finance Statistics yearbook, OECD economic surveys, Bank of England Quarterly Bulletin, National Institute for Economic Revue, Datastream, Duns Europa, Principal International Businesses, Census Data.

Journals
Marketing, Marketing Week, Campaign, Market Research Great Britain, Retail Britain, Marketing in Europe, Mintel, Which?, Bank of England Quarterly, Investors Chronicle, European Economy, UK Economy, Economic Bulletin, Journal of Industrial Economics, British Tax Review, Risk, Overseas Trade, Quarterly Revue of Marketing, European Journal of Marketing, Journal of Marketing Management, Long Range Planning, Euromonitor, Business History, Patent and Design Journal, Designs in View.

CD-ROMs
F & S index (marketing and products), General Business File (850 journals for business and management) BNI (British News Index), ABI/inform (business/management database), BPI (business periodicals index), Emerald (complete journal articles on HRM, quality, info mgt, mgt ops), Anbar (accountancy, business, marketing, management), access personnel management (apm – personnel), infodesk (EC bibliography), European business (bibliography), Hansard (economic forecast), Justis (EU legislation), British Humanities index, general academic index.

On-line
Reuters business briefing, Microexstat (on-line accounts), Microview (on-line share information), economic indicators database, Fame (Financial Analysis Made Easy).

Other sources
British Library Collection (all marketing reports), professional associations, Patent Office, Extel, library keynotes (market share and reports), market research society abstracts,

Companies House, Monopolies and mergers commission reports, Customs & Excise, local development organizations, banks, trade associations, Department of Trade and Industry (DTI), Export Market Research Scheme, Export Market Information Centre, Funding Councils.

Because a large amount of market data can be accumulated it is useful to organize it within a framework. A common way is to sort data under the headings 'Near Environment' for facts that a company can influence and 'Far Environment' for facts that a company has little or no control over. The far environment can be further organized under the headings Social, Technological, Economic and Political data which is referred to as a STEP or PEST analysis (or STEEP if a separate environment section is included).

Example 1.2.4

Example STEP analysis

Social

- Forty per cent of jobs are secure, 30% are insecure, 30% are unemployed.
- U3 unemployment (a 'wide' measure which includes all employable people) is increasing to 4.4 million.
- Twenty-two per cent of people leave new jobs within three months.
- The average person now has ten jobs per life compared to seven, 20 years ago.
- 1998 saw 18 000 engineering jobs lost, a figure predicted to rise.
- There is an increased inequality in earnings differentials between the rich and poor. One-third of children are now in poverty.
- It is estimated that mankind will need to be 20 times more efficient with the earth's resources than is presently the case.

Technical

- Europe (EU) and the United States (US) spend equal amounts on R&D with equal amounts of publication, with Japan shortly behind.
- EU invests less in science and technology than US or Japan (per capita or % GDP).
- European patents are strong in traditional industries such as aerospace, automotive, pharma and chemicals but weak in computers, electronics and instruments. Europe exports 12% of its high tech products compared to 25% for the US and Japan 25%. This is growing.

- The UK/GB R&D spending by top companies has remained average. However, using the measure of intensity (the extent to which sales are reinvested in research), the UK has the lowest ratio of research spending/sales of the G7 countries, which is 2.2% below competitors.
- The most active sectors for patent applications are in telecommunications, followed by machine elements, civil engineering, conveying equipment, measuring/testing, electric circuits and household furniture. Growth areas include genetically created immunogenic viruses, optical data storage, and security devices.
- In the area of design, filings were made in the following areas – electronic apparatus (walkmans, pagers, radios), containers, and measuring instruments. Games and toys, however, saw the biggest area of growth.
- The trend for large companies to outsource is likely to continue, particularly for new technology development.

Economic

- The UK accounts for 14% of world production (manufacturing and services).
- Manufacturing represents 25% of the UK bank account but is in recession. Only 23% of factories are running at full or satisfactory capacity.
- Service industries generate 60% of GDP and make a profit.
- Ninety-seven per cent of all businesses sell less than £1 million per year.
- Seventy-eight per cent sell less than £100 000 per year.
- Small businesses account for 23% of GDP.
- The 3000 largest companies account for 43% of UK turnover.
- Unlisted companies are valued at £57 billion.
- Listed companies are valued at £500 billion.
- Since 1918, stock values have risen 18% on average per annum. National recessions occurred in 1982, 1987 (where a stock market crash reduced company values by one-third) and 1991/92.
- Growing money supply, real earnings and reduced tax will continue to fuel consumer demand for products.
- The strong pound makes British goods expensive and makes it harder to export.
- Britain ranked eighth in world economies (Swiss World Economic Forum).
- Seventy per cent of listed companies are owned by pension funds.

Political

- A Labour government traditionally has policies that are Keynesian. In other words they rely on demand stimulation through low lending rates, high employment, rewarding reinvestment and low exchange rates. Conservative policies have been more monetarist, relying on objectified, self-balancing, individualism, competition, money controls, high interest and exchange rates for low inflation, self-responsibility, inward investment and centralized control. Labour is currently ahead in the polls which would put it in power for the next five years.

- A proposal by Turnbull for management of risk legislation may stifle innovation.
- A new standard (ISO 140001) details environmental management systems. Twenty-five per cent of all products must be recyclable and emissions must be cut by 2010 (EU 8%, Japan 6%, Africa 0, US 7%).
- The Plugs and Sockets Safety Regulations 1994 – from 1.2.95 plugs must be fitted to domestic electrical appliances 13 amp to BS 1363 or euro plug IEC 884–1.

Each of these factors will have a varying degree of influence on the commercial success of your design. For example, you might consider low unemployment would not affect the sales of laser guided cutting equipment – but it might do if you could not recruit skilled people to make the product.

Activity 1.2.2

Rate how you think the following factors will affect sales of the hard ink pen, discussed in Activity 1.2.1, based on the degree of effect, the degree of importance and the time to take an effect.

	Effect 0–10	Importance 0–10	Impact 0–10
The world's first electronic pen that applies charged carbon to paper is announced.			
Marks and Spencer agree to distribute your pens.			
The government bans private motoring on Sundays.			
The Arctic ice cap thins by one-third.			
A breakthrough in thin metal wall technology will make battery power as cheap as mains electricity.			
UK Gross Domestic Product (GDP) is predicted to rise by 4% for the next eight quarters.			
Your distributor only sells within the EU.			
The IMF proposes a single world currency.			
China, Taiwan and Japan agree to form an Economic Asian Union (EAU).			

Sorting through the data to decide which factors are more important than others can also be complex and another framework can be useful. Headings of Strengths, Weaknesses, Opportunities and Threats are known as a SWOT analysis and are useful in weaving together facts to create a picture that will summarize the market demand.

The market research, STEP and SWOT analysis combined are often referred to as an environmental analysis. Environment here refers to the commercial rather than ecological environment.

Assignment 1.2.2

Environmental analysis

Rank the data provided in the STEP example (1.2.4) into three classes; not important, medium importance, very important relative to design of the new hard ink pen (if you prefer, you could use your own concept or the product you derived earlier from a recent scientific discovery assignment (1.2.1)).

Arrange the data to create a SWOT analysis. From this, try to predict what the likelihood of success is, and what you could do to improve the chances of success.

What other information would you need to find in order to increase the confidence in your forecast and to produce a more accurate sales forecast?

Case study

Triumph Motorcycles Ltd

In the mid-1990s there were six major motorcycle companies competing for the UK's power bike market. These were the big four Japanese bike manufacturers of Honda, Suzuki, Yamaha and Kawasaki, the Italian Ducati company and Britain's Triumph Motorcycles. Triumph and Ducati occupied the lower end of production each with sales of around 30 000 bikes per year. Yamaha, Suzuki and Kawasaki each sold up to around 300 000 bikes per year. Honda's total range of bikes' sales was in the region of 3 million. The following might be said of the power bikes that these companies offered at the time:

Yamaha

A range of six bikes, the newest being the FZ 600 and FZ 750 which were considered to be reasonable and faster versions of their old XJ models. The FZR 600 was a dated bike considered to be sound but with weak points. The FZR 600R was faster but less versatile. The FZR 1000 and FJ 1200 had their roots back in the mid-1980s but with regular improvements each year were considered to be fair, fast, low cost tourers.

Suzuki

A range of seven bikes. The GSX 600, GSX 750, GSX R 750, GSX 1100 F and GSX R 1100 all dated back to the 1980s and represented a range of power bikes that were considered fair, cheap, but slower, overweight and twitchy. The RF 600 introduced in 1993 and the RF 900 were good value sports bikes.

Kawasaki
Also had a range of seven bikes, although the GPX 750 was known to be shortly deleted from the range. The GPX 600 was a competent budget sportster introduced in 1994. The ZZR 600 introduced in 1990 was a strong, reliable sports all-rounder although it was slightly overweight. The ZXR 750 dated back to 1989 and had some weak points. The GPZ 900 was a market breaker when introduced in 1984 but was now very dated. The more recent ZX 9R was a much better bike but was too slow for a sports model and too uncomfortable for a tourer. The best bike in the range was the ZZR 1100 introduced in 1990 and was probably the best in class tourer.

Honda
Honda had a range of just four – but what a four. Its CBR 600, CBR 1000 and VFR 750 all dated back to the late 1980s but were generally considered to be the best sports bike, the best tourer and the best all-round bike respectively on the market. The more recently introduced CBR 900 or 'fireblade' was a ground-and-sector breaking sports bike. Prior to its introduction the 750 class was considered to be the size between 600 and 1200. The fireblade was considered by many to be the best motorbike in production.

Ducati
Ducati had two bikes. The 750 and 900 bikes were both introduced in 1992 and comprised high specification parts making them very fast, sporty and distinctive but also making them very expensive.

Triumph
Triumph had a long history in the motorcycle market but had floundered in the 1970s and 1980s at the hands of the Japanese onslaught into the bike market. A change in ownership, factory, and design methods had created a platform of six bikes with which it had started to re-establish itself. The 750 Daytona and Trident, and the 900 Trident and Sprint, were introduced in the early 1990s and thought of as fair general-purpose bikes. The Triple and Super three were 900 and 1200cc bikes respectively introduced in 1994 which were quick but expensive.

The concept for Triumph was a new bike. It was urgently needed to consolidate its position. But which sort? An update of an existing model? Or something completely new? What size? What type? A small design team of around eight people with an input from key engineering, production and marketing staff was allocated to the project. The concept they developed was for a competitive 600cc sports bike.

Key point

If your product was 99.99% perfect and everybody in the UK bought it you would still have 6000 unhappy customers.

Concept definition

The environmental analysis should establish the viability of the concept and help to define the concept so that further design work will be focused in the right direction.

Activity 1.2.3

You are to play the role of a customer representing the consumer market. Sketch a product you have in mind. Try to pick a product that is unusual. If you do not have a product in mind, select one at random from a magazine or catalogue (an inventions type catalogue is good for this).

A colleague will now act the part of the designer who will attempt to translate your concept. They must reproduce the sketch so that it can be developed. However, they must not be allowed to view your sketch and can only ask 20 questions in order to do this. You may only respond with yes/no answers.

Now view the designer's sketch. Is the designer going to be working on a design that is what you wanted? If the answer is no, imagine how much time and money will be wasted.

When finished, reverse the roles.

Key point

The modern definition of a 'quality' product is not a luxury or expensive product but one that meets the customer's requirements.

Poor communication between the designer and customer is often the reason for a poor design. The example in Activity 1.2.3 is contrived precisely because the customer can only answer yes or no so that communication fails. Yet it serves a purpose in demonstrating what happens when designers or companies fail to communicate with customers. It sounds improbable that companies would do this but it is a commonplace event and there are several reasons for this:

- Companies are full of people all of whom have different ideas and different aims.
- Controlling companies and people in them is difficult.
- Companies are not good at listening to customer complaints (which would help them to develop new and better products).
- Designers often do not use their own products *in situ*.
- Companies change personnel and rarely learn from previous mistakes and experiences.

Quality function deployment

Quality function deployment (QFD) is a technique that aims to improve communication between the customer and the designer. It stems from work by the quality 'guru' Juran who proposed a ten step approach to quality planning beginning with the identification of the customer and their needs and ending with the transfer to operations. Starting at the Kobe Shipyard in 1972, the Japanese adopted and developed this idea as a business strategy to plan the quality related aspects of products and coined the term 'quality function deployment'. QFD is hence a tool that ensures customer requirements are built into a design. In focusing on risks and achievables at an early stage it avoids mid-course corrections caused by traditional problems of relying on weak specifications and ill-defined concepts. It has since become a widespread technique through many industries.

QFD is a 'structured method for product planning and development'. Because the stages build into a 'house' like table, it is sometimes referred to as a 'house of quality'.

Key point

By using QFD, Toyota were able to save 60% of new product development costs.

Different companies have different versions of QFD but they all revolve around the same six basic steps. The design of a new boiler control system (© Buta, Douch, Wells, Winborn) is used in the worked example.

(1) *The voice of the customer*

What the customer wants, or the 'voice of the customer' involves listening to customers and analysing what is heard to create a list of customer requirements. Customer requirements can be further categorized into 'basic' or expected features, 'one-dimensional features' which increase customer satisfaction as they improve or 'exciters' which are non-expected features that truly differentiate a product (Kano). Methods of listening to the voice of the customer might include:

- improving unsolicited data collection.
  Only 1 dissatisfied customer in 10 will bother to complain so encourage comments through easy access such as multimedia, help desks and 0800 numbers
  listen to everyone in the company (drivers, cleaners, sales, etc.)
- improve solicited data collection
  Use iterative 'how and why' questioning to gain the emerging picture
  Ask people! Use market research and unbiased consultants
  ensure queries are installed into everyday work routines
  avoid 'don't know' answers (use rating or comparisons instead)
  focus groups (pay your customers to talk)
  overcome respondee prejudices (e.g. use descriptions not specifically related to your product)
- make better use of data
  quantify types of complaint
  use fishbone diagrams that link effects of complaints with causes
  use multivariate analysis (e.g. clustering responses to find the route important parameters)
  prioritize parameters (iterative ranking between groups)
  metrics (should it be smaller, larger or nominal) can help to simplify the process

> **Example 1.2.6**
>
> 'Customer demands for a combustion control system in the biofuel industry were classified under four main categories. These requirements may be seen in the left-hand column of the affinity diagram (Figure 1.2.1). Each customer requirement has been further classified as being a basic, one-dimensional or excitement feature.'

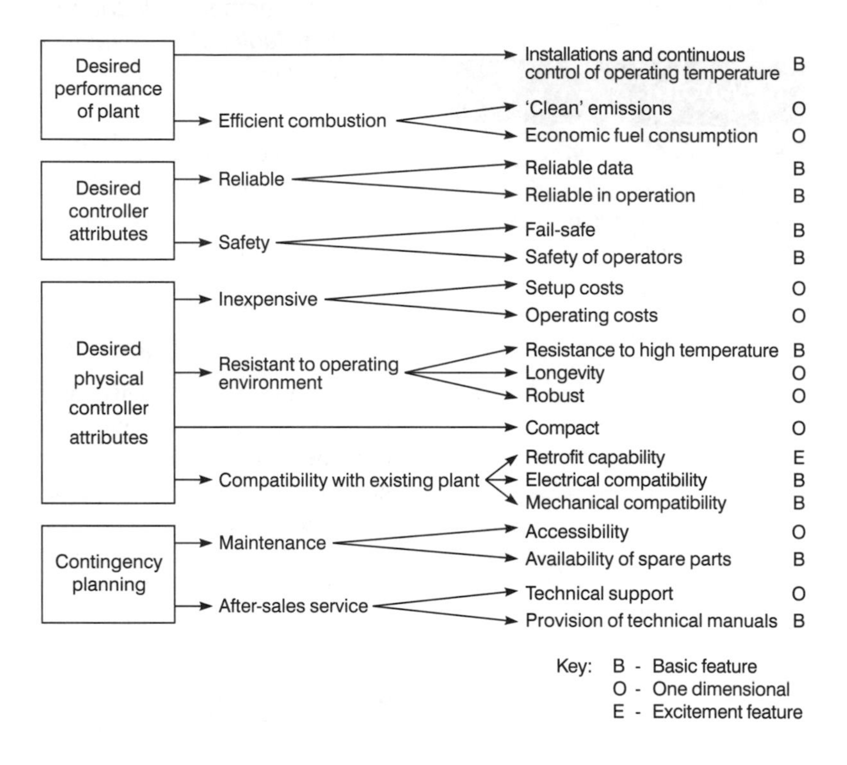

Figure 1.2.1 *Affinity diagram*

(2) *Competitor analysis*

Comparing customer requirements with what competitors currently offer can summarize the competitive scenario, order the customer requirements and identify the key areas for improvement. In the example house of quality we will use 2 competitors (N and E) and 8 analytical columns:

column A is the company's concept
column B is the competition products
column C is the target level of improvement
column D represents the level of improvement required (C/A)
column E is weighted customer needs (1.5 strong, 1.2 potential or 1.0 normal)
column F is customers' rating of importance
column G is a weighting factor of D × E × F
column H is column G expressed as a percentage
columns A and B may also be presented graphically.

Customer requirements	Max, Min, Target	KANO	A: Customer evaluation US	B: Customer evaluation N	B: Customer evaluation E	C: Target	D: Ratio	E: Sales point	F: Importance	G: Weight	H: Normalized %
Constant operating temperature		B	4	2	5	5	1.25		5	9.4	7.4
Clean emissions		O	4	3	5	5	1.25		5	9.4	7.4
Economic fuel consumption		O	5	3	4	5	1.0		5	7.5	5.9
Reliable data		B	4	1	4	4	1.0		4	4.0	3.1
Reliable in operation		B	5	3	4	5	1.0		4	4.0	3.1
Fail-safe		B	1	1	4	5	5.0		5	30.0	23.5
Operator safety		B	5	4	5	5	1.0		4	4.0	3.1
Setup costs		O	4	5	1	4	1.0		3	3.0	2.4
Operating costs		O	4	5	2	4	1.0		3	3.6	2.8
Resistance to temperature		B	3	4	2	4	1.3		4	5.3	4.2
Longevity		O	3	4	2	4	1.3		4	6.4	5.0
Robust		O	4	5	2	4	1.0		4	4.0	3.1
Size (compact)		O	4	5	3	4	1.0		3	3.0	2.4
Retrofit capability		E	5	2	1	5	1.0		3	4.5	3.5
Electrical compatibility		B	5	1	5	5	1.0		5	5.0	3.9
Mechanical compatibility		B	4	5	3	5	1.25		5	6.3	4.9
Accessibility		O	4	4	2	4	1.0		4	4.0	3.1
Spare parts available		B	5	2	3	5	1.0		4	4.0	3.1
Technical support		O	5	3	4	5	1.0		5	6.0	4.7
Technical manuals		B	3	3	3	4	1.3		3	4.0	3.1
									Totals	127.4	99.7

Figure 1.2.2 *Competitive analysis*

'Our concept was rated by our customers in comparison with our two companies identified as the two most likely sources of competition. N are active in the biofuel market and E offer high technology control products (although to a different market sector). Ratings for these companies were based on the assumption that their current product would be adapted to exploit this new market in direct competition with us. Additionally, importance ratings were applied to the customer requirements to create a precedent, enabling strategic product improvement. The figure summarizes the competitive scenario and identifies key areas for improvement. After reviewing the data, the key areas requiring product improvement were identified as:

- Fail-safe: If the control system malfunctions, the plant must always be left in a safe condition. One of our competitor's products rates highly on this criterion, while our own product and that of our other competitor did not rate so highly.

- Constant operating temperature: Whilst our product competes well under this criterion, the importance of controlling this variable is paramount. Thus, functionality in this area should be maximized.
- 'Clean' emissions: Increasingly stringent EC legislation has brought this issue firmly to the forefront of public concern. Our product fully satisfies current standards, but further development would promote a 'best-in-class' situation and accommodate further legislative adjustments.
- Economical fuel consumption: Whilst helping to promote the widespread use of biofuel, our product must also allow efficient combustion of this resource. In doing so, environmental benefits will not be attenuated and the customer will be rewarded with reduced fuel bills. Although presently leading the market, continuous improvement should be employed to maintain this position and maximize customer satisfaction.'

(3) *The voice of the engineer*

Technical responses to customers requirements are referred to as the 'voice of the engineer'. For example, if a customer requirement is for a product to be attractive, the technical response may be an enhanced surface finish. A new product should identify technical features for each customer expectation. A product improvement should identify controls and influencing factors for each expectation.

'The customer's control system requirements are translated into objective engineering parameters. The technical measurements that must be optimized to fulfil customer needs and expectations, along with target values, are:

(1) Temperature – ±20°C.
(2) Flue gases – emissions should comply with current EC legislation.
(3) Fuel consumption – 0.3 kg (kWh).
(4) Accuracy – ±5%.
(5) Downtime – 0.01% (approx. 1 hour per year).
(6) Fail-safe – yes.
(7) Operator safety – should comply with Health and Safety at Work Act.
(8) Purchase and installation costs – £1000.
(9) Running costs – £250 per year.
(10) Maximum operating temperature – 90°C.
(11) Service life – 7 years.
(12) Impact resistance – 10 N.
(13) Volume – 0.008 m^3.
(14) Retrofit – yes.
(15) Electrical compatibility – yes.

(16) Mechanical compatibility – yes.
(17) Accessibility – time to access 10 mins, local clearance 0.0643.
(18) Spare parts – all components stocked.
(19) Technical support – engineering personnel on 24 hr callout.
(20) Technical literature – comprehensive operating manual supplied.'

(4) *Correlations*

The matrix that forms the body of a house of quality displays the extent to which technical responses meet customer requirements. These can be shown by simple ticks but more complex priority ratings using complex symbol and number methods will help to indicate where a design can be improved.

Customer requirements \ Technical responses	Temperature	Flue gases	Fuel consumption	Accuracy	Downtime	Fail-safe	Operator safety	Purchase and installation costs	Running costs	Max operating temperature	Service life	Impact resistance	Volume	Retrofit	Electrical compatibility	Mechanical compatibility	Accessibility	Spare parts	Technical support	Technical literature
Constant operating temperature	⊙	▽	○																	▽
Clean emissions	▽	⊙	▽	▽																▽
Economic fuel consumption	▽	▽	⊙																	▽
Reliable data				⊙						▽	○	○			▽					▽
Reliable in operation					⊙				▽	▽	○	○				▽				▽
Fail-safe					○	⊙	▽													
Operator safety	▽					⊙	⊙										▽			
Setup costs								⊙						▽	○	○	○			
Operating costs					⊙	▽			⊙		▽	▽						○	○	
Resistance to temperature										⊙										
Longevity											⊙	○						▽	▽	▽
Robust					▽				▽	▽	○	⊙								
Size (compact)													⊙	○		○	▽			
Retrofit capability													○	⊙	○	○	○		▽	
Electrical compatibility														○	⊙					
Mechanical compatibility													○	○		⊙	▽			
Accessibility									○				▽	⊙		▽	⊙			
Spare parts available									○									⊙	▽	
Technical support					○				○								▽	▽	⊙	▽
Technical manuals					○			▽												⊙

Key:
⊙ = 9 ○ = 3 ▽ = 1

Figure 1.2.3 *Customer requirements vs technical responses*

Rows without strong, or any, correlations are uncontrolled and should be attended to. Columns without correlations are redundant. A query may require experimental work. The matrix is asymmetric (not necessarily reversible).

> 'Strong, medium and weak relationships between customer requirements and technical responses are depicted by the double circle, single circle and nabia symbols, respectively. The matrix indicates that each of the customer requirements has at least one strong correlation with the engineering measurand. However, it should be noted that the matrix is asymmetric, which indicates that the engineering measurand and customer requirements are sometimes mutually exclusive. For example, the engineering measurand of temperature, used to satisfy the customer requirement of constant operating temperature, also guarantees a degree of operator safety. However, when operator safety is considered as the engineering measurand, this in itself holds no implication for satisfaction of the constant operating temperature requirement. Thus, it is concluded that no areas pertinent to customer satisfaction have been overlooked in the QFD process to date. Therefore, the house of quality will be constructed on firm foundations yielding a product which will fully satisfy customer expectations.'

(5) *Technical comparison*

The technical responses can also be compared between what the actual performance is and the performance of competitor's solutions. Where scores between solutions and competitors vary, an incongruity of customer perception between desired and actual performance can occur.

It can be difficult to monitor the technical performance for a totally new product. Try benchmarking, not necessarily with companies selling the same product, but with those using similar products or processes.

> 'The competition was analysed on the basis that the two closest competitors would slightly redesign their existing products to exploit the biofuel market. Competitor's literature and prior knowledge enabled the rating of our product with its rivals against the technical criteria defined earlier. Once again, a rating of 1 (low) to 5 (high) was used to indicate the performance of each product.
>
> The raw importance figures were obtained by multiplying the value of the correlations ($Q = 9$, $0 = 3$, $V = 1$) by the weight calculated in the competitive analysis and then summed for the individual measurands.
>
> By comparing the customer evaluation in the competitive analysis with the technical evaluation, using the engineering measurands, it was possible to see where customer perception

differed from measured performance. Where the scores differed by two or more points, the category was noted as an incongruity.

Two categories were thus selected: operator safety and technical literature. The operator safety category was identified as requiring a more representative measurement, so that the customer's perceptions and the technical measurements were more consistent with one another. However, in the absence of a more suitable measurement of operator safety, this measurand was retained. The measured performance figure for the technical literature exceeded its respective customer evaluation, thus this discrepancy was considered to be beneficial.

Three critical technical measures were identified for improvement at this stage, based on the importance rating, the competition and our product's comparison to that competition. The categories targeted were: (1) fail-safe, (2) downtime and (3) retrofit. By concentrating on improving a few features, incremental, rather than revolutionary, product improvements were introduced.'

Engineering measurands	Temperature	Flue gases	Fuel consumption	Accuracy	Downtime	Fail-safe	Operator safety	Purchase and installation costs	Running costs	Max operating temperature	Service life	Impact resistance	Volume	Retrofit	Electrical compatibility	Mechanical compatibility	Accessibility	Spare parts	Technical support	Technical literature	Totals
US	4	3	5	4	4	1	3	3	4	3	3	3	4	4	5	4	4	5	5	5	
N	1	1	3	2	5	1	2	4	5	5	4	5	5	5	5	3	5	4	1	1	
E	5	5	4	5	2	4	4	1	2	2	2	1	2	3	4	3	1	4	3	3	
Raw importance	105.8	101.3	105.1	45.4	192.4	309.6	66.0	31.0	82.4	59.7	97.2	82.8	63.4	122.4	71.5	96.2	77.8	59.2	79.7	82.7	1931.5
% importance	5.5	5.3	5.4	2.4	10.0	16	3.4	1.6	4.3	3.1	5.0	4.3	3.3	6.3	3.7	5.0	4.0	3.1	4.1	4.3	100.1

Figure 1.2.4 *Technical evaluation*

(6) *Trade-offs*

Technical responses can also be analysed to determine how an improvement in one might affect others. For example, making a product thicker may make it stronger and safer but more expensive and harder to hold. Again the correlations can be indicated simply as strong or weak and a ± will suffice for QFD as a basic tool. Companies pursuing a more stringent policy may choose to achieve a correlation of ±99% or 95% (a six sigma approach).

> 'Once again, different symbols were used to denote the magnitude of the correlation, whether positive or negative. There were only six negative correlations between technical categories, indicating that the technology embodied in the system formed a cohesive product. It may be seen that most of the negative correlations were concerned with the purchase and installation costs. It is inevitable that design improvements will add cost to the product, so these trade-offs were not considered to be of too great importance. Another negative correlation was the fail-safe category, which would involve another feedback to the system controller and hence increase the complexity of obtaining electrical compatibility. A final negative correlation existed between the volume (or 'size') and the spare parts availability, since miniaturization would make spare parts more complicated to produce. All the negative correlations identified were expected, and not deemed to be significant enough to warrant changing any of the technologies embodied.'

After the six steps are completed the sections are pieced together to form the house of quality.

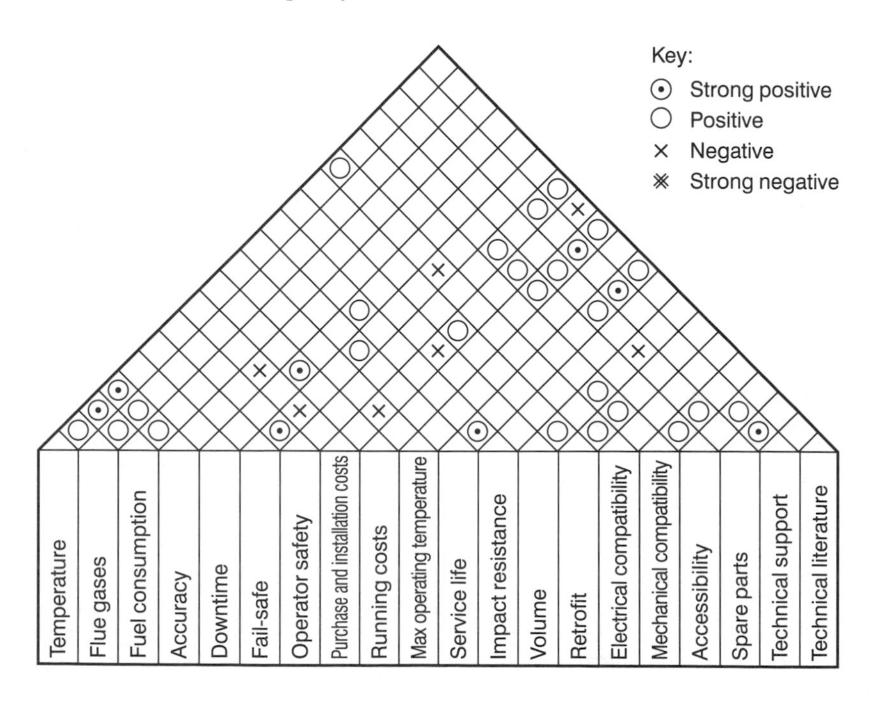

Figure 1.2.5 *Trade-offs*

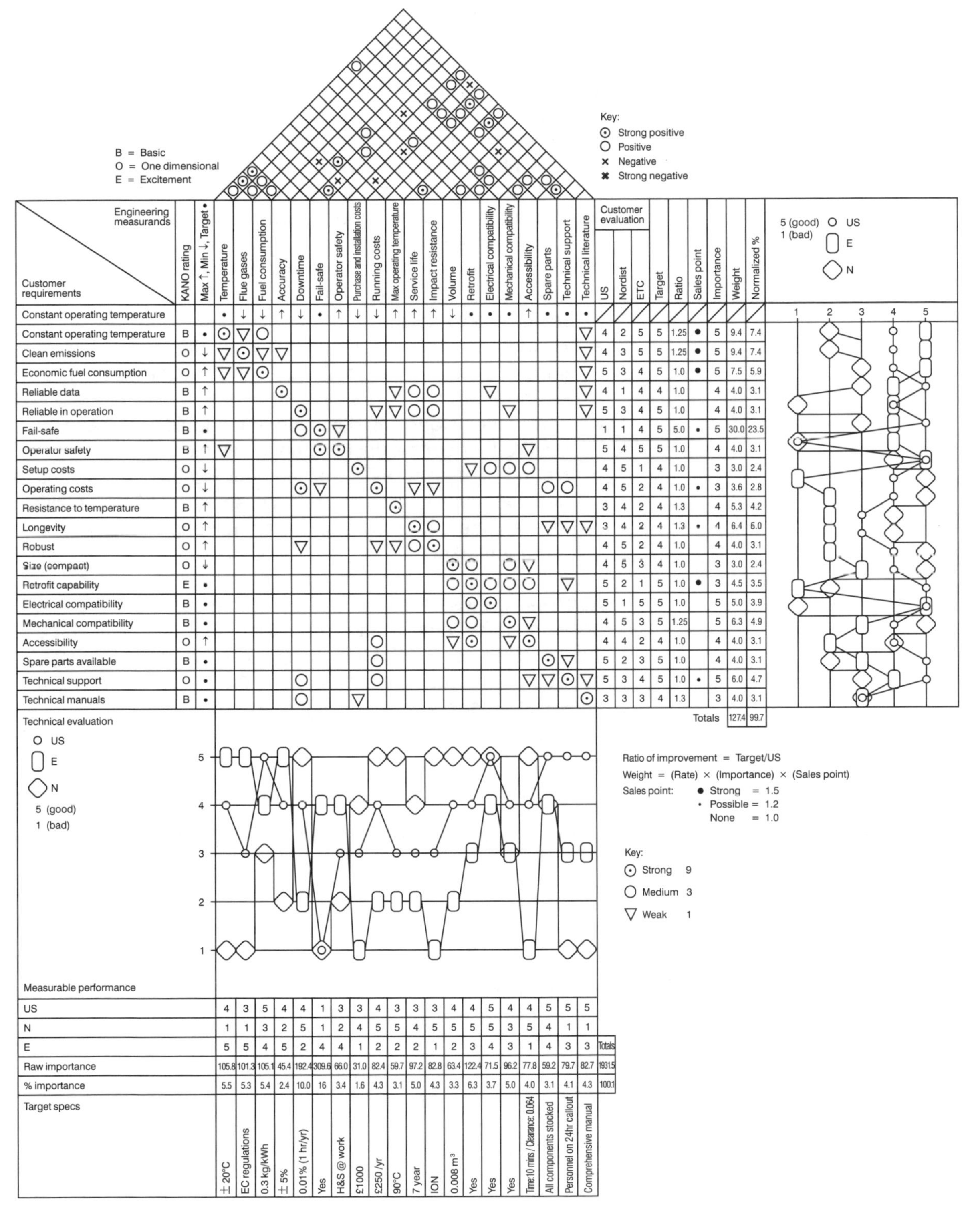

Figure 1.2.6 *The completed QFD*

'The QFD analysis highlighted areas of the design requiring improvement to more fully satisfy the customer requirements and thus remain competitive with the other products in the market. The key categories identified from the competitive analysis and technical comparisons sections as requiring additional or continuous improvement were:

- Fail-safe: This is a critical area for product improvement. It is absolutely imperative that in the event of system failure the plant is always left in a safe condition.
- Constant operating temperature: This is the fundamental customer requirement.
- 'Clean' emissions: Although not of direct concern to our customers, failure to comply with EC legislation could result in prosecution.
- Economical fuel consumption: This enables more efficient use of the biofuel resource and subsequent cost saving.
- Downtime: Constant development should be made to ensure that our product is wholly reliable.
- Retrofit: Identified as an excitement feature, maintained development in this area will serve to horizontally differentiate our product and allow full exploitation of the current market.

Addressing these areas would mean that the product could attain the target specification. At present, our product is rated 'best in class' for 30% of the customer requirement features, but sufficient further development in the areas emphasized above could easily raise this figure to 55%.'

QFD can also be used beyond product development and into process development where it is referred to as QFD II.

Specification

The final act in the concept stage of the design process is to capture the information in the form of a written specification (or 'brief'). The lack of a complete and thorough written specification is now generally accepted as being one of the main reasons for design failure.

There is a debate as to whether the specification should be narrow and descriptive, or broad and vague. One argument is that a truly agile or flexible business should start with specifications that are as wide as possible as they will be imaginative and capable of adapting to new ideas. By reverse logic, businesses that are not flexible should use a narrowly defined specification so that solutions are in keeping with their capabilities.

Figure 1.2.7 *Triumph TT 600. (Reproduced courtesy of Triumph Motorcycles Ltd)*

Example 1.2.7

The specifications for the Triumph concept bike included a combination of high speed performance, precision handling and performance best brakes.

The final solution was the TT 600.

Problems 1.2.1

(1) Define serendipity. Consider how this could be encouraged within an organization.
(2) Complete an analysis of the last product that you purchased using Kano's descriptors.
(3) Undertake a QFD to produce the concept for a bus stop.
(4) Consider whether a QFD is better for creating a completely new product or for improving an existing product.

1.3 Scheme designs

'No ideas come from staring at a drawing board.'
(after James Dyson)

Knowing what you are setting out to design is major step in the design process. The next is to establish some ideas that will enable a solution to be developed. One of the specifications Triumph wanted for their concept of a 600 cc motorbike was speed. They could achieve this in a number of ways. Improving the aerodynamics and rider position, lightening the weight of the bike, increasing the engine power or breathability, or adjusting the drive train. Another concept was precision handling. It is possible to make motorbike frames using aluminium beams, trestled tubular tensile steel, pressed steel or tubular steel spines, each of which will affect handling in different ways.

The array of different solutions to satisfy each of the specification requirements are known as schemes. This section will look at how to develop and select schemes.

Creativity

Activity 1.3.1

A pet food company currently sells its dog food in a traditional can. Customers have complained that opening the can is messy, awkward and can be dangerous. The company has the concept in mind for something new. Develop three scheme designs for the concept in new dog food packaging.

It is rare that the schemes suggested in response to concepts are revolutionary. Most new products represent a continuation or improvement on an earlier design or a copy of another design. As few as 5% of products are completely new to the world.

Key facts

Edison is recognized as the inventor of the light bulb. Yet his specification was to develop softer lighting for a shipping company. He simply enclosed existing filament technology within a standard glass dome inside the ship's existing gas fittings. None of the technology was therefore new.

Similarly a motorbike produced today might have a new shape and performance but as much as 90% is based on old technology. The engine (Daimler 1895), carburettor (Mayback 1893), spark plug (1898), radiator (1896) and tyres (1888) are all substantially unchanged in principle.

However, it is important to strive to be as creative as possible as the more innovative a design is, the more likely it is to appeal to customers. Creativity is also a unique feature that human beings have. One view on human thought is that the brain can be likened to a Swiss army knife where all the 'blades' represent different specialist mental processes such as linguistics, dexterity or social skills. Like all animals we can select and use these tools as and when necessary. However, man alone has the capacity to move from one area (or blade) to another. Making connections between specialist thought processes is known as cognitive fluidity. It is this ability that has enabled man alone to develop intelligence and creativity. Other primates can be taught to talk a few words and to use a typewriter, but they are not able to put the two together to produce a written sentence. Neither would they be able to transfer these dextrous and linguistic skills into social skills. If apes that had been raised in captivity were released into the wild, mankind would not be faced with a 'planet of the apes' scenario at a later date.

Key point

Imagineering is concerned with not relying on a flash of genius to develop novel new designs. It considers that novelty is an entire process that involves preparation, effort, insight, incubation and evaluation. As such it is a process and can therefore be modelled, learned, repeated and improved.

Whilst humans have this ability, it is not, however, always readily available and creative thoughts are not always forthcoming! This is because the brain prefers to work in a set pattern for gathering information and processing it in a logical way. Concept engineering or 'imagineering' is the process aimed at breaking down this logical pattern and increasing the cognitive fluidity so that creativity can flourish. This can be achieved through the use of a number of unstructured or structured methods.

Unstructured methods for improving creativity

(1) *Wider information.*
Be sensitive to surroundings.
Encourage dreaming and freewheeling.
Gather data beyond your own thoughts (literature, research, observation, role play, statistics).

Key point

Brainstorming sessions are easy to run but also easy to run badly. Good sessions should be fast and furious, the wilder the better. There should be no restrictions on time and often the best ideas will only appear after the first 60 to 80 suggestions. There should also be no criticism of ideas and people should later help to develop other people's ideas rather than their own.

(2) *Think laterally.*
Relate previously unrelated things (try free association or random word connectivity forcing an association between the problem and a word chosen at random from a dictionary).
Think on more than one plane (bisociatively).
Switch between activities and hence brain functions.
Use the more creative right hand of the brain.
Brainstorm.

(3) *Avoid being judgemental.*
Rely less on facts (rationality focuses on facts, facts are about the past and the faster things change, the less you can use facts, the more you need imagination).
Allow plenty of thinking time – lack of time and opportunity often leads to premature evaluation (allow free time or total flexitime).
Use the Gordon method (discuss many problems without knowing the key issue).

(4) *Overcome the rational control mechanism (Freud's 'superego').*
Tolerate uncertainty and ambiguity.
Challenge the non-obvious.
Be inquisitive.
Overcome self-imposed limits.
Avoid giving the expected answer.
Avoid belief that there is only one right answer.
Control emotions.
Use the Delphi technique (voting without knowing the opinions of others).

(5) *Create the right environment.*
Motivate positively through rewards (encouragement, financial or merit awards) or negatively through pressure (time or customer visitations).
Avoid comfort and routines (Dyson's 'constant revolution'), reorganize regularly.
Avoid red tape, bureaucracy, efficiency goals, group thinking (encourage solitude), resistance to change and unclear goals.

Structured methods for improving creativity

(1) *Problem analysis*

Many people try to create an improvement, which is difficult. For example, in trying to design a better telephone, they may consider 'what's a good innovation for a telephone?' It's much easier to identify negatives, working on problems rather than needs/wants. For example, asking 'what are the problems of using a telephone?' is a much easier approach.

It is also easier to rank those problems by the degree of irritation that they cause and the number of times they occur so that solutions are found to key issues and not wasted in minor ones. Again, consider the redesign of a public telephone:

	Irritation	*Occurrences*	*Score*	*Rank*
Difficult to hold without hands	0.6	0.9	0.54	1
Accidentally cutting people off	0.5	0.3	0.15	3
Dirty earpiece	0.8	0.4	0.32	2

Failure mode and effect analysis (FMEA) and cause and effect diagrams (C&E) similarly identify potential failures in design but also identify why problems occur and what effects they have. They collate factors on a chart to aid visualization. They too can be considered as problem analysis techniques but are generally used later in the design process to check designs rather than as creative tools.

(2) *Scenario analysis*

- Scenario analysis attempts to picture the future. What will people want from your product in ten years' time? (Use experts and/or mixed discipline panels of assorted experts and laymen.)
- Seed trends: What are the current trends and what will they grow in to?
- Precursor trends: Judging trends based on similar/previous product behaviours. For example, mobile phones have moved through distinct changes; large analogue, small analogue, large digital, small digital, hinged digital, twin hinge, folding. What will the future be (roll-up phones)? Can this trend be applied to other similar products such as portable computers?
- What stage is the current design in the life cycle of the product?

(3) *Attribute analysis*

Attribute analysis involves identifying attributes of a product such as its features, functions and benefits and then changing them, for example by adapting, modifying, reversing, combining, substituting, resizing, rearranging and so forth.

To consider a toothbrush, its features are its bristles, head, handle, grip, colour, density, shape. Its functions are toothpaste carrier, water storer, tooth polisher, food remover, crevice seeker. Its benefits are healthy teeth and gums, fresh breath and an aesthetic bathroom tool.

We could adapt some attributes. Mirror the head to see what is going on in the mouth? Hollow the handle to store foam as it's produced and reduce the amount of liquid in the mouth? Or, for example, we could combine attributes. Such as toothpaste carrier and handle. Could the handle be hollowed to store toothpaste? Could shape and aesthetics be combined to produce a toothbrush that stands up on its own?

(4) *Gap analysis*

By selecting and plotting two attributes on a chart it is possible to spot where there is scope for a new idea. For example, if we plot the attributes of speed against comfort in the lower end of the power motorcycle market, we might identify four gaps for which we could design a new product (see Figure 1.3.1).

(5) *Trade-off analysis*

Also known as conjoint measurement, this technique pits various pairings of attributes against each other. For a toaster 'fast acting and aesthetic' may be preferred to 'controllable and crumb removing'; 'controllable and thickness adjustable' may be preferred to 'lightweight and visible toasting'. Eventually a winning pairing evolves, highlighting the attributes which should be focused on.

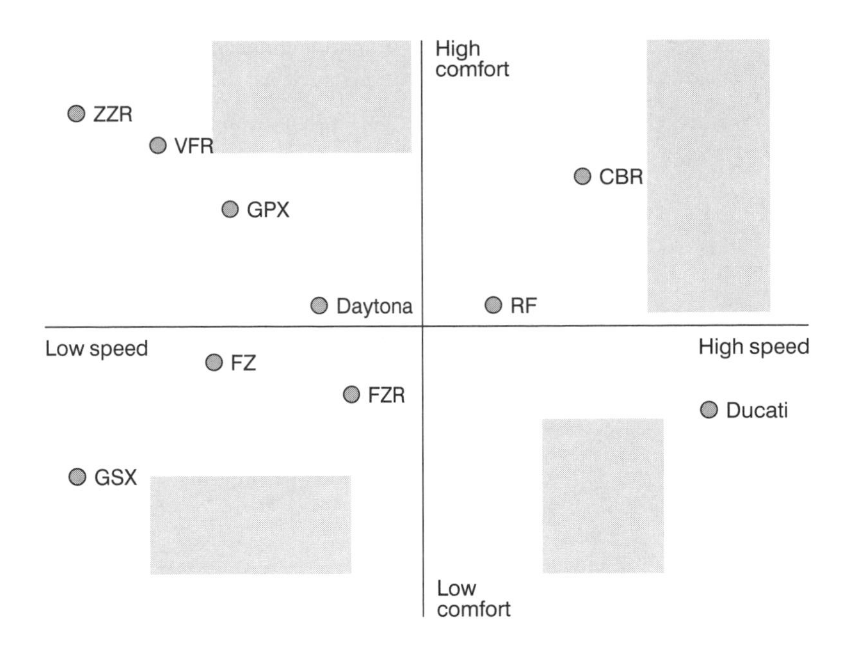

Figure 1.3.1 *Example gap analysis for the powered motorcycle market in the 1990s*

(6) *Relationship analysis*

A two-dimensional matrix is drawn up to force the comparing of attributes. Each comparison is numbered and the cross-sections are analysed afterwards. In the example of a toothbrush, the attributes of cleaning medium and bristles are compared:

	Paste	*Powder*	*Foam*	*Gel*	*Pump*	*Tube*
Nylon filament	1	2	3	4	5	6
Sponge	7	8	9	10	11	12
Short bristles	13	14	15	16	17	18
Long bristles	19	20	21	22	23	24
Hollow bristles	25	26	27	28	29	30

The technique creates combinations which might otherwise be overlooked. In this case, combination numbers 2 and 25 might be worth exploring, perhaps filaments made from toothpaste powder (throw away the brush when it has run out).

(7) *Morphological matrix*

A morphological matrix uses a table to list multiple attributes through which corrections are made at random. Look again at the toothbrush example:

Head	*Medium*	*Handle*	*Storage*	*Cleaning action*
Filament	Paste	Long	Rack	Up and down
Sponge	Powder	Sort	Hang by neck	Sideways
Bristle	Foam	Fat	Free standing	Circular
Hollow	Gel	Textured	Lay down	Vibratory
Short	Pump	Telescopic	Tumbler	Ultrasonic
Long	Tube	Hollow	Magnetic	Jet wash

Random connections drawn through the matrix might produce combinations such as:

long bristle – gel – textured handle – free standing – circular motion

May be a toothbrush with a cone-shaped base for free standing could result. Other combinations might produce a magnetic toothbrush that could be stored on the side of the bathroom mirror, or have ultrasonic stimulation from the bristles.

(8) *Analogy*

In designing a new article, consider comparing by analogy its use with that of another article. For example, when designing a new vacuum cleaner, compare its use with that of a lawn mower. Or in considering methods of securing a house, compare it with securing a car.

Alternative view

The use of structured and unstructured techniques suggest that there is an infinite number of possible schemes available and that these simply need to be 'unlocked'. It is also believed that 'generic' solutions can be found. For example, in trying to find a novel handle for a toothbrush the solutions could apply to any product with a handle or any product needing to be hand held. Analysis of patent applications shows that solutions to problems can be categorized and applied across any range of design problems. Computer programs are available that will generate creative solutions based on this premise.

Activity 1.3.2

Use a mixture of structured and unstructured techniques to find scheme designs for the following concepts:

- Uses of a brick.
- One hundred uses of a yoghurt pot.
- A method of mixing food.

Scheme selection

'Businessmen may be relied upon to make wise, intelligent and statesmanlike decisions – having first exhausted all other alternatives.'
(after Lord Chalfont)

'Decision' comes from the Latin term 'to cut off' (as in the mental process of reducing options) and is defined by dictionaries as the action of selecting or reaching a judgement on options.

Key point

A large company may have as many as 3000 design concepts. These will lead to 11 workable schemes of which only three may be developed, and ultimately one successfully released.

At this stage schemes are merely ideas with little development work and little understanding as to practicality. It is unlikely that any company has all the resources available to develop all of the schemes that are suggested and must therefore select only those that are most likely to succeed.

Selection has to be done right. Firstly, because choosing the wrong scheme is expensive. Forty-six per cent of a manufacturing company's development resources are wasted on schemes that go nowhere. Secondly, because even choosing the right scheme is expensive! Costs up to now have been cheap (ideas, after all, cost nothing) but from here on they begin to get expensive. It is in the manufacturing stage where between 50 and 85% of the entire costs of a new design are incurred and it is only at the manufacturing stage that these costs begin to materialize. Whilst scheme selection is therefore necessary and important, it is not always easy.

Activity 1.3.3

Select one scheme from the following list:

Product concept A

Expected profit/unit	£2
Market size	2 million
Product life cycle	5 years
Technical difficulty	high
Manufacturing fit	good
Competition	fair
Investment	£0.5 million
Originality	fair

Product concept B

Expected profit/unit	£1
Market size	2 million
Product life cycle	2 years
Technical difficulty	low
Manufacturing fit	excellent
Competition	fair
Investment	£0.1 million
Originality	high

Product concept C

Expected profit/unit	£1.5
Market size	5 million
Product life cycle	2 years
Technical difficulty	fair
Manufacturing fit	poor
Competition	fair
Investment	£0.5 million
Originality	fair

Product concept D

Expected profit/unit	£3
Market size	10 million
Product life cycle	5 years
Technical difficulty	fair
Manufacturing fit	fair
Competition	high
Investment	£0.8 million
Originality	low

Product concept E

Expected profit/unit	50p
Market size	5 million
Product life cycle	3 years
Technical difficulty	low
Manufacturing fit	excellent
Competition	high
Investment	£0.2 million
Originality	low

Product concept F

Expected profit/unit	£2
Market size	5 million
Product life cycle	3 years
Technical difficulty	high
Manufacturing fit	good
Competition	low
Investment	£0.4 million
Originality	high

One of the difficulties in making decisions is that the process requires logical analysis. It was stated in the section on creativity that the brain has a logical thought pattern. Making decisions should then be easy. Unfortunately this is not the case. The brain's logicality is also prone to undue emotional influences that can affect the outcomes.

Chance is one of the key emotions in affecting decisions. If you toss a coin ten times and it comes down heads each time, what side do you think will land on the next spin? Instinct may have told you that it should be a tail whereas in reality the odds are still 50:50. If the outcome of an event has no causal laws then chance is the perception of an event happening. Chance includes concepts of frequency and credibility rather than a statistical probability. Frequency is the perceptive probability that two equal outcomes must be maintained over a period of time and credibility is the acceptance of the number of times. People also perceive chance as being individually controllable.

Judgement also affects decisions. Judgement is influenced by utility outcomes affecting the assessment of likelihood. In other words, if people think that the outcome from a decision is almost certain, or extremely important, then they will behave differently than if the outcome was 50:50 and the effect was of interest only. Utility functions are not linear – there is bias at high and low probabilities and high and low values. People may make a decision because the act is risky, safe or neutral (depending on the personality) rather than considering the decision itself. For example, a manager should develop product A because it would do his company well. He may, however, choose product B because it would hinder a rival company which he dislikes.

A whole range of other personal factors can also affect judgement. Cognitive ability, motivation, personal goals (physical, biological, social, cultural), the person's role and the objectives of the decision. For example, a manager, shareholder, machinist, designer and customer may all decide that a different scheme is preferable (but may change their decisions if they swapped roles).

Key points

- People can make poor decisions if there is poor clarity over what the outcome of the decision is intended to achieve. Is the outcome of scheme selection to select a design that is innovative, stylish, low cost, profitable, bulk selling or any number of other criteria? The Mini and Concorde for example are both considered to be design classics yet both failed to yield a profit. Do you think they were successful designs or not? Whether they are successful designs or not depends on the criteria for which they were selected.
- Quality can be used as a criteria. If the design fulfils the customer's requirements then it can be considered to be successful. However, this requires customer requirements to be clearly understood, clearly stated in the specification and for the requirements to be met. Even this will not then guarantee commercial success.
- A 'mission statement' helps a company to ensure that decisions made throughout an organisation have greater clarity and a common purpose.

Activity 1.3.4

What are your top six new products of all time excluding the last 50 years?
What are your top six new products of the last 50 years?
What criteria did you use to select your best products?
Compare your lists with that of your colleagues. The lists are probably very different and this is because people use different criteria based on their own goals to judge a successful product.

Decision models

It is also possible to practise techniques which help to avoid humanistic flaws in decision making. Using groups can help to overcome problems of chance and judgement (but beware now of illusions of vulnerability, collective rationalizations and pressure on dissenters). Becoming 'focused' on the problem also helps. Entering a highly focused concentration zone (Japanese 'mushin') in much the same way that a sportsman does is thought to increase alpha waves and use of the central cerelebrum resulting in more inductive decisions and actions.

Decision models are also useful tools. Models are representations that help to clarify, objectivize the subjective and aid straight thinking. In the coin-flipping example of spinning 10 heads in a row, a mathematical model may overcome your 'chance instinct' by reassuring you that the odds in the next spin are always 50:50.

Activity 1.3.5

What is the likelihood that two players on a football pitch with two teams will share the same birthday? Take an estimate. Now find a mathematical model that will prove it.

There are several types of decision models. Simple models clarify and simplify thoughts, benefit models compare thoughts against preset standards, comparative models compare options with each other and risk models which aim to quantify the dangers of different decisions.

Decision models are not foolproof. The model itself may be wrong, the cost of gathering the information and creating the model may be more than the product is worth itself and the information used may also be wrong. Computerized models may suffer from a lack of familiarity of users with these systems, overcomplexity, and a tendency to use the model to make the decision. Comparative models may also select the best scheme from a group but may not indicate that all of the schemes are in fact weak. However, models do aid the decision maker in collecting, collating, analysing and presenting data and can be helpful if designed and used properly.

Key point

It is important to remember that models are intended to be aids to straight thinking. They should not make the decision for you.

(1) *Simple models*

Simple models might be expressed in a variety of ways, including diagrams, mathematical equations or templates. An example of a simple model that has been developed to explain the design-making process itself is given in Figure 1.3.2.

Activity 1.3.6

A packaging firm uses fixed lengths of standard width card to construct its cardboard boxes. Inevitably there is a certain amount of waste material. However, the company is

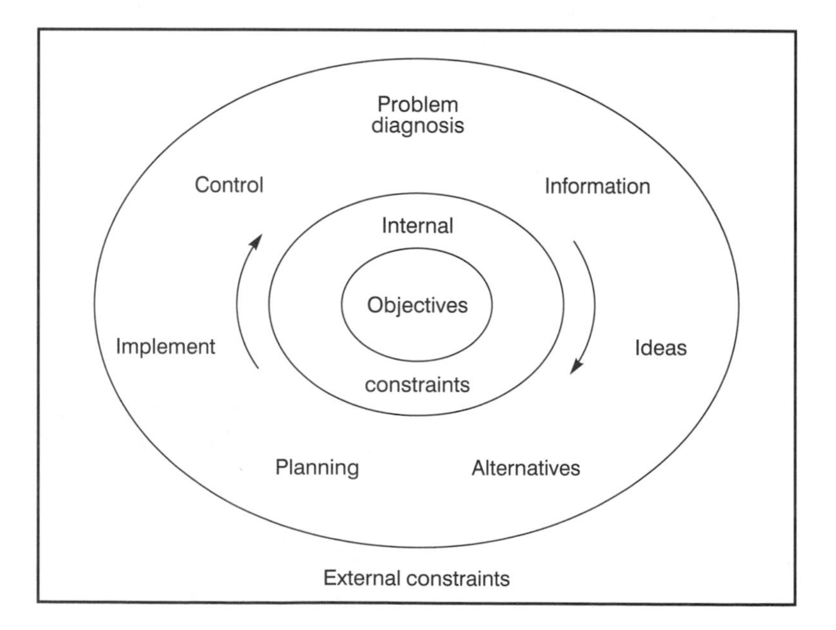

Figure 1.3.2 *A typical decision-making model*

determined to utilize the waste from the end of each length of card. It has decided that this waste material is to be used to construct variable sized containers. The containers are to have a rectangular cross-section. The tops will be made from other waste material from elsewhere in the plant. The price the company can charge for these non-standard packages depends upon the volume of the container. The larger the volume the more it can charge. Width of standard card = 40 cm. All packaging produced by the company is constructed from one piece of card using a simple folded net. The edges are joined using normal industrial tape.

Construct a simple model to help the company decide what size of container it should construct from a piece of waste standard width card to maximize its profits? Produce the simplest decision aid for use by the worker manufacturing these boxes.

(2) *Pay-off matrices*

These models are based on statistical decision theory. Consider the decision of whether to redesign a product or not. An important factor in the decision may be whether the economy will be prosperous or in recession. The matrix in Figure 1.3.3 models this decision with an expected value per product shown.

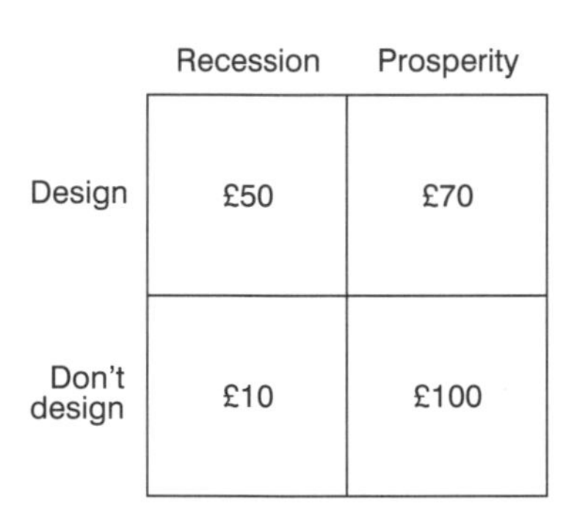

	Recession	Prosperity
Design	£50	£70
Don't design	£10	£100

Figure 1.3.3 *Pay-off matrix*

'Utility' is the return expected from a certain decision. It is normally calculated as the estimated pay-off for a given event multiplied by the probability of the event occurring. When utility is calculated in monetary terms it is expressed as 'expected value'.

'Probability' is the value given to the likelihood that an event will occur. The value must be between 0 and 1.

	Recession (0.3)	Prosperity (0.7)
Design	35	21
Don't design	7	30

Figure 1.3.4 *Pay-off matrix incorporating probabilities*

The same model can be used to determine the expected pay-offs when probabilities are applied. If, for example, it were considered that the probability of a recession was 70% (or 0.7) then the revised utilities would change (Figure 1.3.4). Note that the probability of a recession will be 0.3 as the sum of all probabilities must equal 1.

(3) *Decision trees*

Information can be represented in the form of a diagram known as a decision tree. The design/don't design is modelled in Figure 1.3.5.

The advantages of the decision tree are that the layout is visual and the number of events (factors) and decisions can be built up into several stages.

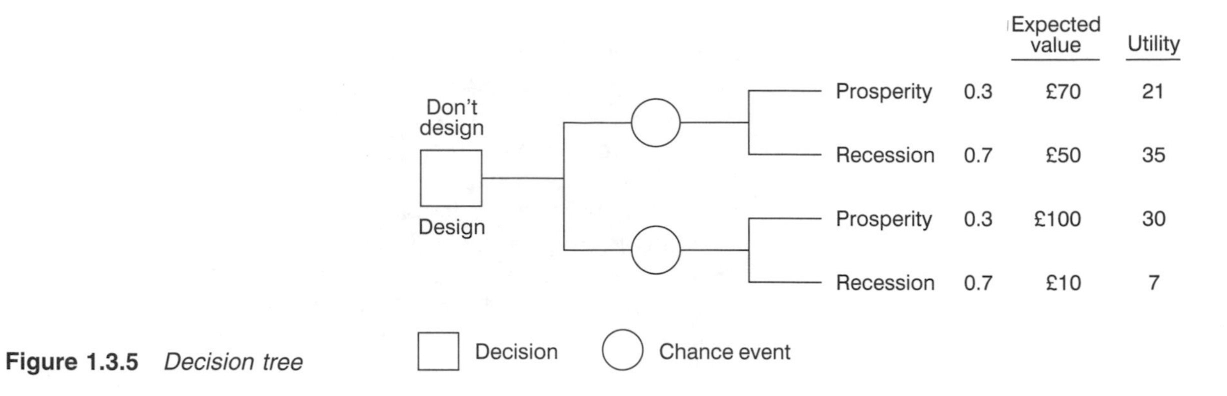

Figure 1.3.5 *Decision tree*

(4) *Product scoring models*

Companies can build up their models based on their previous knowledge of new designs. The following ATR (awareness – trial – repeat) model is such an example:

Buying units	3 million
% aware of new product	40
% of those who will try	20
% of stockists	40
% of repeat buys	50
Units per user per year	2
Revenue	2
Cost	0.65

Profit = $3 \times 0.4 \times 0.2 \times 0.4 \times 0.5 \times (2 - 0.65)$ = £129 600

(5) *Profiling or screening*

Companies should measure (profile) key factors to decide whether a scheme will be successful or not. Research suggests that the following factors are key to the success of a design.

- Market need
- Product superiority
- Economic advantage to the end user
- Newness to firm
- Compatibility with technological resources
- Compatibility with management resources
- Market competitiveness
- Capable of mass production
- Constant market

Activity 1.3.7

Companies are able to rank the factors and give them 'weightings' to increase or decrease their importance. What order of importance do you think the factors should be in?

(6) *Concept testing*

Similar to scenario testing for creative ideas, this model uses a mixture of individuals, panels, experts and laymen to evaluate ideas. Techniques include verbal, drawing and prototype examination. Q sorting is similar to Delphi in that schemes are sorted anonymously into order.

(7) *Investment appraisal*

Investment appraisal is another type of benefit model involving the evaluation of financial benefits including payback periods, internal rates of return (IRR) and net present values (NPV). Details of these techniques would be included in most books on finance.

(8) *Risk models*

Bayes' theorem assumes knowledge of prior probabilities and postulates that when nothing to the contrary is known, the probabilities should be assumed to be equal. Prior probability is the assessment of success based on current information. Posterior probability is proportional to the prior probability multiplied by a likelihood. Likelihood is the probability of an outcome given a certain event. For example, the probability of succeeding with a design is 0.6, and therefore the probability of failing is 0.4. This is the prior probability. The company is considering undertaking additional research. The likelihood of the research improving a successful design is 0.9. The likelihood of the research improving a failing design is 0.2. This is the posterior probability.

The probability of success is now

$$\frac{0.6 \times 0.9}{(0.6 \times 0.9) + (0.4 \times 0.2)} = 0.87$$

If the company estimates the costs of the design and the expected profits, it can make a decision on how much extra to invest in research based on an improvement of 0.6 to 0.87.

> *A 'risk' is the degree of loss that might be sustained. It has an uncertainty to which a probability can be assigned.*

Risk models also make use of the concept of randomness. Randomness and random numbers are derived from modern theories on evolution and statistics. Its use as a conceptual tool for explaining causes is valuable for explaining increasing order at the expense of unconscious knowledge and primitive guesswork. It finds uses in fields of science, chaos, deviance control as well as innovation.

(9) *Game theory models*

Game theory models bring emotional conflict to an intellectual level by focusing more on the effect decisions have on other people. It teaches not just what's known, but what's missing and identifies 'saddle points' (best/worst points) where the decisions are equal for both sides. For

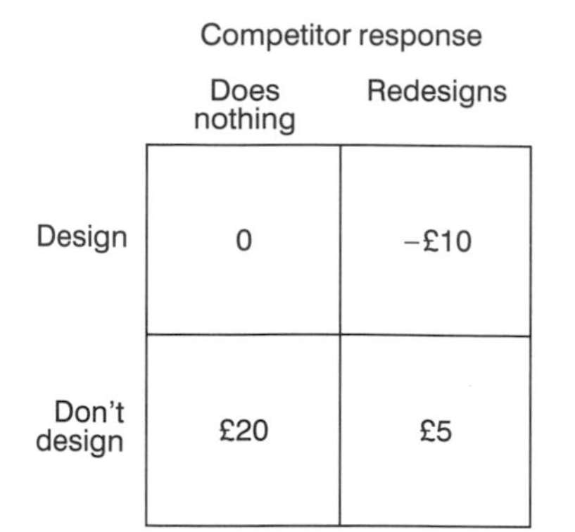

Figure 1.3.6 *Game theory model*

example, in game theory, a decision normally made by comparing scheme A with scheme B might be made instead by comparing what effect choosing scheme A or scheme B will have on a competitor. Game theory models should therefore include ethics, norms and principles. They are valuable in avoiding consistent and therefore predictable strategies.

We might reconsider our earlier design/don't design problem in this light (Figure 1.3.6).

(10) *Decision support system*

Decision support systems (DSS) are computerized models that form a sub-branch of management information systems (MIS). There is no common or acceptable definition for these systems but in reality they are a series of integrated computer software and computer-based information systems that helps decision makers with semi-structured and unstructured tasks.

Assignment 1.3.1

Resource allocation decision support system

A local authority sends lorries to collect quantities of waste from six different areas which it places into three major tips. The costs of transporting from the area to each of the three tips is given in the body of the matrix below. The total quantity that the tip can process each day is also given. Use a spreadsheet to produce a DSS that will help the council produce a low cost schedule for his operation.

Sites	1	2	3	Amount of refuse to move
Area 1	2	1	2	120
Area 2	3	2	3	231
Area 3	2	1	2	323
Area 4	4	3	1	241
Area 5	6	2	4	200
Area 6	3	3	5	100
Site capacity	500	400	400	

Scheme development

One of the problems with decision making in scheme selection is knowing when to make the decision. An early shakeout may save costs but may lose valuable ideas. A later decision may gain more data but is more expensive and in the delay your competitors may beat you to the market. Larger companies have formal gateway committees or regular review periods. Even when the choice is made, there are still a number of questions that must be asked before moving on to the detailed design phase:

Key point

In-house design is thought to be more effective. Increasing the tiny in-house design budget by one-third creates growth in the much larger manufacturing budget of 0.3%.

- *Is the chosen scheme worth doing?*
  How much will it cost?
  How much will it make?
  Can it be protected with patents and similar intellectual property rights?
- *Can we do it?*
  Do we have the resources?
  Is it technically achievable?
  Is it legally feasible?
  Should it be done in-house or should it be done using an external consultancy.
- *Should we be doing it?*
  Has it been done before? Prior art should be checked to avoid reinventing the wheel. It is estimated that 75% of all European funded research (and 30% of research in general) is wasted because innovation, and related technological information, is already available.

  If it has been done before who owns it? Most inventions represent an improvement on previous concepts and due diligence searches should establish if and who it belongs to, where rights start, any chains of ownership and any infringements. If these original concepts are patented then licensing is required (and enforced by US anti-trust laws). If you use somebody else's background intellectual property ensure that they are actually entitled to that background and be wary if your source of background has a variety of funding sources.

  Are you actually sure that your company will own the design work, particularly if you are subcontracting? Generally ownership belongs to the originator but this will depend on the type of work and any contractual arrangements. For example, unless stipulated otherwise, consultants may own patent rights and copyrights (but not design rights). What about your employees too – will they or you own the development work?

Key point

Better decision making has helped the success rate of newly launched products increase from 10% in the 1980s to around 60–65% today.

Example 1.3.1

'X' was a cycling expert and 'Y' a supplier of bicycle stabilisers. Y commissioned X to assist in reviewing and updating his stabiliser range. X agreed and stated that after the review he would provide Y with a design which Y could use if he paid a further commission. X conducted his review and concluded that a number of amendments were required. He subsequently produced a schematic drawing of a redesigned stabiliser. Y later produced the stabiliser using X's schematic drawings.

X claimed that the design was produced not as part of the commission which was for reviewing and updating. He stated that the design was intended for Y but was a separate project carried out by X on his own behalf which Y would be entitled to exploit or develop only if it had placed a further commission which it had not. X therefore complained that the stabiliser or certain parts of it constituted infringement of the design right in those designs, ownership of which he claimed to be vested in himself. His alternative claims were for infringement of copyright or misuse of confidential information.

In court it was determined that a design for bicycle stabilisers had been produced in accordance with the original commission and therefore the designer would not claim ownership or infringement rights. (Apps v (1) Weldtite Products Ltd (2) Wharmby Associates Ltd, 2000)

Problems 1.3.1

(1) Find ten schemes that might solve the concept of keeping dry when waiting for a bus.
(2) Where would the Gordon method be of help to you and why?
(3) Analyse your last important decision and explain it in terms of a decision-making model.

1.4 Chapter review

Review questions

(1) How many active UK businesses are there in the UK?
(2) What do the following acronyms stand for?
GDP
SME
NPD
DTI
BNI
MIS
(3) In what fields are the following experts?
Juran
Bayes
Ansoff
De Bono
Bavelas
Bell
Stephenson
Edison
(4) Match these library classmarks with their subjects:
620.0042
658.575
745.2
728
(5) What is the study of demographics concerned with?

Test paper questions

(6) Explain to a lay person the difference between designing and inventing.
(7) What problems would you need to be aware of if you were introducing QFD to a company?
(8) Analyse and explain why companies do not spend more on R&D if it yields improved profitability.
(9) Discuss how you would embark on, and what you would include in, a business plan aimed at raising finance from a bank for the design of a new folding bicycle.
(10) Describe three structured methods for breaking down the methodical pattern of the mind and explain how they work.

Assignments

(11) Write a report comparing the use of gut feelings with logical analysis in decision making.
(12) Make an in-depth study of a designer or engineer of your choice. Analyse their successes, failures and personality.
(13) Identify and justify an entrepreneurial opportunity.
(14) Write a report on why engineering design companies fail.
(15) Implement a '20/20/20' improvement programme on a product of your choice. You should aim to achieve 20% fewer parts, 20% improved efficiency and 20% lower costs.

Sources

Bibliography

Merle Crawford, C. *New Products Management*. Irwin 4th ed., 1996
Pugh. *Creating Innovative Products*. Addison, 1996
Shigley. *Mechanical Engineering Design*. McGraw Hill, 1996

Reading

Chell, E. *The Entrepreneurial Personality*. Routledge, 1991
Deakin, D. *Entrepreneurship and Small Business*. McGraw Hill, 1999
Eppinger. *Product Development in the Manufacturing Firm*. MIT Press, 1993
Eppinger & Ulrich. *Product Design & Development*. McGraw Hill, 1995
Ertas & Jones. *Engineering Design Processes*. Wiley, 1993
Flurscheim. *Industrial Design in Engineering*. Design Council, 1983
Green, J. *Starting your own Business*. How to Books, 1999
Morrison, A. *Entrepreneurship*. Oxford, 1998

Journals

Design
Design Week
Design Studies
Entrepreneurship, Theory and Practice. University of Texas, El Paso
International Journal of Entrepreneurship & Innovation. IP Publishing
Innovation & Technology Transfer
Journal of Engineering Manufacture
Mechanical Engineering Science
Professional Engineer
Solutions
Stuff
Technovation
T3

2 Creative design practice

Summary

This chapter gives a range of views, discussions and techniques that relate ideas, various types of creative research and development, and thinking about design and your own activities into the creative role. The aim is to relate these aspects together to sustain the practice of design, to provide a basis for professional activity.

One of the main weaknesses to be avoided in the study of design is the proposal of 'aesthetic' or 'functional' objects or products in isolation, or with no context. Informed and responsible design practice has to see products and artefacts as part of a story or as part of a set of relationships. The extent and nature of these relationships has to be researched, appreciated and reflected in both the procedures used in design development, and in the resulting design that is subsequently used, relied upon, enjoyed and paid for (if it succeeds!).

Objectives

At the end of this chapter, you should be able to:

- consider an overall or 'holistic' view of design practice;
- relate aspects of decision making in design to the human aspects of visual and tactile aesthetics, and to the perceptual issues that underpin these;
- develop a 'questioning' approach to design in order to make the product connect with the user or the context to which it relates;
- consider the development and role of the design 'brief' and the impact of this on the design process.

2.1 Objects, messages and experience

Thought question: ergonomics – what is it?

The first thing in considering the relationships impacted upon by designed objects is to ask the question: What messages do people get from this situation/artefact?' Messages can be visual, tactile, experienced before you touch something, while you are using it, or after you have used it. Such messages can touch upon emotion, psychology or pragmatics. For example, a working chair may have ergonomic styling – it may look as though ergonomics have been considered – but after a

Figure 2.1.1

short period of use you may discover that it is badly considered and in fact irritating to use. In this case the initial 'message' given out by the visual appearance has failed the test of experience.

It is more satisfying when the experience of using something is close to what you expected – but there is a range of ideas and values here that can be explored playfully in order to see what it means. For example, we may say there is a spectrum from those artefacts where it is vital that there is a match between perception and expectation (such as tools, anything with health and safety aspects or where performance is relied upon) to those in which elements of surprise, delight and discovery are achieved by deliberately challenging the way something moves or acts in relation to what you may expect. The designer's awareness of this range is vital if the dull uniformity of conventionalized 'schemes' of design is to be avoided. See Exercise 2.1.1a below.

The question of messages given out by objects, and our psychological relationships with the whole field of visual 'language' of form, is a vast subject and runs deep in human behaviour. In relation to the purchase of products, buying activity is indicative of the human need of expressing change, aspiration or development, and is the subject of detailed research and financial investment for the designer. However, the designer's main activity is the bringing together of seemingly conflicting types of information, ideas and requirements, and synthesizing a vision and a model that unifies these ingredients into something that other people, other specialists, and clients can recognize as being a solution or a 'model' idea.

Key point

In cognitive terms, visual information has meaning only when confirmed by physical experience. This is how illusions work; what you 'thought' you saw was based on what you 'thought' happened, not what actually happened.

In this way design practice is potentially a key interdisciplinary skill in the working environment. In contemporary society many specialists in various fields have difficulty communicating clearly with each other – but given well-researched and informed visual expressions of concepts, such visual 'models' can successfully catalyse a good design process. Some of the reasons for this are indicated by what follows.

What messages do we get from objects?

Back to the question 'What messages do we get from objects?' First, one thing everyone shares to a greater or lesser extent is the vast amount of physical and visual experience we have all received and grown up with since birth in the development of cognitive and physical 'body' skills. Although in one sense 'basic' (such as perceptions of hot/cold soft/hard shiny/rough wet/dry, etc.), the means by which we have acquired this experience and the ways that our knowledge is underpinned by such concepts are individually and elaborately detailed, extensive, delicate and exact.

Secondly, it is not possible to gain all this experience independently of specific features of our environment as we experience it; so, for example, that hot water was in a blue metal saucepan with the wooden handle. The glue I got stuck to the table with came from that red coloured squashed tube with a black top. The bicycle that fell on me was the black one with the gold stripes that my auntie had. I'm sure you can think of your own examples – and you can see that experiences of heat, stickiness, hardness, etc. are in fact embedded within a rich emotive and experiential language that is value and memory laden. Such visually associated experience is common to us all with the exception of those without sight from birth, in which case the 'haptic' territory of memory is mapped with the other senses.

Next, it's a short step to linking this idea with such things as object 'types' ('my' things and 'other people's' things, 'dull' things and 'interesting' things, etc.). Advertising and marketing work with these associations, manipulate them and build them into brand identity for the purpose of priming or maintaining our perception of a particular product. Some types of product are less dependent on this, such as a chair you can try out by sitting in it before you buy, and some are most dependent, such as over-the-counter pharmaceuticals. See Exercise 2.1.1b below.

Product architecture – how to think about structure

Messages from objects are also related to ideas or implications of 'typical' artefacts, often seen as associated with a particular period in time (for example, jet plane, framed painting in a gallery, clamshell, skeleton). Referring to familiar 'types' is used often to describe 'product architecture' independently of any other attributes. In this way we can have a 'clamshell' moulding to encase a product, a 'modular' structure put together in interchangeable sections or a 'spine' around which other parts are attached. Development proceeds by analogy. The language of ideas takes what we know and abstracts or extracts features we can use to increase our understanding, which is done by deconstructing and reconstructing our experiences. In this way a 'language' of design forms becomes transmitted through different times, combining the 'new' with the familiar.

Messages from objects range from the factual to the transitory, for example:

- If numbers are used make an indication on a control, does the user know what the numbers mean? Is what the user thinks they mean the same as the designer's intention? (Practicality)
- Currently it can be extremely important to get colour combinations right in the area of fashion – but this will change soon. Colour prediction in fashion design is a major business activity. (Fashion)
- Colour rendition of visual displays is an important area of interaction design. (Perception)

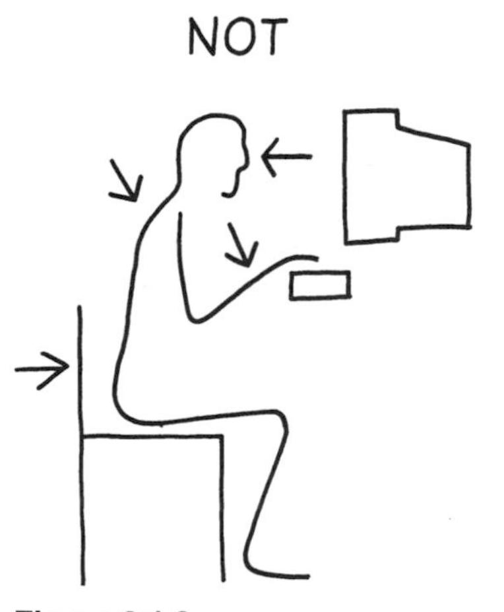

Figure 2.1.2

Example 2.1.1

'Sideboard' is an Anglo-Saxon term for an item of domestic furniture. It is a typically pragmatic term meaning a wooden board placed at the side of the room for putting food and drink on before serving. The more formal term is 'credenza', a Latin sourced word for a 'credence table'. It was for offering food for checking its credibility, or testing, before bringing to the table (a table has an entirely different social function). There is a religious layer of meaning to 'credence table' associated with the intermediate positioning of an offering, to be checked or approved before it is used. See Exercise 2.1.1c below.

So let us expand our ideas about messages we get from objects by thinking of as many ways as possible that this happens. The following three exercises explore this.

Exercise 2.1.1a

Object criticism exercise
Consider an artefact you do, or do not, like to use or own, and list up to ten features you do, or do not, appreciate about it, including the perceived messages or signals given out by the selected objects. This can be done on your own, or by bringing examples into a 'group critique' situation in class. (Unless you are practised in doing this you will find it harder to do well than you might assume. Ideas and perceptions need introducing or putting in context.) Critical appraisal is a fundamental skill of a design team, and the way such review processes are carried out will determine just how constructive they are in practice. This exercise can illustrate how our appreciation of key issues can be expanded and shared in order to establish a basis for the next stage of work.

Exercise 2.1.1b

Brand image exercise
Think of a range of examples of product design covering the spectrum from 'most dependent' (upon independently established brand image) to 'least dependent'.

Exercise 2.1.1c

Object messages exercise
Using your own terms and associations, draw up an extensive list of all types and forms that messages in designed objects can be about, in your experience (and indeed in 'found' objects for comparison). There are no right or wrong answers here, it is a matter of extending our vocabulary and appreciation of how we interpret our own experiences in connection with objects we own or use. Start with a range of products and extend your own list: washing machine, bicycle, door handle, shower gel, etc.

2.2 Ergonomics: analysing use and experience

Ergonomic awareness

The field of ergonomics expands this issue into our experience, over time, as we use and live with designed artefacts. It is about understanding the detail of our interaction with things. Ergonomic awareness starts with critically evaluating what you do, how you do it, what supports or inhibits an activity or process. Can you close your suitcase? Do you know when something is turned off? Does your chair give you a neck-ache or did you sit in it too long? What does 'too long'

Figure 2.2.1

mean? What does using a VDU for four hours non-stop make your eyes do? Do you know? What is the lighting like where I am right now? Do I trust this piece of cutting equipment? Can I interact with a cash machine in the sunshine? How do I get my bike onto this train?

A fallacy underlying some early attempts at 'ergonomic' design was an underlying assumption that by studying variations one could arrive at an 'ideal', which could be universally applied. There are many situations where an ideal doesn't work, and where it is more helpful to establish exactly what degree of compromise fits a given need. Design situations will cover a spectrum from 'no compromise' (racing car driver cockpit fits one named driver) to examples where a wide compromise is possible (pillow) but not exclusively so (orthopedic pillow).

Stepping back to take a wider view for a moment, for the designer, because of the many factors involved, the total design and product development process itself has to be a compromise. In this case, compromise between sometimes competing factors in one design – such as development time, manufacturing methods, cost-benefit analysis of different components and compatibility – has to be an active balance in order to get a result. By 'active balance' we mean something you consider and are aware of from the outset, not something that happens at the end.

Although it is possible to get reference works on ergonomics topics covering specialist areas (physical body data, vision – related ergonomics, sports performance ergonomics) you can start simply by increasing your own awareness and critical skills.

Do you know the dimensions of the chair you are sitting in? A kitchen is probably one of the most familiar 'workshop' situations for most people. Have you thought critically about what it is like to work in 'your' kitchen? What about all the 'input' and 'output' of the kitchen? Remember to think of it as an eventful story (not in isolation like a photo in a magazine). A study of 'A day in the life of my kitchen' would be a good start to ergonomic awareness. Many things we do involve repetitive activities – just how much of such activity is advisable or desirable in a given situation is an important ergonomic study. Bad ergonomics can lead to discomfort or irritation in some cases and possible serious long-term health difficulties or even life threatening effects in others. An example of the latter is repetitive strain injury (RSI) which is an increasingly common and serious injury associated with the users of certain machines and notably computer keyboards and screens. Although not unique to the industrial age ('tennis elbow' is related to RSI) its incidence is much more commonly noticed now.

Figure 2.2.2

Early ergonomic study tended to use static measurement, but to be effective, ergonomics really has to be about movement, including such concepts as degrees of change, range of events, different values or types of measurement working together. See Exercise 2.2.1a below.

Example 2.2.1

Example from the history of the design of prosthetics

The earliest knee replacement designs were shown to produce serious pain, difficulty and damage to the recipients. Why should this be so? It's only a hinge isn't it?

The fact is that the whole picture of body movements in walking and running, for example, shows complex and variable types and dimensions of movement that are much too challenging for a hinge. The knee has rotational, sliding, rolling and rocking actions and the many associated muscular contractions and extensions work together in harmony with these. In researching the design of effective knee replacements, It took many three-dimensional animated studies of natural knee movement in a variety of active conditions to begin to *conceptualize* what was really going on, rather than what had been *assumed* was going on. In other words, a more comprehensive conceptual model was developed, which showed up the crudeness of the initial quasi-static model. These studies in turn allowed for the production of a more intelligent design with much reduced negative side effects on the recipients.

How do you know what something is?

Sometimes messages from objects are hard to see, and sometimes they are all we've got to go on, in terms of understanding a product. An obvious example of messages given by design exists 'outside' the product – for example, with foods, pharmaceuticals and many materials. In these cases we have no idea what things are if we tip the contents out of the container, or separate them from their label. We rely upon a carefully orchestrated mixture of such text, graphic design, colour and brand image projected previously into our perceptual world through communications media. These elements working together establish a 'perception' of the product that predates our experience of it. A friend once purchased shampoo from a Japanese department store and had started using it before discovering it was household bleach. Next time you buy a product such as washing powder for your clothes, see if you can find anywhere on the package where it says in plain language what the contents are actually for! You may have little trouble finding what you need in a familiar marketing environment, but when you travel to a different culture all of the programmed 'signals' we rely upon are either different or absent.

Visual research

Whilst open to such manipulation, our perceptions of designed visual form, colour, texture, weight, etc. nevertheless do connect with our own inner library of formal and aesthetic association. It is vital for the student of design to explore this repertoire for themselves by designing and making things that allow an intuitive or personal dialogue with ideas and expressions through a 'doing' process, i.e. by drawing, visualizing, making models or structures (in a playful way and not necessarily with a target in mind) as a preparatory or background activity. The fact that in the contemporary commercial culture we are constantly bombarded with manipulated visual information has to be acknowledged – and unless a personal language of 'visual research' is developed, design and creative work becomes simply a recycling of received imagery. In poor design this is why so much has a tired or second rate identity to it.

How to develop personal visual skills is discussed in more depth in the Section 2.4, 'Teaching and learning design'.

So, whilst analysis and a well-defined brief are essential for successful product design, equally important are those activities that underpin your own personal way of informing and developing your own creative process as a working designer, which process is 'fed' by some of the means described above. These means include:

- Curiosity: Allowing curiosity to work its way throughout the process – by asking questions even if seemingly irrelevant.
- Getting more information: Gather information 'around' the subject or the idea, not just directly 'at' it, or what you and others may assume is the subject.
- Listening: Be a good listener; notice what colleagues, clients, users actually say especially when presented with anything they can react to, for example with models, presentations, interviews. (As opposed to what you want or expect them to say.)
- Visual study: Draw what you see, or model your responses to perceptions and information. Give a visual/tactile dimension to your researches.

Alternative view

Two quotations concerning perception

'Ah, but Watson,' said Sherlock Holmes, 'people *see*, but they do not *observe*'.

(From *The Adventures of Sherlock Holmes* (1892). Sherlock Holmes is a fictional amateur detective character created by Arthur Conan Doyle in England in the nineteenth century.)

'How strange it is to see with so much Passion
People see things only in their own fashion!'
(Molière, *The School for Wives* (1662).)

Remember Sherlock Holmes' advice about the necessity to 'observe', which is different from 'seeing'. Observing is an active process which is a form of work. One of the most powerful forms of observation for the designer is to draw what you are looking at. Whether you are 'good' at drawing is irrelevant. The act of drawing is a physical, cognitive link, and it allows many forms of knowledge to be either developed, made explicit, or revealed. Drawing connects us with our 'body' of knowledge, only a small part of which is easily expressed through language. This concept of course extends to all forms of visual construction, whether using pencil and paper, wire, macquette or model materials, or 2D + 3D computer software. Such different methods will suit individual preferences. It is fundamental to regard these as processes that must be gone through in order to experience and realize development. As more experience is gained, these activities can become more directly a 'statement' of mental concepts, but at the early stages of design study the importance of continued reflective play with such processes cannot be overemphasized. Drawing and modelling is a two-way process – both informing and revealing.

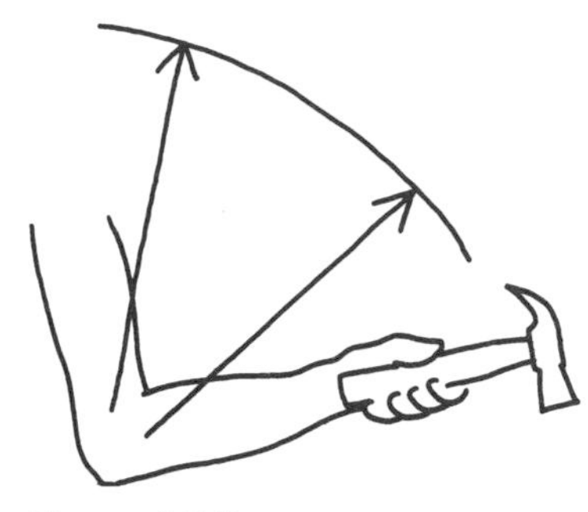

Figure 2.2.3

Exercise 2.2.1a

Ergonomic study exercise
Think of a situation involving movement in either the interaction with or the workings of a simple product such as a corkscrew or a hammer. Devise a way of illustrating or representing the various features of the movements so that the object itself can be deduced from your depiction of the movements.

Exercise 2.2.1b

Drawing and visual research exercise
Look again at 'Product architecture', page 56. Identify one of the given types of structure (skeletal, modular, shell, etc.) or think of another type of structure of your own choice, and go out and find an example of this, either from an industrial or a natural source. Make drawings or macquettes based on the chosen structure. These should explore and/or imaginatively reinvent its basic features. If a drawing, imagine you have X-ray vision and do an 'investigative' drawing, not just a superficial picture of the surface. If a macquette, be selective in what aspect of your chosen structure you are interested. See also 'Appropriate Modelling', page 70.

2.3 How to make and use a 'brief'

What is a brief?

The term 'brief' was used above – how does it work and what is it? The context is that product design always involves many people, usually from different educational or professional disciplines working together to achieve an end result. The usual cause of difficulty and sometimes even disaster is the making of assumptions at an early stage, either about what people mean, or what they want or what they actually do. One of the essential skills in teamwork is that of ensuring that all parties share a similar understanding of what is going on at different stages. This is much harder than it may seem. There is not space here to go into group dynamics or role play, but you can understand that there must be some form of reference for everything that happens. Your ability to listen, speak clearly, sum up, be prepared to change your viewpoint with ease, and present information and ideas simply and effectively is central to performing an effective role.

The function of a 'brief' is to catalyse a process. An example of an initial brief may be a client stating an objective or a requirement, or a company outlining some research it has done or a direction it wants to move in, and asking for ideas/products to be developed to effect this. It can be seen from what has been said previously that it is essential for the designer to respond to this with creative insight of her/his own, not just be led by what is given. By considering the information and requirements and the context for this brief, the designer can creatively restate the brief – and in so doing strengthen, improve, or in some cases totally change it. In this way the 'brief' has to be a product of rigorous and tested dialogue to be of any use. When it has been developed in this way a brief becomes the departure point for various types of development work, acting as a reference and a tool to link different activities. It often forms a contractual element to maintain a focus on what work is done and on the timescales involved.

A notable feature of a brief is that it should identify what 'is' and what 'is not' known in order to make progress – in other words, to identify areas of research that have yet to be done and to make provision for this to be reviewed, and further decision-making times considered. In larger interdisciplinary working, periodic meetings to review the brief and the impacts of work in different areas are needed. With the increasing use of information technology particularly in design, the sharing of complex information, concepts, and data, is now done as often in a 'distributed' way (i.e. geographically distant) as it is through personal contact, and this too has its own particular challenges. In this way the design of systems and the representation of processes have become rapidly developing fields for designers in their own right.

Design, knowledge and choice

Many of the challenges presented by the processes outlined above have to be dealt with through 'visual' language – through the representations of ideas and processes via a visual and tactile interface. This brings us back to a basic premise of this chapter – which concerns those practices and activities which, for the designer, help develop an active dialogue with our visual life; drawing, making and visualizing any ideas that interest us, from the imagination, from observation, from other sources, exploring our curiosity and establishing a personal repertoire of visual language, the 'meanings' of which we are prepared to test in the public domain.

In the same way that you cannot design things without some knowledge of materials of which they will be made, you cannot design for people or for users unless you acquire knowledge of them, and of their circumstances. This work can be generalized, such as with issues of human shape, behaviour, size, movement, senses or sensitivities, or the knowledge may become specific, such as how to prevent a certain age child accessing something potentially dangerous, or how to allow for people with identified difficulties resulting from disability or illness, to use or access something despite the problems. In this latter case general experience is not enough and further study is essential. Even in general situations, never assume you know basic things. Study the circumstances over time in order to avoid your preconceptions which may be fallible, and extend your understanding.

Alternative view

If you are interested in buying an apartment or house to live in, although much is influenced by first impressions, it is wise to revisit the place at different times of the day and of the week – to find out more about it over a cycle of time. Then you have a better picture and are less likely to get unpleasant surprises that arise through assuming that the image you have in your head is the true story.

This principle applies to design concept development. It was discussed earlier that failures in design often occur through the design of an object or product as just that – a 'thing', with insufficient consideration to what happens over time. Apart from being at the root of the misuse of materials and negative environmental impacts, this failure in thinking also ensures a poor 'user experience'.

Example 2.3.1

In his book *How Buildings Learn*, one of the phenomena the author Stewart Brand illustrates is the fundamental differences between the image of an architectural design as it may be presented in magazines (i.e. like a decontextualized fashion photo), and the reality as experienced by individuals who have to use, maintain, and live with the built environment. The book addresses conceptual topics that are essential in any practice of design that is not to be superficial.

Figure 2.3.1

By looking at products as a time-based process involved in an adventure of usage, maintenance, possible recycling and for recovery, we can establish a more informed and responsible platform for decision making concerning product design. Have you noticed that many products spend a greater proportion of their life in a slightly broken state than in the pristine state? It is interesting that a common feature of many indigenous or folk artefacts, some of which display remarkable and advanced design, is that the 'well-used' condition is the stable or dominant condition for the artefact, because the materials' understanding, making technique, and use, are evolved into a complete set of relationships. Think of a wooden 'Windsor' chair, where it is obvious that over time some flexing or looseness will set in. The design allows this without catastrophic failure – the failure is progressive and delayed. Failure is a feature of the design. Continual repair – in other words the 'involvement of the user' – is another feature of much indigenous clothing, tools, furniture and dwellings. The modern analogy with this concerns the matched performance and characteristics of different parts of a product, again demanding that the designer understand relationships both within and outside the product.

All the designers I know stockpile bits of information from everywhere. Samples, materials, connections, fixings, tests, finishes, specialists, etc. Two of the most influential European designers of our age, both famous for their collections, are Achille Castiglioni and Ettore Sottsass. In Castiglioni's case, he is known particularly for his collections of objects which are unselfconscious designs from the past hundred years and more that tell us about function. Sottsass is known for his extensive photography from a lifetime of travel and observation of detail, of materials, of the unexpected, the humorous, the curious, and so on.

Decision and communication

Even if you work only with 'virtual' ideas, at some point either you or someone else has to evaluate and make decisions concerning the physical design and assembly of products. With most things, assembly has to be designed as part of the whole scheme, in other words the product itself must be thought of as a 'process'.

In carrying through a design, the designer will work with production engineers to evolve product or component designs in detail. This specialist field of production engineering is a part of the feedback loop in the overall development process closest to the designer, as a concept is taken through to implementation. This often represents a major portion of the work, since as the commitment and investment of time, equipment and personnel increases there must be a very clear focus on what is being aimed at and specifications are agreed upon in detail. The scope for creative play or turning ideas on their head will have been severely reduced at this point. It's the creative team's responsibility that such creative orientation is done at the appropriate time and has been fed into the brief; not left too late or indulged in as a delayed reaction to what has happened elsewhere or later on in the process.

Stepping back from an individual case, and despite the points above, the design–development–production process is of course a continuous learning exercise for all, and the importance of reciprocal feedback, monitoring and review practices cannot be overstated, since more can be learnt in any one cycle than may be realized in any one product.

2.4 Teaching and learning design

Experience and knowledge

If one of the main aims in teaching is to build confidence, and confidence is acquired as a result of experience, then it follows that for any teaching process to be effective it has to be about promoting and extending opportunities to *experience* the subject. This may sound obvious, but reflection will show the significance of this.

It is tempting to say that this is particularly so in the creative fields, but since any endeavour in any specialist field requires creativity this holds true for any educational programme that is intended to extend the field. Design is particularly enmeshed with research, being concerned with synthesis, vision and generally bringing 'new' formulations of objectified ideas into being. But research lies at the very first steps in design teaching, and links seamlessly with the cognitive development that occurs in childhood, and continues into adulthood. The basis of this was discussed earlier in Section 2.1.

Many actions that we learn and take for granted (like using your hand, or riding a bike) are in fact very complex in terms of muscular and perceptual co-ordination, so much so that only a small portion of the control/sensory activity lies at the conscious level. For example, there are at least nine degrees of freedom of the human thumb alone. Put this into combination with the rest of your hand, and the computations for movement go 'off our screen' very quickly! Although all of these actions are 'known' to our sensory and control system, we do not experience all this complexity mentally at a conscious level. This is significant because throughout cognitive development our experiences are embedded with this degree of complexity, again much of it below the conscious level. It is this reservoir of rich 'haptic' knowledge that is drawn upon and lent endless variety of expression, nuance and recombination in the process of struggle for creative expression. ('Haptic' means of, or pertaining to, touch.)

It is in view of this extreme complexity that a learning strategy used in cognitive development, for example in the throwing of a ball to hit a target, is that of trying things more or less at random and simply repeating those attempts that work. This approach is used in the 'AI' (artificial intelligence) aspects of robotics design in the 'learning' of complicated actions or responses. It appears to be a quicker route to success than a process of analysis followed by an 'instruction' to move. In the 'trial and repeat' approach, the 'repeat' response is based upon experience that has been evaluated – there is a feedback loop (like the 'trial and improvement' technique in GCSE maths). The significance of this in a learning environment is that so-called 'failure' is a principal building block of confidence in that it is an important result of trying something, and a significant link in that chain of events known to include what is being sought. Fear of failure is a block to growth. It follows that all effective teaching and learning (in design especially) must embrace both failure, and effective responses to it, as one of the primary ways forward.

Meaning and experience

We describe things by using analogy and metaphor – these analogies and metaphors are sourced from and referenced to our 'body' of knowledge and experience. The dimensions of this body of knowledge that we all share (and which occasionally cross-references with others) are as varied as all the ways we ever experience or express anything. But in the process of expression, we convert or translate between *dimensions* and this is where the concept of *meaning* becomes elaborated. We can attempt to think of a framework for this to help us be clear about the language we use concerning matters of aesthetics. So, for example:

- When we read a particularly satisfying sentence or line of poetry we can feel that it has a particular 'shape'. (FORM)
- When we see certain combinations of colours we can experience different 'energies'. (ACTION)

- When we pick up something that 'fits' the hand, it has those properties of sensitivity that reflect the complexity that we 'know'. (KNOWLEDGE)
- We say that an argument can 'leave a bad taste', or that someone was 'blue'. (SENSATION)
- Or we may say they had an 'uplifting' experience. (MOTION)

Alternative view

'Form, in the narrow sense, is nothing but the separating line between surfaces of colour. That is its outer meaning. But it also has an inner meaning, of varying intensity, and properly speaking, *form is the outward expression of this inner meaning.*'

(Wassily Kandinsky, from *Concerning the Spiritual in Art*, Dover Publications Inc., New York (1977).)

Although this is an extensive subject, it is important that the principles of this are appreciated and acknowledged in teaching through experience.

The foregoing should go some way to explain why design cannot be taught as solely a linear or information giving process, since it is concerned primarily with discovery, rediscovery, and reformulation with the touchstone of common human experience and aspiration. Much of engineering and its associated technologies rely upon obtaining an optimized solution to a problem or objective. The practice of design involves an understanding of the subjective and the objective in an effective synthesis. A designer, whether of a child's toy or a flight control system, needs to be prepared to have their assumptions and views challenged, and further, to know how to challenge such views effectively in the advancement of a design scheme.

Figure 2.4.1

Example 2.4.1

Teaching example; the group critique

The group critique, or shared critical review, is a common learning strategy used in art and design education, as it is in design practice. Experiences of this for the participants range from the seriously negative to the inspirationally positive, depending entirely upon how and on what basis such group critiques are managed. In the critique, listening to a range of personal opinions in an unstructured format is usually to be avoided unless the group involved already knows each other and has agreed the remit of the session.

A more creative strategy for a group review is to concentrate upon the intentions of the designer, and to give one of the group a role of excluding opinions that digress from this. Another function for such a group review is to consider any sources of information or influence and how such influences could be related, checked, or researched. A plan of work to develop further these researches or responses can be agreed to carry the project forward to the next review. In a study situation, it is important that members of a critique group have a sense of trust in the process and what it is for, and take time to review and agree these aspects before engaging in a process that is intended to be positively challenging.

So how do you learn and teach design?

- Skills need to be acquired – gain experience of placing ideas outside yourself for others to appreciate, and practise expressing what is significant to you about them. Don't be scared of rejecting ideas – you'll get another. Some ideas are right first time but others (most) need working on. So-called basic skills such as drawing, illustration, modelling of any kind, writing clearly, learning how to do research are in reality not so basic. These skills will continue to develop and will always be your main vehicle for interaction.
- Promote the ability and fluency of translating ideas into different media and of communicating a relationship between what is intended and what is actually being done. For example, if in a presentation you are showing a clay model of a steel tool, have some steel and/or another means of helping your audience connect what you are actually showing them with what you are trying to communicate.
- Promote an awareness of process – i.e. that it is normal not to be able to see the end result from the starting point, and that by experiencing and trying out different types of process you give your faculties and skills some breathing space to contribute to the work. The ability to share views on this is improved by experience.
- Develop the insight that just because you know what you are doing in a particular situation, it doesn't mean that this is at all clear to other people who think in a different way. Get used to simplifying and communicating aims, activities and objectives as part of how you work and at its various stages. See Exercise 2.4.1b below.
- Gain experience of specifying components or artefacts that have to be made by someone else – get used to what this is like, its potentials and shortcomings. See Exercise 2.4.1a below
- Gain experience of encapsulating key issues or features of a situation, or context, for a design activity, by way of illustration, visual communication, simple text and with a sense of organization. How information is organized is equally important to the content, since this will either enable or obscure a person's access to it. See Exercise 2.4.1b below.
- How do you know when something is finished or complete? Reflecting upon this feature to creative work will open up a greater appreciation of what you are doing in terms of its scope, aspirations and limits.

Alternative view

Sometimes you will hear people say:

> 'You can't teach creativity. . .'

In some respects this may be true, but you *can* teach ways and means and techniques to allow creative work to flourish and reach expression. Through expression and interaction, and the confirmation received from others through these means, confidence in creative work can grow, and your abilities will be supported and developed.

Knowledge, expectations, intuitions

You cannot find out enough about materials solely by reading about them. To build up appreciation of materials and the complexities of their behaviours it is necessary to understand some fundamental concepts about them such as the differences between polymers, ceramics and metals, to handle them (where possible!), to see them being processed and preferably to experiment with them first hand. Talk with those specialists who do work with specific materials as their main activity. This will improve your understanding of the key issues as well as revealing any inaccurate preconceptions you may have.

It is a safe assumption that you will know very little detail about, for example, wood, (complex natural polymer) ceramics, synthetic polymers or metals unless you research them as suggested in addition to finding out 'facts' about them. This is despite a rich body of experience and memory you *do already have* through contact with made objects in your environment. This relates back to the discussion on 'haptic' or tactile knowledge of which we all have a significant amount. It is necessary to continue to develop the links between factual or articulated knowledge and this store of experiential knowledge. Also of fundamental importance is the need to be open to the idea that your intuitive understanding of something, while being your main gift, may also be entirely wrong when wrongly applied, or put in an inappropriate context, since it may not be based on appropriate experiences.

The example given in Section 2.2 above of the design failures of the first artificial knee replacements is a good example of 'counter-intuitive truth'. Intuition said the knee was a 'hinge' – a perfectly adequate mental model for a knee until you try to replace it. Further study took the designers to a different level of appreciation which was necessary for an effective design. Who could have guessed that knees actually rotate in the y-axis? Mostly when we talk about knees we are not concerned with the complexity – we take it for granted since we have other concerns. The expertise of a designer to be able to research beyond the superficial is, professionally, what people pay for.

You will notice that in exhibitions or shops people naturally wish to pick up or touch anything they are seriously interested in (remember that visual information has little meaning until 'confirmed' by physical experience). 'Please don't squeeze the peaches' is a sign you'll see at the vegetable counter – because we can't resist the urge. This principle extends in a particularly critical way to the evaluation of products in the design stage. The tactile dimension, the 'use' and 'performance' dimensions, in fact everything about the reality of a product is affected by a complex grouping of stimuli and input via all senses. Hence the importance of model making, mock-ups and prototypes.

Against this, a person's expectations of a product are fuelled by the full weight of his or her own accumulated experience. If we are to match or exceed these expectations it becomes obvious that the richness of sources of material information, material properties, strengths and weaknesses, etc. are fully reflected in the input to the design process. It follows that the 'feel' (what could broadly be termed the 'functioning' of any object) must form an essential part of a designer's background experience, and ongoing research.

The 'feel' of a product may be remote, such as in using a VDU or an instrument readout, or it may be very direct, such as in where a surgeon uses an item of specialized surgical equipment such as a cranio-blade (used in brain surgery). Product 'feel' extends to the full sensory range including memory and 'past–present–future' indications.

User feedback can even become a more significant design challenge than the actual product, since there are situations where what guides a person's 'use' of something which requires skill is the mental model they carry of the process rather than the thing itself. The most obvious example is riding a bike, where no matter how carefully you steer or pedal, riding it is not possible until you get the 'idea' of riding a bike. Other examples include the operation of machinery or using computer software, or using tools. Paradoxically, you can only learn how to do something by doing it first.

Exercise 2.4.1a

Design and make exercise

Make a drawing of a simple object that needs constructing or fabricating and then get someone else to make 'it' from your drawing without asking you for any explanation. See if the two things bear any relationship to each other. When the maker has finished, designer and maker should discuss the outcome and the processes that each went through.

Exercise 2.4.1b

Presentation exercise

Choose a designer or manufacturer and an example of their work which interests you and prepare a 10-minute presentation about your topic to a group audience. Identify features in your presentation that will help your audience understand the points you wish to put across and why you have chosen the example.

2.5 Models and design tools

Model, mock-up or prototype?

- A 'model' represents a concept in some way.
- A 'mock-up' is a quick approximation of some aspect of a design for evaluation.
- A 'prototype' is a first version of a finished product prior to the production process being 'ramped up', and it defines the processes as well as the product.

Appropriate modelling

A 'macquette' really means a quick 3D sketch study, to help think about an idea. 'Model' has many meanings for the designer, from a small-scale 3D representation to a full-size one-off. Scale models are unbeatable for allowing someone else their own experience of, say, an interior design such as an exhibition. Accurate visual models or 'appearance models' can be used to evaluate only the appearance of a product. A 'functional model' may operate as intended but not deal with the appearance. A common example of this is with consumer electronics where the first prototype is a tangle of wires, microchips and bits of keyboard and display components, the interrelationships must be sorted out before a rationalized and integrated product is designed. From the buyer's point of view they see a product. From the point of view of the manufacturer it may be 2763 components. The designer can model any, or some, or all of the attributes of a product for evaluation purposes. Indeed, it is usually vital to be able to model different product attributes separately in order to optimize a design or indeed to understand the issues involved. This understanding of appropriate modelling will increase the designer's and engineer's insight and provide a source of creative input in its own right.

For example, the form of a hand-held drinking vessel may be 'modelled' the same but translated into five different materials such as aluminium, glass, wood, ceramic, and polystyrene. Although the form is outwardly the same in each case, the tactile experiences of using each of these is dramatically different and nobody will have any idea about the sensory correlation between the form, the use and the material until these examples have been tried.

Prototyping is an advanced stage of modelling, and one which establishes or fixes the product attributes (aesthetic and technical), as well as the processes by which it will be produced in volume. Prototyping is always actual size, and commonly requires a high level of skill to manipulate materials to a very high standard of accuracy. Traditionally, and commonly at the present time, prototyping relies upon a combination of craft and design skills of the highest order, at the point of synthesis of design intention and material possibility.

Rapid prototyping

Rapid prototyping (RP) technology has revolutionized the business of pre-production evaluation of many components and products, since various ways now exist to manufacture one-off complex forms direct from 3D computer models in a variety of materials or as waxes for casting. All RP techniques rely on encoding a file of three-dimensional data derived from a virtual prototype. The '3D' (i.e. 3D illusion of a form projected onto a flat screen) image can be regarded as one type of output from the mathematical model, and the RP object another. Techniques for processing such a model into tangible materials all rely upon the concept of slicing the object into layers, defining the layer, then putting them together in sequence. The 'cheap and cheerful' approach uses layers of paper of the same thickness as the sampled slice literally stacked together to make a laminated 'object'. Imagine an apple sliced into a thousand layers, each slice being printed as an individual outline on a sheet of paper. With a thousand such printed layers cut out and stuck back together in the correct sequence, we have recreated an apple form but now in solid paper, albeit with a slightly 'jagged' surface which is a function of the thickness of the sampled layer. With 40 000 layers you probably would not notice this resolution effect. Other RP processes use ways of curing progressively a resin object out of a liquid resin bath, or of sintering materials in granular or powder form. 'Sintering' means binding together. 'CAM' or computer aided manufacture is a closely related technology for making full-size components with computer operated process machines such as milling machines, routers, welding robots, etc., working in two or three dimensions of space. An interesting contrast with the 'sampling' approach of a mathematical data set in the case of RP is where, for example, a skilled spray painter can have his or her expert movements mapped in 3D space by wearing a data glove. The movement sequence thus mapped can be then transferred to robot spray arms.

A consequence of rapid prototyping technology is that since the digital 'file' of information can be exchanged anywhere, the 3D 'printer' can be remote from the computer that holds the file. This will lend a new dimension to distance collaboration, as a designer can 'produce' an object on a remote device.

Design tools for communication

All forms of idea generation involve various forms of communication, both 'communicating to yourself' and 'communicating with others'. The tools used to do this cover the entire range of what is available for presentation purposes, from a chalk or pencil through to full-scale real-time simulation on computers and full-size models or prototypes of products or their parts. Such idea communication can include either separately or together things from the following list:

- Freehand drawings.
- Measured drawings.
- Working drawings or drawings which specify details.
- Two-dimensional or three-dimensional drawings created on a computer.
- Mock-ups.
- Sketch models or macquettes in many materials including clay, wood, paper, plastics and plastics foams.
- Virtual models, i.e. 3D models or renderings generated on computer.
- Photography.
- Presentations of images supported by text or voice.
- Briefing papers, messages or documents.
- Illustrated presentations which can be mailed.

Conclusion: learning through paradox or, doing more with less

The maxim 'doing more with less', although seemingly admirable, can lead to absurdities if not considered in depth. For example, by making functional items such as furniture out of less and less material a point can be reached where such items become regarded as rubbish by the users – the 'perceived value' has been degraded to a stage that encourages an attitude of contempt in the user, which further shortens the life of such an object. This can happen when the conventional appearance of one technology or value system is carried over uncritically to new processes; for example, fibreboard furniture covered in 'pictures' of real wood looking like watered down Chippendale. (Chippendale was an influential furniture maker of the eighteenth century whose name became synonymous with a certain elaborate decorative style.) If we looked at this collection of ingredients afresh and took its elements as value-free processes, i.e. reconstituted natural polymer, photographic imaging, functional object, we could get an entirely fresh starting point for design. Probably the most influential European designer who showed the way forward with this approach at the beginning of the twentieth century was Gerrit Rietveld. Rietveld created furniture using (for example) riveted aluminium airframe techniques being researched as replacement for the wood construction used in biplanes. In this way he created domestic objects with a transfer of aesthetic values fifty years ahead of the time. We now take this approach for granted.

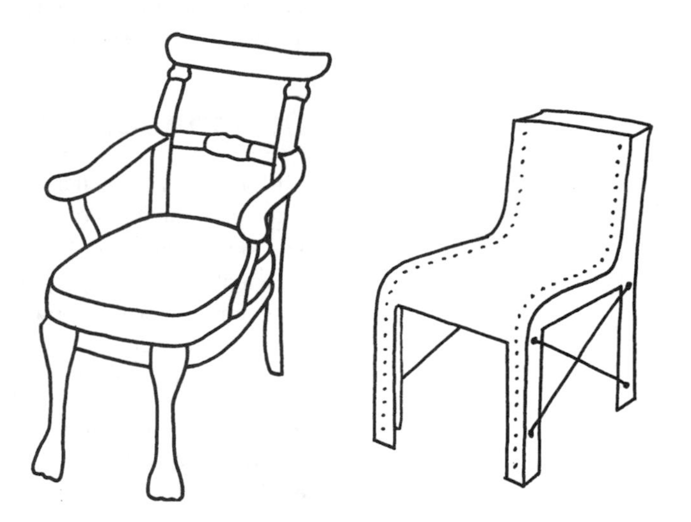

Figure 2.5.1

A good example of 'doing more with less' concerns the use of flax (a harvested natural fibre related to cotton) to make moulded automobile components. It is an example of using a natural polymer to do the job of what was previously that of a synthetic polymer in the production of interior components in vehicles. In this example the 'less' referred to concerns the energy consumption involved in manufacture, and the bad environmental impact of the disposal of synthetic polymer materials. The 'more' refers to the regenerative effect of developing a renewable resource, in this case a natural polymer sourced from plant fibre, that boosts effective land use and provides employment in an area of work undergoing rapid changes (agriculture). Do you mind if your car interior is formed from flax or glass reinforced polyester resin?

Instead of masquerading as things they are not, it is arguable that low-grade products should be and look what they are, and celebrate the fact. This is the only way that a level of honesty can be achieved with product design which addresses our consumer culture head on, rather than perpetrating the pretences which are a hangover from an age which did its best to emulate 'luxury' products with mass processes. 'Doing more with less' can be better interpreted as improving our understanding and appreciation of different types of product existing in different contexts and with widely varying attributes as is appropriate.

One of the paradoxes thrown up by industrial culture which has now reached crisis proportions, is that many of our long-term use products like buildings and furniture are more readily recyclable than short-term products. This is acutely so with very short-term products such as computers and food products, i.e. processed foods dependent upon packaging.

The fact that a so-called disposable container can last longer than a house is a feature of our consuming culture that can no longer be ignored, and it is in a revised relationship between design, business structures, and we as users, that the way out of this can be developed. Materials themselves are a story involving energy, long- and short-term consequences, and design development concerning material properties. There is an ecology for everything – a bottle, a hospital, a bird, and ourselves. By extending our awareness to the circumstances and the context for what is being designed, any contrasts between what we assume, and what could be, can be explored and new knowledge brought to bear. This is one of the essences of design, and what it indeed signifies to the user. The designer, in reconstructing this mixture of processes, aspirations, and messages, acts as a powerful reinventor of our material culture and what it signifies for us.

Problems 2.1.1

(1) Identify five ways in which we interact with products.
(2) What is the most effective way to think of ergonomics?
(3) What are some of the key aspects of a design 'brief'?
(4) Why are aesthetics significant?
(5) Why can't you design things effectively only from a visual point of view?
(6) What is the purpose of a critical review, or critique?

3 Properties of materials

'Good design works. Excellent design gives pleasure.'
(M. F. Ashby)

Summary

The materials that are available and their properties are fundamental to the design and development of all products. In this chapter, we will begin by looking at the periodic table in Section 3.1. We can then compare this arrangement of all elements against the nature of materials we are familiar with. In Section 3.2, we will proceed to look at the nature of bonds between atoms and molecules, which lead from the periodic table, and hence their qualities. The arrangements that form will determine how the materials will behave in particular environments. In Section 3.3, we look at specific factors such as the mixing of different elements and the reasons for disorder in real materials. Finally in Section 3.4, we examine some of the qualities or properties of materials, as they are determined, such as density and electrical and mechanical behaviour.

Materials and design

Every object that is either man-made or existing in nature is dependent on materials and the way they behave in their environment. Materials and the forces of nature combine to give us the structures. Materials and energy combine to give us different forms of matter.

In this chapter we will look at how materials are put together, their classes, the properties they exhibit and how the designer is able to use these properties to make components. Whether it is a crane hook designed to carry heavy loads or ink particles deposited on a toffee wrapper, the designer needs to understand what is expected of the materials and how different materials behave before an effective choice can be made.

The natural world is full of examples of excellent design with materials. The wings of a bird provide lift and are flexible enough to change shape and propel the bird forward. They can be rigid also to allow the bird to glide. The first attempts to make a man-made flying machine began with the wing of a bird. Much refinement led eventually to the first powered flight by the Wright brothers. Daedalus and Icarus in Greek mythology are said to have used wings held together with wax to fly successfully. As the story goes, Icarus, in his enthusiasm, flew too close to the sun and the wax melted, sending him plunging to his death without his wings. This shows that all materials have their limits and design must take this into account.

Just as in the case of a bird's wing, the single materials can rarely meet the needs of modern engineering applications. It is most likely that metals will be in the form of alloys such as steel or even composites where at least two different materials are combined to give the right properties.

Objectives

At the end of this chapter, you should be able to:

- show how the materials we recognize have come into existence from the arrangement of elements called the periodic table;
- show how the forces between parts of atoms determine the arrangement of atoms into structures and the types of structures that can form;
- explain how we come to define properties and how they are determined;
- explain how materials behave when subjected to external loads.

3.1 Classes of materials

Solid materials used in the manufacture of products from buckets and bottles to cars and buildings can be classified broadly into metals, ceramics (including glasses) and polymers. We could add advanced materials such as coatings (thin layers applied to solid surfaces), semiconductors (with special electrical properties), adhesives (materials for joining different solids) and composites (materials made up in selected combinations to give the best combination of properties) to this list. The classification is based on their nature and the way they are processed. All materials have to be created by refining either from ores or oil extracted from the earth or from plants and animals. This process can involve a great deal of energy and invariably, most materials tend to return to their natural state if they are allowed to.

Ceramics are hard and strong but brittle. They can also take very high temperatures. They can conduct heat but are excellent electrical insulators. Examples in nature include minerals and products such as stone and even bone. Most do not melt and so cannot change from solid to liquid as the temperature is increased. Instead, as the temperature is increased, they react chemically with the atmosphere and degrade, losing their properties. Glass is a subset of ceramics but is often treated separately especially since glass appears to melt. Strictly speaking, glass does not have a melting temperature since it never actually switches from solid to liquid. Instead, as the temperature is decreased, its viscosity increases until it appears to resist flow. At this point, it is regarded as a solid. Modern ceramics such as silicon carbide are used where components have to operate at very high temperatures and where low friction and wear as well as chemical stability are important.

Metals are neither as hard nor as strong as ceramics and are generally ductile. They have a melting temperature, and so change from the solid state to a liquid state when the temperature is raised to a precise level. This level is generally not as high as the temperature at which ceramics degrade. Metals are excellent conductors of heat and electricity. Ferrous metals generally make very good permanent magnets. Metals need to be extracted from naturally occurring ores as they react with the atmosphere relatively easily. Once refined and converted into products, they generally need to be protected from the environment, for example by coating. The most important metal is steel, which is used in structural applications. Other metals such as copper and aluminium are widely used for their electrical and thermal properties. In more specialized aerospace, biomedical and superconductor applications, where low density, good biochemical compatibility or low electrical resistance are required, materials based on titanium, niobium and yttrium are used. The high cost of refining metals makes their recovery at the end of life of a component very important.

Polymers have the lowest hardness, strength and stiffness when compared with metals and ceramics. They are again subdivided into thermoplastics (that melt as glass would), thermosetting polymers (that do not melt like general ceramics) and elastomers or rubbers (that may or may not melt but are elastic to very high extension). Thermoplastic polymers are ductile and have low melting temperatures. Thermosetting polymers are stronger, but brittle and degrade like ceramics but at much lower temperatures. All polymers make very good insulators against heat and electricity. Naturally occurring polymers include wood and fibres such as cotton and proteins. Modern industrial polymers are largely derived from crude oil and hence can generally be burnt to recover energy.

Many naturally occurring components contain combinations of materials designed to satisfy different functions. In the same way, many materials have been combined. For example, steel is a combination of iron and carbon, with a few other additions. Such a combination is called an alloy. Hence steel is an iron alloy. The resulting structure has the ductility of iron but also the strength that comes from the small additions. Oddly enough, the strength comes from disrupting the perfectly ordered structure of the iron atoms. Compare this with copper where the perfectly ordered structure has low strength but very high ductility allowing copper bar to be drawn, without breaking into very fine wire. If tin is added to copper, its ductility decreases, but the resulting alloy, a bronze, was strong enough to advance civilization into the so-called Bronze Age. Polymers contain fillers that can improve strength and toughness, the appearance and the ability to mix and flow. For example, you will have come across 'uPVC' window frames. This is a rigid material, which with aluminium reinforcement is able to support glass and provide an effective barrier against the weather (and intruders) for twenty years or so. The chemical name for the polymer is polyvinyl chloride. The 'u' refers to the fact that it is unplasticized. When a plasticizer is added to PVC, it softens the material and makes it very flexible. Hence with a large amount of plasticizer, the same PVC becomes ideal for the sleeve of all electrical cables found in the home. In this case, it is the electrical insulation and the flexibility of the material that are important. All PVC used to be black since carbon black (carbon in the form of a very fine powder) was added to prevent the material degrading in sunlight. Today, other stabilizers are added to allow the material to take on any colour – even brilliant white as in the case of window frames. Ceramics are combinations of basic materials held together by strong forces. The combining of different materials is taken to its limit in the class of materials known as composites. They could be combinations of polymers, metals or ceramics but consist essentially of a fibre and a binder. The fibre provides the key properties and the binder helps to compensate for weaknesses in the fibre such as toughness. An example would be glass fibre reinforced polyester. The glass has excellent strength and stiffness but is brittle and susceptible to mechanical damage of the surface. The polyester, which is a thermosetting polymer, helps to increase the toughness and to create the shape of the component. Composites are truly engineered materials, tailor-made to meet the particular needs of modern applications.

As engineered materials come to the fore in different industries, new classes are being defined. There are semiconductors and superconductors with particular electrical characteristics. There are shape memory alloys that, as the name implies, change their shape under particular conditions, and piezoelectric materials that convert mechanical stimuli into electrical responses. There are also specific material needs for joining of components.

Basis of materials

When faced with the huge range of materials and their properties, the engineering designer may retreat to the safety of relying on an expert to make choices. This is just like referring to a professional translator when having a conversation with, for example, a French speaker. The problem with this approach is that some information can always be lost in the translation process. There comes a point when, to make progress, you need to be able to understand precisely how the chosen material for an application behaves and how that behaviour may be improved. Also as with any foreign language, a little insight into how things work can improve confidence in getting the right answers.

Fortunately, all materials can be traced to elements that can be arranged in a neat order in what is known as the Periodic Table. This is shown in Figure 3.1.1. You would have seen the arrangement before but we are now going to use this to work out precisely why materials behave as they do and indeed what the properties tell us.

All materials are created from elements in the *periodic table*. You will recognize a large number of the elements from their nature but each element has the same basic shape and consists of the same material as you see in Figure 3.1.2. Each element is identified by the number of *protons* in the nucleus of its atom. This is the atomic number, Z, of the element. For example, carbon has an atomic number of 6 and therefore 6 protons in its nucleus. Protons are positively charged particles and so to achieve balance, the atom has to have an equal number of negatively charged *electrons*. Although the electron has a much smaller mass than the proton, its electric charge has exactly the same magnitude as that of the proton. Hence the balanced or neutral carbon atom has 6 electrons. These exist outside the nucleus of the atom in so-called shells that are orbits describing the probable position of each electron. A third particle, the *neutron* also occupies the nucleus of the atom. The neutron does not carry a charge but provides additional mass and forms part of the nucleus of the atom.

There are at least as many neutrons as protons in all elements except the smallest one, hydrogen. Only the number of protons uniquely defines the atom so that if there is one proton in the nucleus, the element will be hydrogen. If there are two protons in the nucleus, the element will be helium and so on. An element can exist with different numbers

	1	2	3	4	5	6	7	8	9	10	11	12	13	14	15	16	17	18
1	1 H Hydrogen 1																	2 He Helium 4
2	3 Li Lithium 7	4 Be Beryllium 9											5 B Boron 11	6 C Carbon 12	7 N Nitrogen 14	8 O Oxygen 16	9 F Fluorine 19	10 Ne Neon 20
3	11 Na Sodium 23	12 Mg Magnesium 24											13 Al Aluminium 27	14 Si Silicon 28	15 P Phosphorus 31	16 S Sulphur 32	17 Cl Chlorine 35	18 Ar Argon 40
4	19 K Potassium 39	20 Ca Calcium 40	21 Sc Scandium 45	22 Ti Titanium 48	23 V Vanadium 51	24 Cr Chromium 52	25 Mn Manganese 55	26 Fe Iron 56	27 Co Cobalt 59	28 Ni Nickel 59	29 Cu Copper 64	30 Zn Zinc 65	31 Ga Gallium 70	32 Ge Germanium 73	33 As Arsenic 75	34 Se Selenium 79	35 Br Bromine 80	36 Kr Krypton 84
5	37 Rb Rubidium 86	38 Sr Strontium 88	39 Y Yttrium 89	40 Zr Zirconium 91	41 Nb Niobium 93	42 Mo Molybdenum 96	43 Te Technetium 99	44 Ru Ruthenium 101	45 Rh Rhodium 103	46 Pd Palladium 106	47 Ag Silver 108	48 Cd Cadmium 112	49 In Indium 115	50 Sn Tin 119	51 Sb Antimony 122	52 Te Tellurium 128	53 I Iodine 127	54 Xe Xenon 131
6	55 Cs Caesium 133	56 Ba Barium 137	71 Lu Lutetium 175	72 Hf Hafnium 179	73 Ta Tantalum 181	74 W Tungsten 184	75 Re Rhenium 186	76 Os Osmium 190	77 Ir Iridium 192	78 Pt Platinum 195	79 Au Gold 197	80 Hg Mercury 201	81 Tl Thallium 204	82 Pb Lead 207	83 Bi Bismuth 209	84 Po Polonium 210	85 At Astatine 210	86 Rn Radon 222
7	87 Fr Francium 223	88 Ra Radium 226																

57 La Lanthanum 139	58 Ce Cerium 140	59 Pr Praseodymium 141	60 Nd Neodymium 144	61 Pm Promethium 147	62 Sm Samarium 150	63 Eu Europium 152	64 Gd Gadolinium 157	65 Tb Terbium 159	66 Dy Dysprosium 163	67 Ho Holmium 165	68 Er Erbium 167	69 Tm Thulium 169	70 Yb Ytterbium 173
89 Ac Actinium 227	90 Th Thorium 232	91 Pa Protactinium 231	92 U uranium 238	93 Np Neptunium 237	94 Pu Plutonium 239								

Figure 3.1.1 *Periodic table of elements*

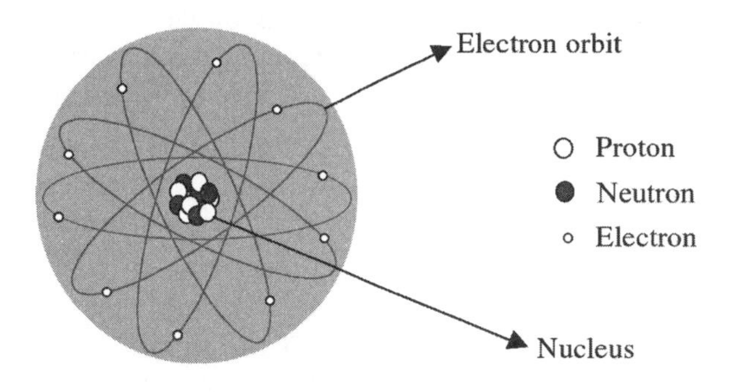

Figure 3.1.2 *Structure of the atom*

of neutrons. These forms of the same element differing in the number of neutrons are called *isotopes*. For example, hydrogen has one proton and no neutrons. There are two other forms of hydrogen called deuterium (with one additional neutron) and tritium (with two additional neutrons). These forms are radioactive, a property of the element which makes them unstable. This is a property that allows one element to change into another by releasing radiation in three forms called alpha (α), beta (β) and gamma (γ). α is a particle consisting of two protons and two neutrons. It is therefore the same as the nucleus of a helium atom. The β is just an electron and the γ is high-energy electromagnetic radiation of the same sort of frequency and power as X-rays. Not all the forms occur in any given reaction, but together, they help to transform the nucleus of one element into the nucleus of another. The α and the β help to adjust the number of protons and neutrons while the γ helps to adjust the energy level of the nucleus. It is interesting to note that the electron acts within the nucleus by converting a proton into a neutron or vice versa. The total number of protons and neutrons in the nucleus gives a measure of the mass number, A. Radioactivity is outside the scope of this book. It is, however, important in areas such as medical physics and nuclear power generation.

The periodic table

The arrangement of elements into a periodic table, shown in Figure 3.1.1, was made long before all the elements were known. However, many elements had also been known by their chemical properties long before the periodic table was first proposed in the nineteenth century. In this section, we will look at how the elements come to be arranged as they are. This is worth studying since, the position of elements will give us clues as to why they have particular properties.

In the arrangement, elements with similar chemical properties are shown in columns, called a group. The atomic number increases along each row, called a period. Most importantly, the number of electrons in the outermost shell increases from one in group IA at the left-hand end to the maximum possible in Group 0. As you can see, the first period has just two elements, the second and third have 8 each and the fourth and fifth have 18 elements each. In the sixth period, there are 32 elements although the table is drawn to show 18 slots, simply because the table would become much too wide to display. The seventh period also has 32 elements although the last 11 do not occur in nature and are generally very unstable. They quickly change, or decay into other more stable forms by the radioactive reactions described in the previous section. From this arrangement, we can see that the elements of the first period can have up to two electrons.

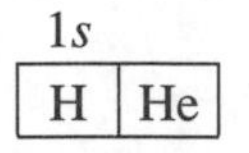

1*s*	
H	He

They form the first shell, designated by the principal quantum number, $n = 1$. Elements in the second period contain a second shell ($n = 2$) which can hold up to eight electrons. Each electron has a different state and so the electrons are organized in pairs in different orbits or subshells. These subshells are labelled *s* (for spherical) and *p* (for polar).

2*s*			2*p*					
Li	Be		B	C	N	O	F	Ne
Na	Mg		Al	Si	P	S	Cl	Ar

Hence in period 2, electrons of lithium (Li) and beryllium (Be) occupy the 1*s* and 2*s* orbits filling the 1*s* state before the 2*s* state. In beryllium, the two electrons in each orbit, or subshell are distinguished by opposite spin. The third element in period 2, boron, will have its first four electrons in 1*s* and 2*s* subshells and the fifth electron in the 2*p* subshell. The *p* subshell is further split into three subshells. Within each, a pair of electrons with opposite spins exists. The last element in the period, neon, will have a full set of electrons in its second shell. This outer shell is very important when determining how atoms react with each other.

Once the second shell is completed, a third shell is filled according to the same pattern as the second shell with sodium (Na) having just one electron in its third shell and argon completing the period with 8 electrons in its third shell. Period 4 represents the filling of the fourth shell. With the filling of the fourth shell, the table widens further as another subshell (*d*) is added. The order in which the subshells are filled and therefore the place in the periodic table is governed by some simple rules. Remembering the notation also helps to identify the electrons that will take part in any reactions described later in the chapter.

The subshells and shells are filled according to the pattern shown in Figure 3.1.3 where the arrows indicate the sequence. The order is therefore,

1*s*, 2*s*, 2*p*, 3*s*, 3*p*, 4*s*, 3*d*, 4*p*, 5*s*, 4*d*, 5*p*, 6*s*, 4*f*, 5*d*, 6*p*, 5*f* and 6*d*.

The subshells filled in any period are shown in Figure 3.1.4 by the 'necklaces'. Each period begins with the filling of the *s* subshell and ends with the filling of the *p* subshell.

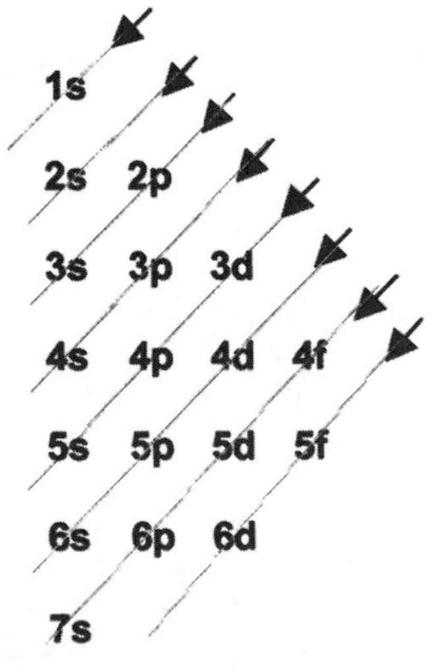

Figure 3.1.3 *Sequence of filling of the electron shells*

Activity 3.1.1

(1) How many electrons will there be when all the subshells up to 6*d* have been filled?
(2) Write down the electron arrangement in the element titanium.

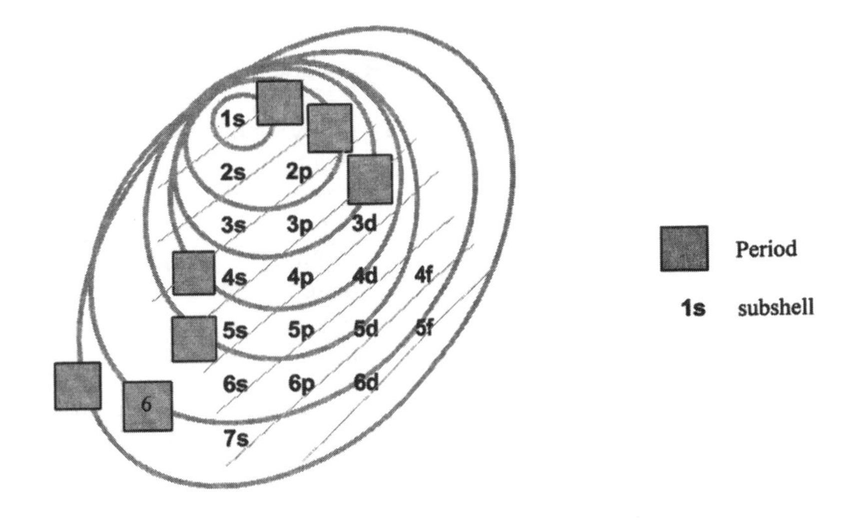

Figure 3.1.4 *Relationship between periods and the filling of electron shells*

Chemical and physical properties

As can be seen now, the elements in the right-hand column (group 18) of Fig. 3.1.1 all have complete shells of electrons. They are all gases that do not react with any other materials, called noble or inert gases. At the left-hand side of the table, elements in group 1 all have just one electron in their outermost shell, existing in the *s* subshell. They are all metals apart from hydrogen. In fact hydrogen is only placed in group 1 because of its single electron. All the other metals in group 1 react very easily in water, producing hydrogen and so are called *alkali metals*. They also have low melting points.

Elements in group 2 have two electrons in their outermost shell, completing the *s* subshell. They react to a lesser extent with water but can produce hydrogen from the reaction and are generally called *alkaline earth metals*.

Groups 3 to 12 begin with the period 4. In this case, the number of electrons increases as the 4*s* and 3*d* subshells fill. The same groups fill in period 5 with the filling of the 5*s* and 4*d* subshells. Groups 3 to 11 contain the transition metals, so called because they form the bridge between the reactive metals in the first two groups and the relatively stable metals in groups 12 to 14. The *d* subshell is being filled as from groups 3 to 11 and completed by group 12.

Groups 13 to 18 are filled with the corresponding *p* subshell electrons. Elements in group 17 are the *halogens* which are reactive gases or liquids. In some cases the next *s* subshell and the *d* subshell exchange electrons, showing that they are very close together in terms of their energy. Many of the elements in groups 12 to 14, such as silicon, have semiconductor properties that make them ideal substrates for the electronic industry.

Review – the importance of the electron

Our review of the periodic table is aimed at understanding where the properties of materials come from. Having established the order in which the electrons are arranged in the atom as we progress through the groups and periods, we come across some basic principles.

- Every electron has a different position within the atom.
- Electrons are arranged in shells and subshells.
- The maximum number of electrons in each shell and subshell is fixed.
- Atoms with complete shells are the least reactive and exist as gases under ambient conditions.
- Atoms with just one or two electrons in their outermost shell react easily in water. They are all metals with the exception of hydrogen.
- Atoms with an outer shell that is one electron short of being full also react easily. They are mostly either gases or liquids under ambient conditions.

3.2 Bonding in solids

As we have established earlier, atoms are neutral when the number of electrons and protons are the same. There is no net electrical charge and since their mass is very small, they could only attract by gravity when the numbers of atoms are as great as we would find in astronomic objects such as our sun. Under ideal conditions there would be no other reactions and all the forms of material we are familiar with could not exist and therefore life forms would not exist.

Fortunately for us, as we have already implied, certain materials are very stable gases and others are highly reactive. This is because the tendency to have complete outer shells is another driving force that all atoms have. We should now also ask ourselves why it is that electrons orbit the nucleus anyway. Imagine swinging a ball on a string about you. The tension in the string represents the inward pull or the centripetal force acting on the ball. If this is removed by cutting the string then the out-of-balance centrifugal force will cause the ball to shoot off away from you.

In the atom, the negative charge of the electron is balanced by the positive charge of the proton. We can consider the electron similarly remaining in orbit (an imaginary shell) around the nucleus. As the number of electrons increases, they will themselves repel each other and so need to be stacked into subshells and shells where the repulsive force between them is balanced by the attraction from the protons in the nucleus. The force of attraction and repulsion is dependent on the distance between the two particles according to the *inverse square law of distance*. Accordingly, the force, F, is related to the distance, r, between the particles by,

$$F = \frac{k}{r^2}$$

where k is a constant.

Therefore the electrons further out from the nucleus will be less well attracted to the nucleus. They are also further shielded by electrons nearer the nucleus. They are therefore more easily removed from the atom. This balanced state is one where the potential energy of the atom has a minimum value. To change this state, energy will need to be given to the atom.

In the same way, when two atoms are brought together well away from any other, it is possible for the negative electrons of one to attract the positive nucleus of another since the electrons are free to move about their 'home' atom but in designated shells. This is illustrated in Figure 3.2.1.

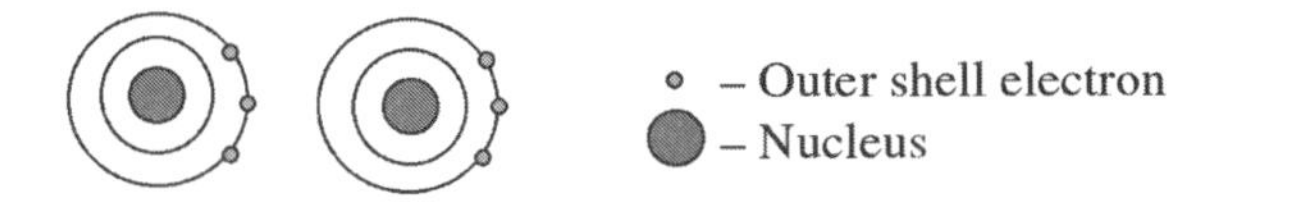

Figure 3.2.1 *Attraction between atoms*

This force of attraction is responsible for the relatively weak forces that hold a liquid together. When we combine the tendency for electron shells to be completed with this electrostatic force between atoms, we come upon a range of very important bonding mechanisms that give rise to the huge range of interesting materials that we are familiar with.

To form a solid, atoms must be attracted to each other in such a way that they lock each other into place just as the electrons are locked into place around the positive nucleus of an atom. The basic unit in the solid could be an atom as in a metal or compounds, which are combinations of atoms. The forces that form the solid metal and the compounds all involve the behaviour of the outermost electrons in each atom. We will now look at the different ways they react and the bonding that will result in each case.

Ionic bonding

Sodium chloride (NaCl) is often given as an example of ionic bonding. We recognize this better as table salt. It will dissolve completely in water and yet it is a solid when water is removed, forming highly organized or crystalline structures. It consists of a regular arrangement of sodium and chlorine. This is possible because, the single electron in the outermost shell of the sodium atom can be transferred to the chlorine atom, which is one electron short of a complete outer shell. The transfer takes place by mutual agreement so that both atoms end up with complete outer shells. The sodium mimics neon and the chlorine mimics argon in terms of complete electron shells, although there is now an imbalance as far as their electrical charges are concerned. The sodium atom is now one electron short and so has a net positive charge. It is therefore referred to as a positive ion and is written as Na^+. Likewise the chlorine atom now has one electron too many and becomes a negative ion, Cl^-. To achieve electrostatic balance, these sodium and chlorine ions arrange themselves so that there is no net charge in the solid in an orderly, crystalline manner. Therefore each sodium ion will be surrounded by a chlorine ion and each chlorine ion will be surrounded by a sodium ion. A two-dimensional picture of the structure is illustrated in Figure 3.2.2. The net energy of each ion is now less than that of the original atom because it is shielded by ions of the opposite charge and so the structure is more stable. This is called ionic bonding. This type of bonding is non-directional; that is to say, it has the same strength in all directions. That is to say, it has the same strength in any direction around the ion. We say that the material is isotropic because it has the same

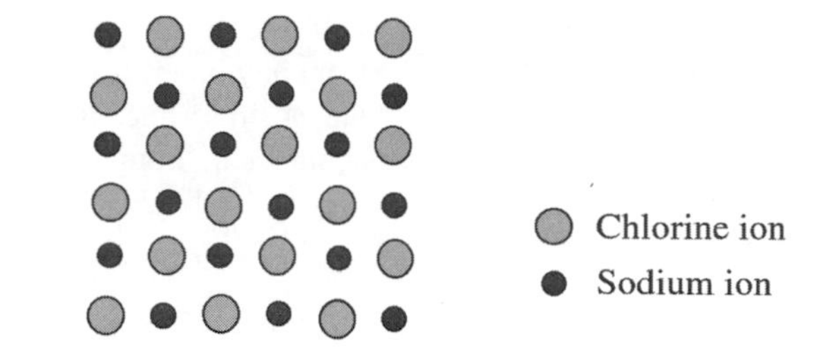

Figure 3.2.2 *Arrangement of atoms in ionic solids such as sodium chloride*

value for any property in all directions. For this reason, the structure that forms will look the same in all directions. The atoms always exist as ions.

All group 1 atoms can form positive ions and will combine with negative ions of group 17 elements since the transfer of one electron is relatively easy. Group 2 elements can also contribute to ionic solids by the transfer of their two outermost electrons. Group 16 elements can do the same by taking on two electrons. In all cases, a positive metal ion combines with a negative non-metallic ion. Ionic bonding is also possible with group 3 and group 15 elements with the transfer of three electrons. However, as we progress inwards into the periodic table from group 1 and group 17, the energy needed to extract an electron increases and so electrons are likely to remain in their host atoms.

Activity 3.2.1

Write down the ions formed and their charge in the bonding of the following pairs of elements: (a) Li and F, (b) K and Cl, (c) Be and Cl, (d) Ca and Br, (e) Mg and S.

Covalent bonding

When two atoms of hydrogen come close together, their valence shells can merge and their respective electrons will effectively attract both nuclei. They become trapped between the nuclei and pair up to complete the shells of each atom as illustrated in Figure 3.2.3. This process needs less energy than the formation of ions as neither electron is completely extracted from its atom. The resulting bond between the atoms is called a covalent bond. It is a strong bond and highly directional since the electrons will be locked between the atoms. The electrons of the two atoms have effectively become shared between them.

It is important for elements involving hydrogen and the non-metallic elements in groups 14 to 17. Gases such as hydrogen, oxygen, nitrogen,

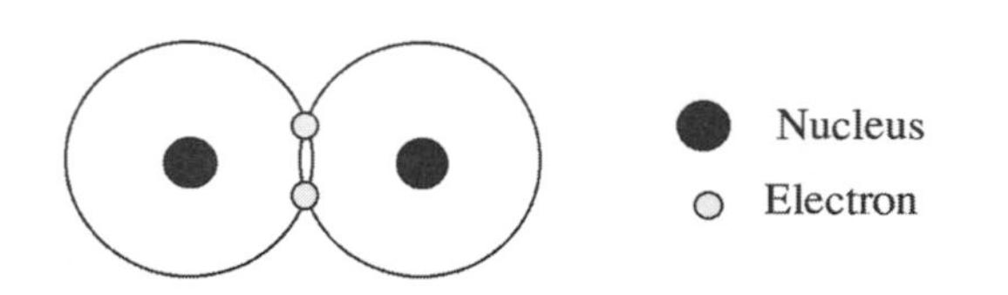

Figure 3.2.3 *Covalent bonding in H_2*

chlorine and fluorine exist as molecules formed by covalent bonding. Hydrogen gas, for example, is formed from molecules made up of pairs of atoms, designated as H_2.

The water molecule H_2O is also an example of covalent bonding. In this case, an electron from each of the hydrogen atoms pairs up with an electron from the oxygen atom to form two strong directional bonds. The direction of the bonds and so the angle between them is fixed by the balance of the electrostatic forces in the molecule. The molecule therefore has a precise shape. Each of the hydrogen atoms now has a full valence shell (Figure 3.2.4). The oxygen atom also has a complete valence shell now.

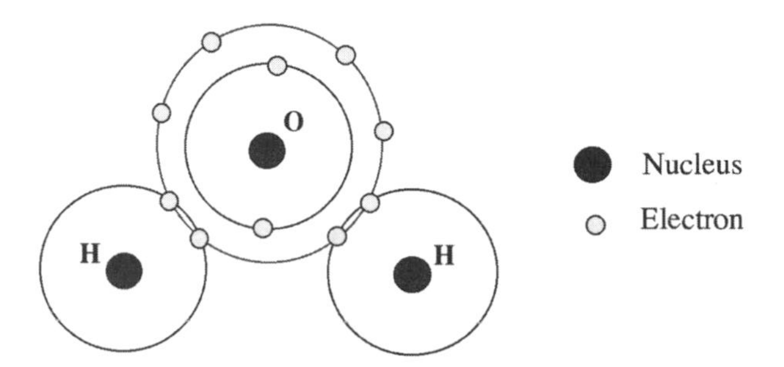

Figure 3.2.4 *Covalent bonding in the water molecule*

Many more complex molecules in nature form by covalent bonding. One example is the molecule methane (CH_4) which is the simplest of the hydrocarbons. In this case covalent bonds are formed between the carbon atom and each of the four hydrogen atoms, completing the valence shell of all the atoms involved. Again the bond direction is fixed. In the next hydrocarbon, ethane (C_2H_6), each carbon atom shares bonds with three hydrogen atoms. Since each carbon needs to take on four electrons to complete its outer shell, the only way that such a structure can be stable is if the carbon atoms share a pair of electrons. Hydrocarbons are very important for the formation of polymers as we will discuss in 'Structure of polymers', page 92.

Activity 3.2.2

Sketch the structures that form in the following examples using the example of Figure 3.2.4: (a) methane (CH_4), (b) ethane (C_2H_6), (c) ethene (C_2H_4) (d), formaldehyde (H_2CO).

Metallic bonding

Most elements in the periodic table are metals by nature. Their valence electrons are loosely bound to the atom so that when a solid is formed with a very regular arrangement of atoms, they become free to drift between atoms. In effect, the valence shells of each atom merge to form one continuous shell across the metal. The electrons in the shell are therefore free to move in any direction within the solid. The remaining positive core of each atom provides an attractive force for the electrons. Hence the valence electrons move freely as a 'sea' or 'cloud' that loosely binds their cores together (Figure 3.2.5). The bonds are non-directional and atoms can be organized into compact structures giving high densities. The freedom of the valence electrons also gives rise to electrical conduction. When an electrical potential is applied across a piece of copper wire, the electrons move in an organized way to give an electrical current. They are also responsible for the metallic lustre seen in reflected light since the electrons in the metal surface will respond to the electromagnetic oscillations that make up the light beam and radiate back the light that surface receives. Because the bonds can form in any direction, metals can maintain their bonds even under extreme deformation as happens when a pure copper bar is drawn into fine electrical wire.

The bonding mechanisms can be mapped onto the periodic table as illustrated in Figure 3.2.6. Whereas most of the elements produce metals, those shaded in groups 1, 2 and 13 have a tendency to give away their small number of valence electrons, and those shaded in groups 14, 15, 16 and 17 tend to accept free electrons to complete their valence shells. Shaded elements outside the metallic envelope can share electrons to a limited extent to form covalent bonds.

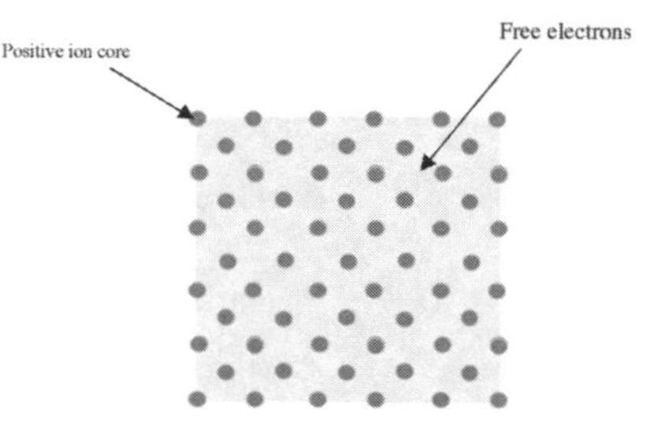

Figure 3.2.5 *Metallic bonding*

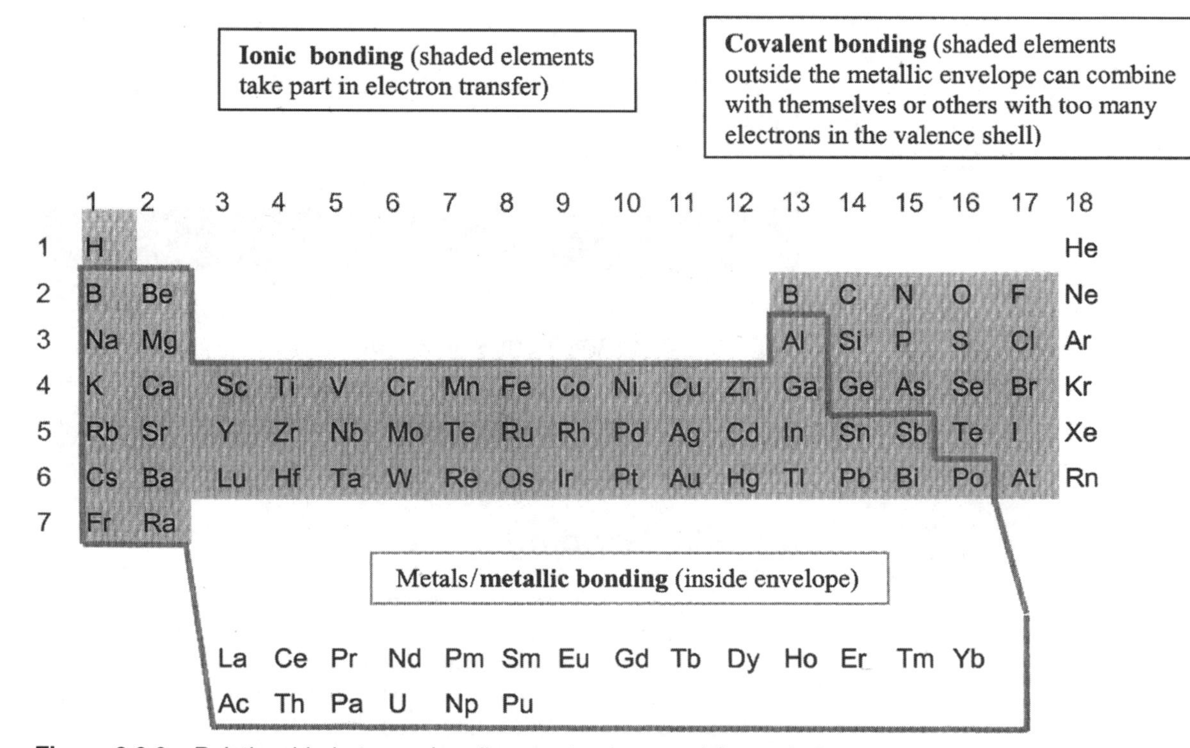

Figure 3.2.6 *Relationship between bonding mechanisms and the periodic table*

Secondary bonding mechanisms

Both ionic bonding and metallic bonding lead to structures such as ceramics and metals respectively. However, a vast number of other materials depend on weaker secondary bonding mechanisms. Water becomes ice, proteins are formed, plants and trees form and man-made polymers are created. These weaker bonds are due to the force of attraction between the positive nucleus of one atom and the electrons of another. The electrons of one atom are never in one fixed place and their position can be altered by the presence of another atom. A good way of looking at this is by considering a spring (Figure 3.2.7). Without a load applied, the free end of the spring will be at one position. If a weight is applied to the spring, its free end will have a new distorted position. If the spring is set to oscillate, the position of the weight will vary between two limits.

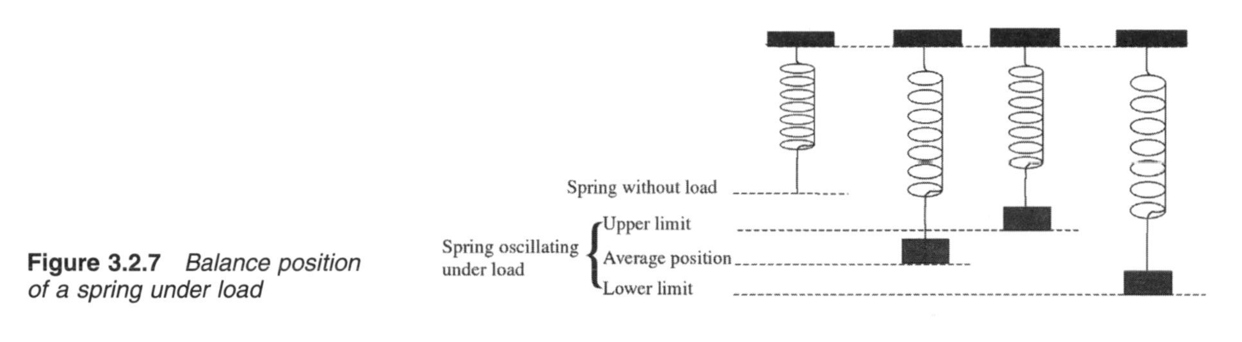

Figure 3.2.7 *Balance position of a spring under load*

In the same way, the presence of the nucleus of a neighbouring atom will cause the electron shell to distort to a new balanced position. This allows electrostatic attraction between adjacent atoms. Such weak bonds are generally called van der Waals bonds.

The example shown in Figure 3.2.8 is a dipole where each atom has a temporary positive pole (dominated by the nucleus) and a temporary negative pole (dominated by the electrons). Many molecules have permanent dipoles because of the distribution of electrons when covalent bonds form. A special case of this is hydrogen bonding. For example, in the case of the water molecule (Figure 3.2.9), the sole electron of the hydrogen atom is locked in position between the oxygen and the hydrogen atoms.

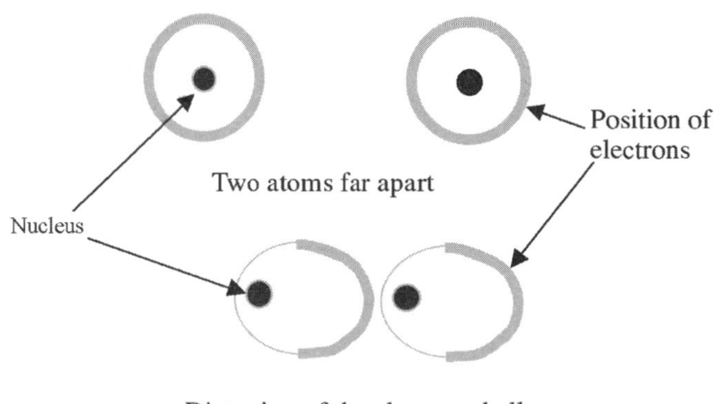

Figure 3.2.8 *Balance position of two atoms*

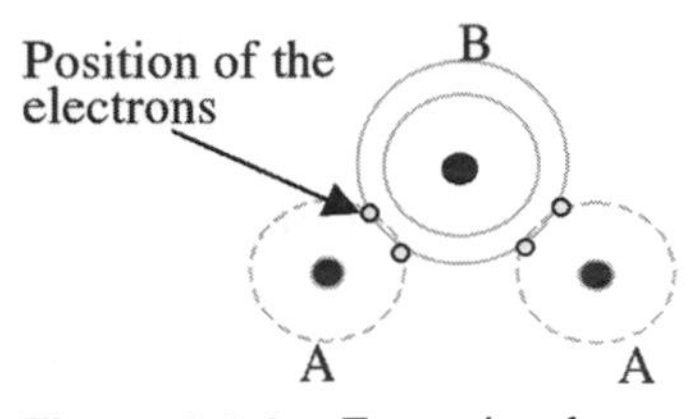

Figure 3.2.9 *Example of hydrogen bonding*

Therefore position A in the molecule will be positive and position B will be more negative (Figure 3.2.9). Since the hydrogen atom only has one electron, other atoms can now approach much closer to the positive nucleus than would be possible when electrons remain to shield the nucleus. Since the force of attraction increases according to the inverse square law, the hydrogen bond is particularly strong. Hydrogen exists naturally in molecular form, where each molecule consists of two covalently bonded atoms. At ambient temperature, it is a gas and remains so down to a very low temperature. This is not surprising since the electrons will be trapped between the positive cores and the exposed positive cores will repel each other. However, hydrogen combusts easily because the positive cores readily react given sufficient energy!

The strength of the bond between atoms or molecules is a measure of the resistance to separate the atoms or molecules. We can see this best in terms of properties such as the melting point and stiffness of a material, and its tendency to react with the environment. Polymers have melting temperatures of less than about 200°C in most cases and ice melts at 0°C since only secondary bonds need to be broken to change the material from solid to liquid. In the case of metals, the metallic bond is a primary bond and so melting temperatures are mostly in excess of 900°C whereas in the case of ceramics, they are even higher.

The strength of bonds, whether primary or secondary, depends on the distance between the positive and negative parts of atoms or molecules. Additionally, the flexibility of the bond and the shape of the molecule will determine how ordered the structure will be. For example, in metallic bonding, the attraction between atoms has the same strength in all directions and so a metal structure can be highly ordered. Such a structure is said to be crystalline. Other solid structures such as silica glass in the form of a window pane have very little order. They are called amorphous solids. Solid thermoplastic polymers such as polypropylene have both ordered (crystalline) and disordered (amorphous) parts.

Crystalline solids

In a crystalline solid, the atoms or molecules form a regular three-dimensional pattern in space. They are said to have *long range order* since the structure will look largely the same throughout a piece of metal. When the material is cooled from a temperature where it is a fluid to one where it is solid, a substantial amount of energy is released. This property is referred to as the *latent heat of fusion* for the material. Its effect can be seen by measuring the change in temperature of the material as it cools as illustrated in Figure 3.2.10. At the point of transition from liquid to solid, the temperature remains constant over a period of time. This temperature, T_m, is the melting temperature of the material. All crystalline materials have an associated melting temperature.

The energy released reduces the kinetic energy of each atom or molecule so that in the solid, there is no translational energy, that is to say they cannot move freely within the solid. Instead the constituents (atoms or molecules) vibrate about an average position. Since they cannot move about freely anymore, the distance between the

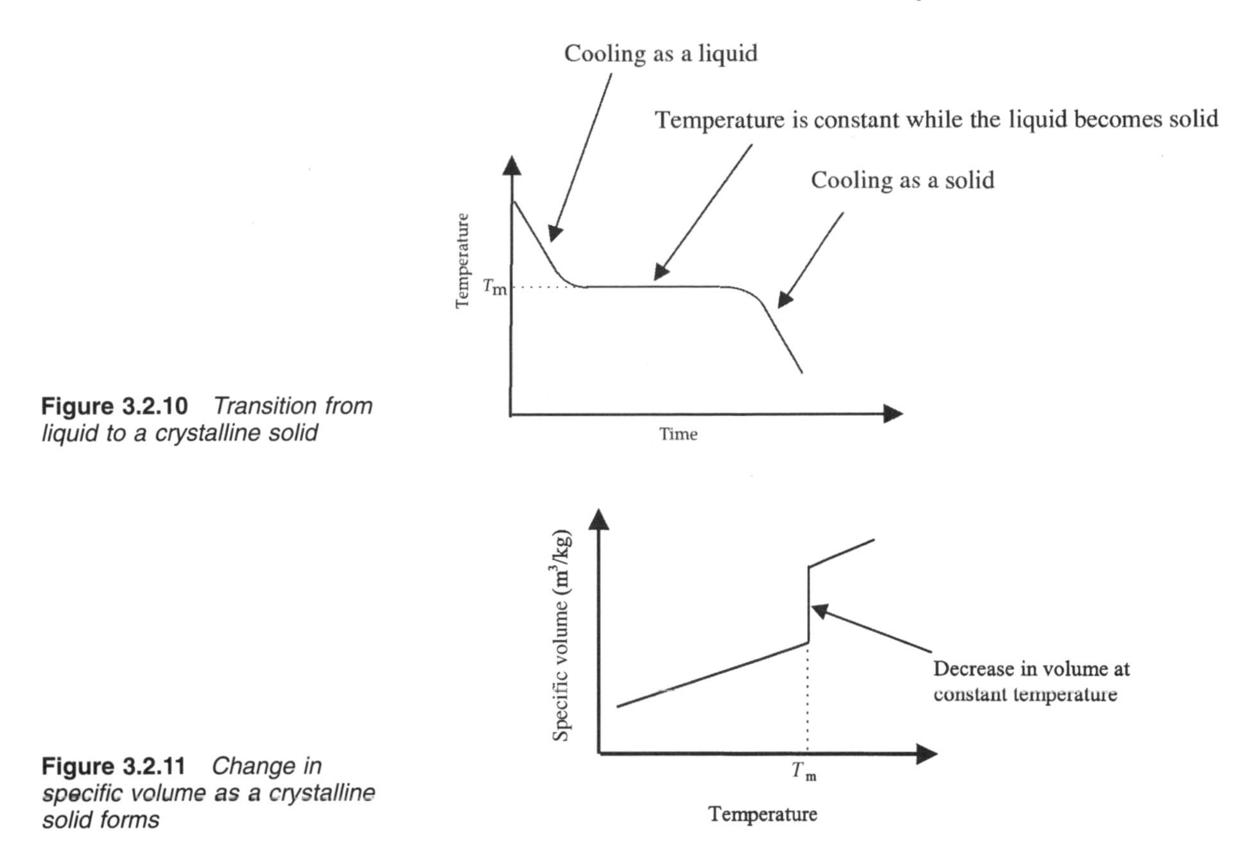

Figure 3.2.10 *Transition from liquid to a crystalline solid*

Figure 3.2.11 *Change in specific volume as a crystalline solid forms*

constituents reduces considerably when the material solidifies. This is measurable as a decrease in the volume of a given quantity of material, or its specific volume as illustrated in Figure 3.2.11.

Since, specific volume = 1/density, there is an increase in density when a material crystallizes.

Types of crystal

The arrangement of atoms or molecules in a crystalline solid can be studied using X-ray crystallography. X-rays of wavelength less than 10^{-9} m will penetrate the space between atoms and from their reflections, the dimensions of the crystal can be established. The smallest repeating unit of the crystal is called a *unit cell*. The knowledge of its structure therefore helps to identify the structure of the material as a whole.

For example, the simplest crystal structure is called simple cubic and is illustrated in Figure 3.2.12(a). In the lattice where the repeating unit is simple cubic, there is an atom at each corner (vertex) of the cube.

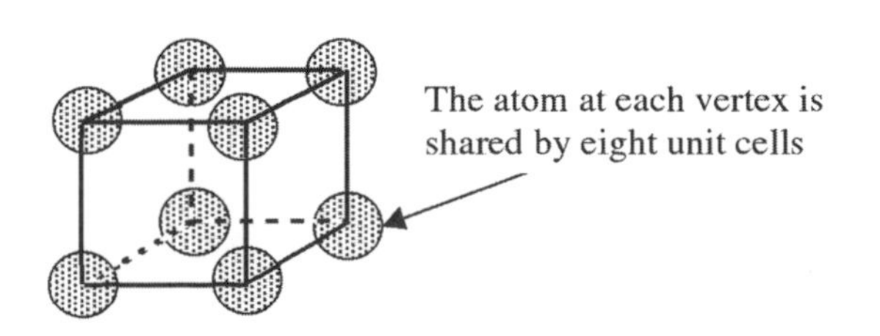

Figure 3.2.12(a) *Simple cubic structure*

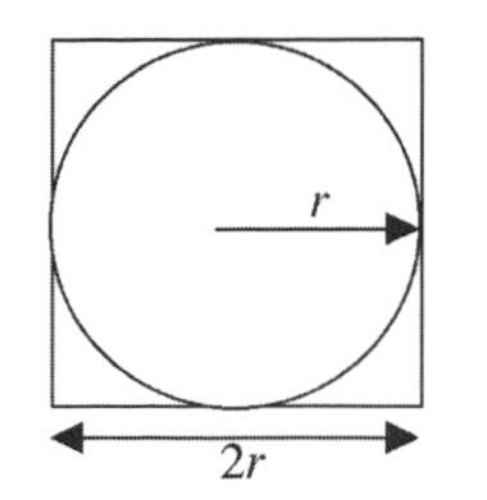

Figure 3.2.12(b) *Determining the packing factor*

Every face of the cube is shared with an adjacent unit cell. Hence only one-eighth of each atom is within the unit cell shown. Thus, the unit cell contains the equivalent of one atom. We can also define a packing factor by comparing the volume of the cube with the volume of the atoms occupying it.

You can see from Figure 3.2.12(b) that in the simple cubic structure,

Volume of the cube, $V_c = (2r)^3$

where r is the nominal radius of the atom and assuming atoms are tightly packed, each side of the cube would have a length = $2r$,

Volume of the atom, $V_a = 4\pi r^3/3$

We can therefore determine the proportion of the cube occupied by the atom by taking the ratio of the volume of the atom to the volume of the smallest cube needed to fit it. This is called the packing factor, (pf). Hence,

$$\text{pf} = V_a/V_c = (4\pi r^3/3)/(2r)^3 = \pi/6 = 0.52$$

That is to say, only 52% of the unit cell is made up of atoms.

Metals form crystals easily because:

(1) Metals are made up of atoms as opposed to molecules. Atoms have simple, regular shapes and therefore can produce tight packing.
(2) The metallic bond is non-directional and therefore the attraction is equally strong in all directions.
(3) Metals are good conductors of heat and so the latent heat removed on crystallization is given up quickly.

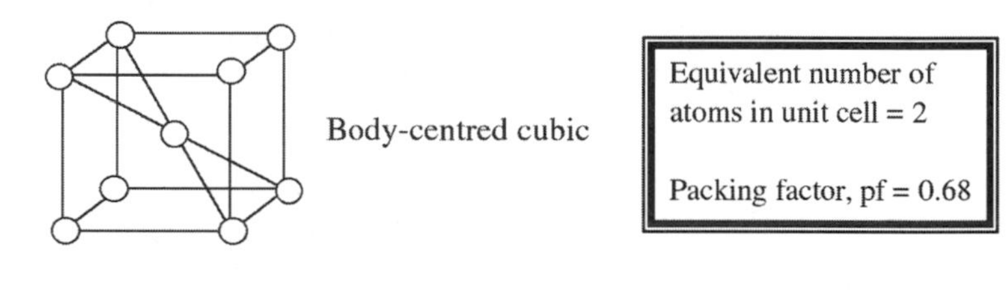

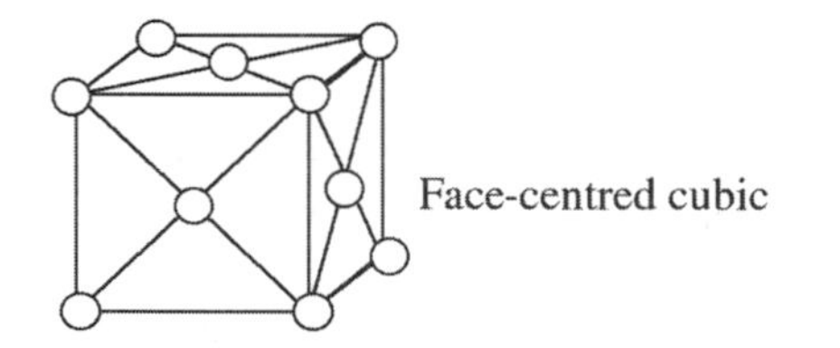

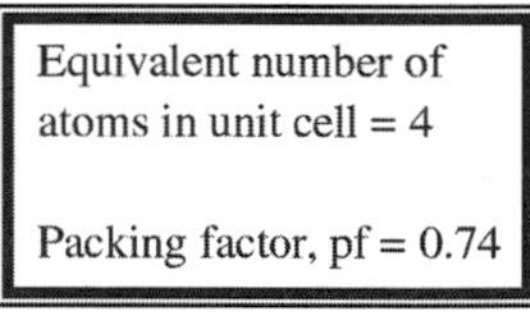

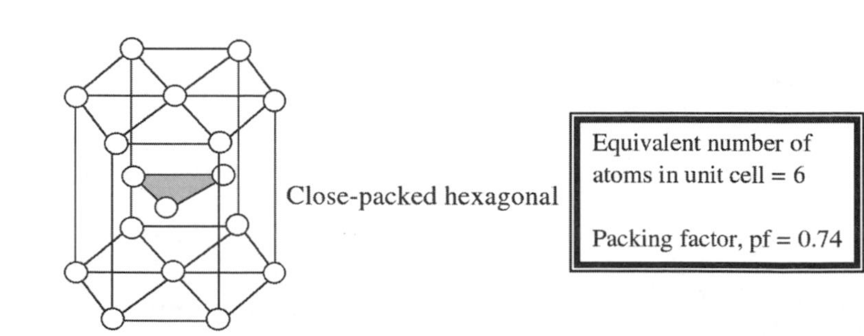

Figure 3.2.13 *Types of crystals and their packing factors*

Most metals form one of three crystal structures or lattices as shown in Figure 3.2.13. In the body-centred cubic (BCC) form, the unit cell consists of one atom at the centre and atoms at each vertex of the cube. Iron, chromium, molybdenum and tungsten all form BCC structures at ambient temperature. In the face-centred cubic (FCC) form, the unit cell is physically larger and contains the equivalent of four atoms. Aluminium, copper, gold, lead and other very ductile metals form FCC structures. The close-packed hexagonal (CPH – also called hexagonal close packed) form has the largest unit cell consisting of two hexagonal faces with seven atoms on each and a triangle of atoms sandwiched between them. The equivalent number of atoms in the unit cell is six. Cadmium, magnesium and zinc are examples of CPH metals. The packing factor for FCC and CPH is the same at 74% but their properties are quite different.

Polymorphism

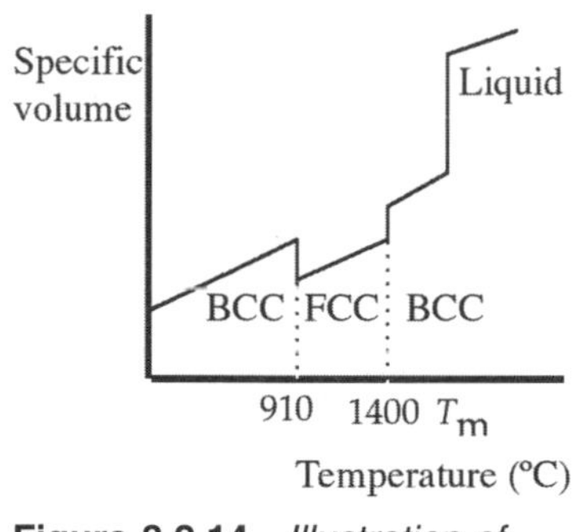

Figure 3.2.14 *Illustration of polymorphism in iron*

A metal is usually identified with one form of crystal structure only. However, some can exist in more than one crystal form, changing their form with temperature. A good example of this is iron which takes the BCC form from ambient temperature to 910°C, FCC from 910°C to 1400°C and BCC again from 1400°C to 1535°C at which point the metal melts. As we have seen from the value of packing factor, FCC lattices are more closely packed than BCC lattices and so the specific volume of iron changes with temperature. This is illustrated in Figure 3.2.14.

As the temperature decreases, the specific volume makes a stepwise drop at the melting temperature, and again at 1400°C and at 910°C. The change at 910°C is particularly important since the volume needs to increase within the solid. This change is exploited to great effect in the heat treatment of steels to control their properties. The change in volume has an impact on the nature of the structures that form in steels. The FCC form can dissolve more carbon than the BCC form when steel is made. The BCC form can become brittle in some conditions whereas the FCC form does not become brittle. The BCC form is magnetic whereas the FCC form is not.

Crystal properties

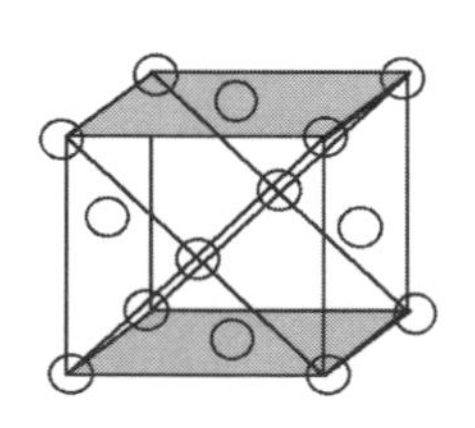

The surface between each pair of parallel lines represents some of the planes of symmetry in the crystal

Figure 3.2.15 *Planes of symmetry in FCC crystal*

The properties are dependent on the freedom that atoms have to move within the structure when under load. Imagine a pack of cards. If the pack was pulled parallel to the face of the cards, they would withstand a high force. The easiest way to deform them would be to slide cards relative to one another and parallel to their faces. If each face represented a plane or sheet of atoms within the crystal, then under load, we can imagine that they could slide past each other easily and thus deform the structure of the piece of material. In a crystal, these sheets are imaginary but can be formed by looking for planes along which the atoms are regularly arranged. The FCC structure has the most number of these planes of symmetry – parallel to each face of the unit cell as well as across the diagonals of the unit cell. This is illustrated in Figure 3.2.15. FCC structures are therefore easily shaped.

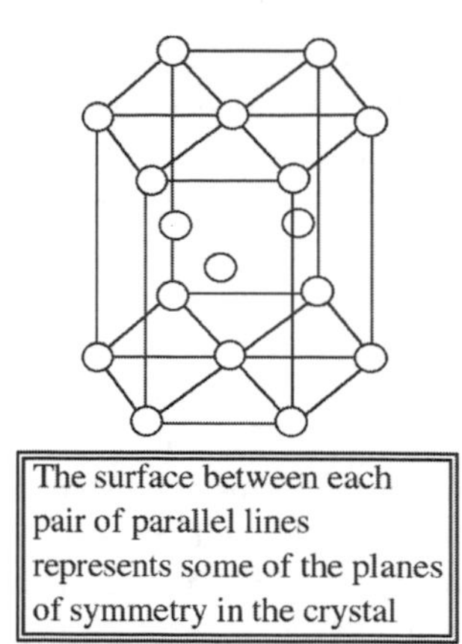

The surface between each pair of parallel lines represents some of the planes of symmetry in the crystal

Figure 3.2.16 *Planes of symmetry in the CPH crystal*

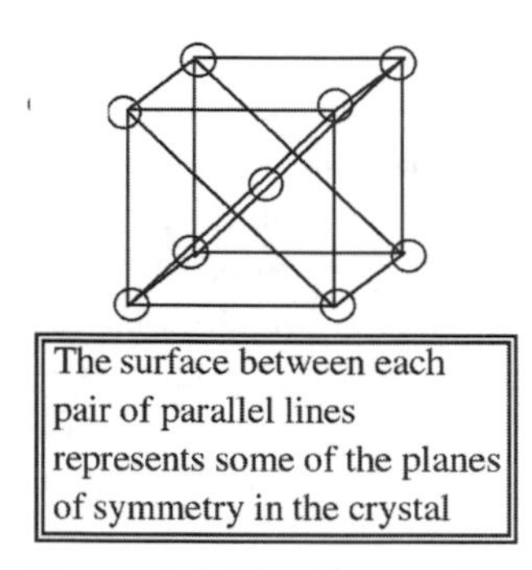

The surface between each pair of parallel lines represents some of the planes of symmetry in the crystal

Figure 3.2.17 *Planes of symmetry in the BCC crystal*

In CPH structures, the base plane is close packed. In other directions, the atoms are less well packed (Figure 3.2.16) and so the CPH crystal is very anisotropic. That is to say, its properties vary with the direction along which they are measured.

The BCC structure (Figure 3.2.17) has almost as many planes as the FCC structure although the planes are less closely packed than in either CPH or FCC structures.

The FCC structure allows deformation most easily while the CPH structure provides most resistance. From an electrical point of view, the FCC structure again provides the greatest number of pathways for electrons to move. Hence, FCC structures form the best electrical and thermal conductors.

Within a single crystal of a metal, the planes of symmetry are defined. However, a given piece of metal may contain hundreds of crystals, within each of which the pattern would be identical, but which are randomly orientated with respect to each other. Because of this, any anisotropy within a single crystal will be masked and the metal will probably have properties which are the same in all directions, i.e. the metal is isotropic. The process of manufacturing solid metals can also introduce anisotropy as discussed in Chapter 4.

Amorphous solids

In an amorphous solid, there is no long-range order as atoms or molecules arrange themselves randomly as the material cools from the fluid state to the solid state. In fact, the solid state is better described as a highly viscous state or a super-cooled liquid. The basic units of the material (atoms or molecules) are unable to move freely because the frictional forces between the units are sufficient to balance the tendency to move about. If a mechanical load is applied, there may be sufficient energy to overcome the friction and then the material will flow. One very important mechanical property, creep, is a measure of this viscosity. We will explore to this topic under 'Creep', page 128. In thermoplastic polymers, the presence of amorphous material dominates the material characteristics and so such polymers can creep at ambient temperatures unless the stress is limited. As the temperature of the environment changes over a summer's day, the change in mechanical performance of a thermoplastic could be noticeable.

Structure of polymers

Polymers are made up of very large molecules that could be described better as long chains. The name polymer is derived from the Greek for many (*poly*) units (*mer*). Typically, a chain will consist of units of a few atoms repeated tens of thousands of times as illustrated in Figure 3.2.18. A chain of polyethylene is illustrated in this figure. These basic repeating units are the monomer (elemental building blocks) from which the polymer is made. If scaled up, each chain would be the equivalent of a piece of cotton thread of the order of one or two metres long. The atoms within each molecule are bonded by strong covalent bonds. The attraction between molecules is dependent on weaker

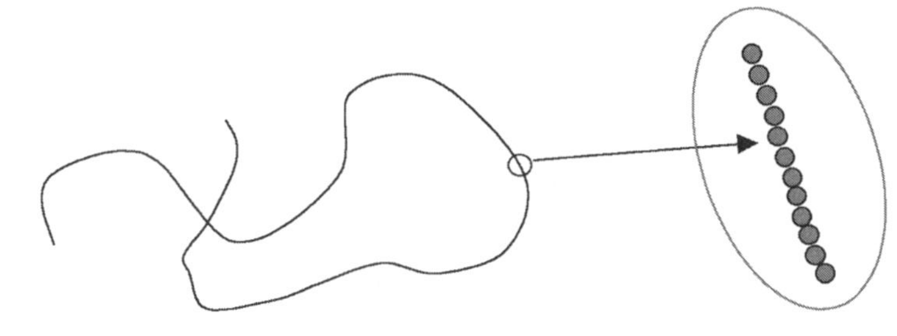

Figure 3.2.18 *Representation of the chain of a polymer such as polyethylene*

secondary bonds. As a fluid, above the 'melting temperature', each molecule has sufficient energy to overcome this attraction caused by the secondary bonds. The strong covalent bonds are not broken significantly. The chains remain coiled and entangle with each other. On cooling, these entanglements can prevent the forming of an ordered crystalline structure.

A useful analogy for the structure of a polymer would be to think of the chains interacting just like freshly cooked spaghetti might in a large bowl. Each piece of spaghetti will slide past each other although remaining attached. If the spaghetti is long enough, the individual pieces could also entangle. For the analogy to work properly, we would need to imagine that each strand of spaghetti was about 10 metres long and that we had enough to fill a small pond! As the spaghetti cools, the pieces become increasingly sticky and eventually set together in a random arrangement. In some areas, several pieces may join together to give parallel strands. In fact, in polymers it is far easier for one chain to fold back and forth to create a small packet than for several chains to come together as parallel strands. This would be the equivalent of an ordered crystalline region (also called phase). In the rest, the pieces will be disordered and entangled. This is the equivalent of the amorphous region or phase.

The crystalline phase should form at the crystalline melting temperature (T_m). Many polymers such as high density polyethylene (HDPE), polypropylene (PP) and polyamide (nylon) actually do form crystals. However, the degree to which crystals form can be restricted because the molecules entangle easily. The shape of the molecules is also important. If the basic unit is small, there is a better chance of forming crystalline regions. The degree of crystallinity will also vary with the length of the polymer chains (longer chains would be more entangled and so give less crystallinity) and on the cooling rate (faster cooling means less time to unentangle and so less crystallinity). The most crystalline polymers such as HDPE could be up to 96% crystalline, the remaining 4% being made up of chain segments linking the crystalline regions. This is because polyethylene has a relatively simple shape as illustrated in Figure 3.2.19(a). The repeating unit consists of a

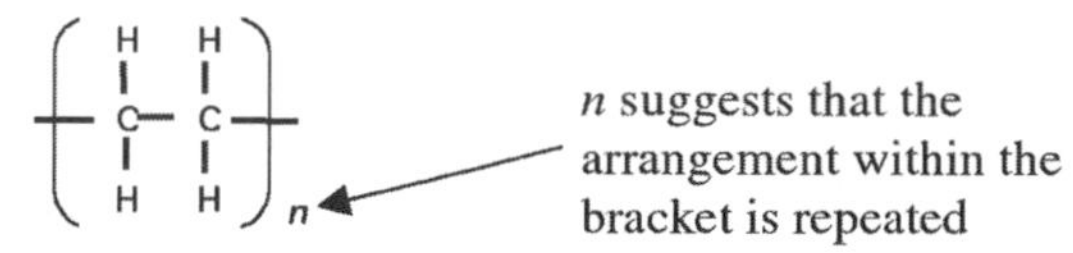

Figure 3.2.19(a) *The repeating unit of polyethylene*

carbon-based backbone with two hydrogen atoms coming off each carbon atom. In this description, each line represents one pair of electrons shared between the two atoms being joined by the line.

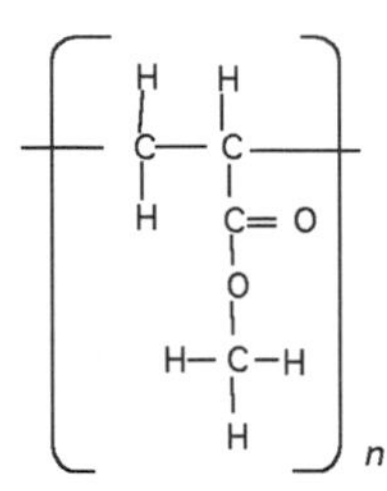

Figure 3.2.19(b) *The repeating unit of polymethyl methacrylate (PMMA) such as Perspex*

Other thermoplastics such as polypropylene (PP) can be up to 85% crystalline. No polymer can be fully crystalline and so such materials are said to be semicrystalline. Thermoplastic polymers such as polymethyl methacrylate (PMMA) and polystyrene (PS) never form crystalline regions and are said to be fully amorphous. Incidentally, PMMA is better known by trade names such as Perspex and Acrylic. Like glass, such materials have relatively large spacing between chains in the solid state and so are optically transparent. The chemical structure of PMMA is illustrated in Figure 3.2.19(b). It differs from polyethylene in that one hydrogen atom is replaced by a larger molecule. This molecule is referred to as a side group, as it 'hangs off' the main chain (Figure 3.2.20). Because this side group is relatively large, it makes it difficult for such chains to bond together to form a crystal.

The amorphous phase does not have a melting point like that of the crystalline phase. The solid is only held together by the frictional forces. The chains remain flexible and the polymer can deform significantly. We call this behaviour *rubbery.* However, as the temperature falls further, parts of the chain begin to 'freeze' as the amount of kinetic energy, as indicated by the temperature, falls below the level needed to initiate their movement. The most important movement involves the combined rotation of several units of the chain. This is like the crankshaft of a car engine and so is called the 'crankshaft motion'. This is also like the motion of pedals on a bicycle. As you step down and push slightly forward on each pedal in turn, the pedal assembly turns in a circular motion about the centre, which is the hub of the bicycle wheel (Figure 3.2.21).

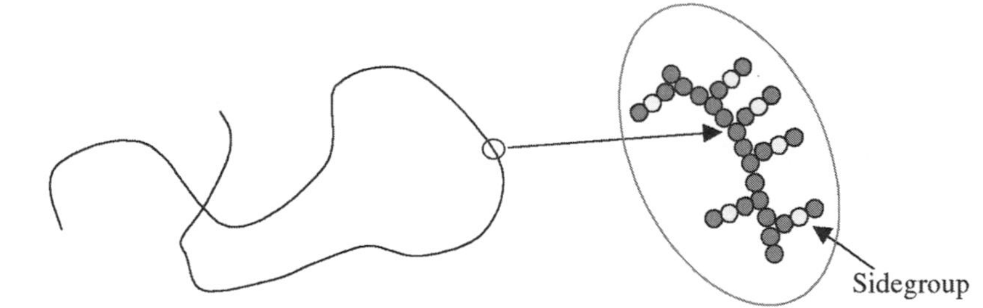

Figure 3.2.20 *Representation of the chain of an asymmetrical polymer such as polymethyl methacrylate*

Axis of rotation through the hub of the wheel

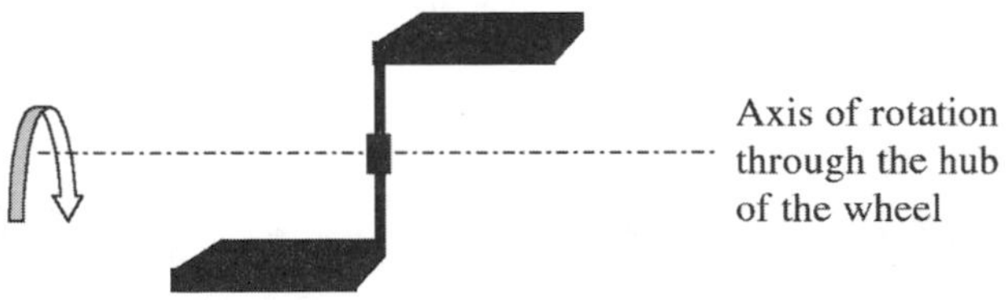

Figure 3.2.21 *Motion of a bicycle pedal*

Without this motion, chains cannot uncoil and so the polymer loses its flexibility. Returning to the analogy of the cooked spaghetti, this is like spaghetti left out after cooking for a day or so. Without its moisture, it becomes brittle. In the same way, the polymer at this lower temperature becomes brittle. It is termed *glassy* and the temperature at which this transition takes place is called the glass transition temperature, T_G. It is therefore important to know the T_G of a polymer before selecting it for an application. For example, HDPE has a T_G of –10°C whereas PP has a T_G of –27°C. Milk crates and small beer barrels are made from polypropylene rather than HDPE because these products are likely to be knocked about during handling. The chains of PP will remain flexible during an exceptionally cold British winter whereas the chains of HDPE may lose their flexibility and an HDPE crate or barrel may crack. As the barrel or crate is meant to last for some time without maintenance, PP proves a better choice of material.

Table 3.1.1 Some properties of materials

Material	*Density* (kg/m^3)	*Tensile strength* (MN/m^2)	*Modulus of elasticity* (GN/m^2)	*Coefficient of linear expansion* ($\times10^{-6}$/°C)	*Specific heat capacity* (J/kg°C)	*Thermal conductivity* (W/m°C)	*Melting or softening point* (°C)	*Electrical resistivity* ($\times10^{-8}$ Ωm)
Platinum	21 450	350	17.0	9.0	136	69.0	1770	11.0
Gold	19 300	120	77.0	14.0	132	296.0	1340	2.4
Lead	11 300	15		29.0	126	35.0	600	21.0
Silver	10 500	15	75.0	19.0	235	419.0	1230	1.6
Copper	8900	215	115.0	17.0	386	385.0	1350	1.7
Phosphor bronze	8820	400	100.0	18.0	379	70.0	1000	9.5
70/30 brass	8530	325	110.0	20.0	379	117.0	935	9.5
Nickel alloy (Inconel)	8440	930	207.0	12.8	410	9.8		129.0
Mild steel	7800	450	208.0	15.0	463	47.0	1495	16.0
Malleable iron	7350	450	168.0	11.0	520	46.0	1100	
Grey iron	7200	350	100.0	11.0	400	48.0	1100	10.0
SG iron	7150	600	165.0	11.0	460	34.0	1100	
Zinc alloy	6700	270		27.0	418	113.0	400	5.9
Aluminium alloy	2790	260	71.0	22.0	965	150.0	600	4.0
Aluminium	2700	80	69.0	24.0	965	240.0	660	2.6
Alumina	3800	350	350.0	9.0	775	29.0		10^{20}
Magnesium oxide	3600		205.0	9.0				10^{11}
Silicon carbide	3170	450	300.0	4.5	630	12.0		10^{6}
Quartz	2650		310.0			12.0		10^{20}
Concrete	2500		14.0	13.0	950	1.0		10^{17}
Plate glass	2500	70	70.0	9.0	800	0.8		10^{20}
Building brick	2300			9.0		0.6		
Silica glass	2200	70	70.0	0.5	750	1.2		10^{26}
Polyvinyl chloride	1390	58	3.0	50.0	1000	0.1	82	10^{22}
PMMA	1200	64	3.0	70.0	1470	0.2	89	10^{20}
Polyamide 6,6	1140	66	2.8	125.0	1680	0.2	75	10^{20}
ABS	1100	45	2.3	100.0	1530	0.2	85	10^{21}
Polystrene	1055	45	2.7	70.0	1340	0.1	92	10^{19}
High density polyethylene	960	35	1.0	130.0	2220	0.5	130	10^{22}
Low density polyethylene	925	12	0.3	170.0	2300	0.3	85	10^{22}
Polypropylene	900	34	1.3	110.0	1930	0.1	150	10^{22}
Ash (air dried)	689		10.1					
Douglas fir (air dried)	545		12.7		2900	0.1		10^{23}
Balsa (air dried)	176		3.2					

Activity 3.2.3

(1) Pour freshly cooked spaghetti onto a flat tray and observe the amorphous and crystalline regions that can form. What can you say about the nature of the material? It is a shame to waste the spaghetti, but in the name of science, leave it to dry for 24 hours. What can you say about the properties of the material now?
(2) Look at properties in Table 3.1.1 and see if there is any relationship between the type of bonding and the properties of common materials.

3.3 Alloys

Types of alloys

While pure materials have some uses, the majority of engineering applications involve mixtures of materials. For most applications, the properties of a given material need to be modified to enhance their properties in some way. In the case of metals, alloying involves the addition of other elements usually in small proportions to a host element. In some cases, two elements may each have a major role in the new mix. The mixture in all cases is called an alloy.

Alloys are usually made by mixing the constituent elements in the liquid state. We will assume that the constituents are completely soluble in the liquid state, i.e. they form a solution. When the solution is cooled and solidifies, the constituents may be:

(a) completely soluble in each other;
(b) completely insoluble in each other;
(c) partially soluble in each other.

In the case of metals, complete solubility can occur in some cases but complete insolubility does not happen. Most metals will show solubility to some degree. Hence we will consider solubility in metals for the two cases (a) and (c) in more detail in the following sections.

Metals that are completely soluble in the solid state

They form a solid solution where the atoms of the two metals mix together uniformly to form a single crystal lattice. The major constituent of a solution is called the solvent and the minor constituent, the solute. The two constituents of the solid solution will be indistinguishable when viewed using an optical microscope. The structure will only show the natural boundaries between crystals as in a pure metal.

This is a *single phase* structure because only one substance can be seen. Such solid solutions can form in three ways as described in Figure 3.3.1. The solute atom may be similar in size to the solvent atom and so could substitute for a solvent atom. This substitution could be either ordered to form a regular pattern or disordered. In a third form, very

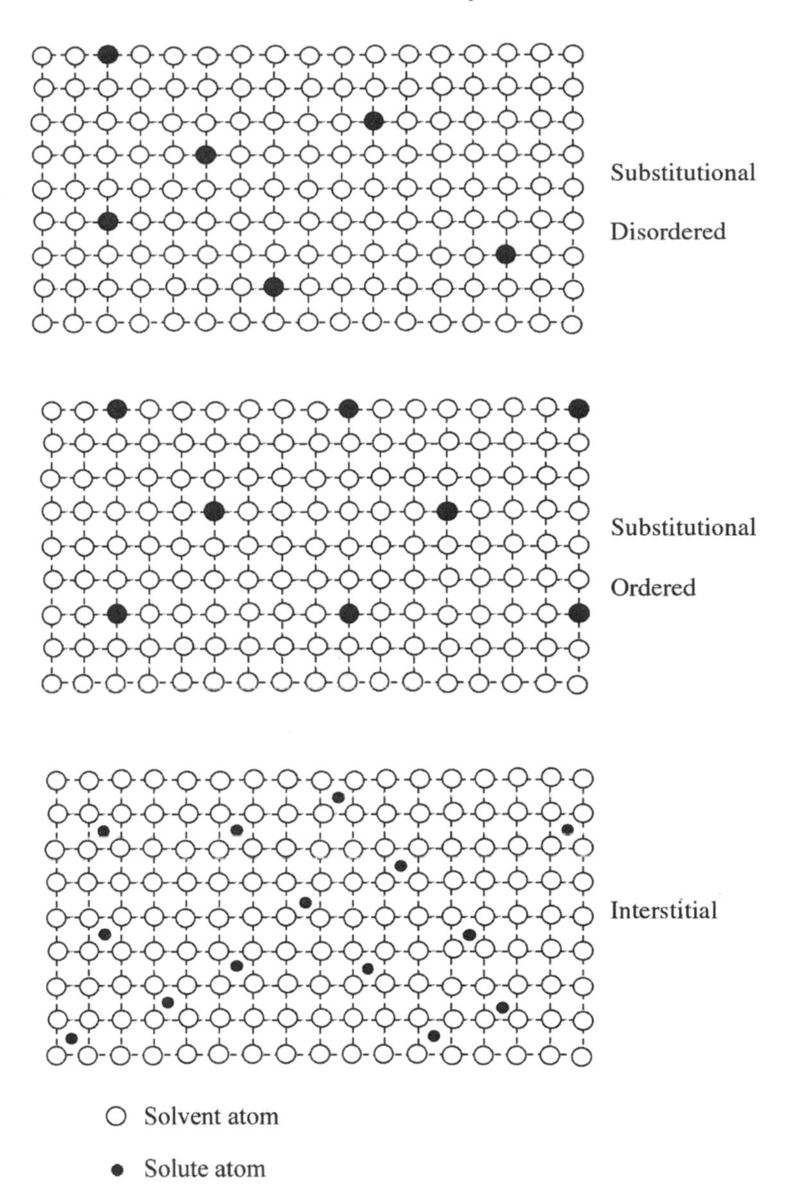

Figure 3.3.1 *Types of solid solutions*

small atoms could be located in the gaps or interstices between the solvent atoms. This is called an interstitial solution.

Since the size of an atom of one element will be different from that of another, only elements close to each other in the periodic table will form substitutional solids easily. For example, nickel and copper can combine to give an important alloy called Monel metal. This is used where good chemical resistance is needed. If the size of the atoms is very different, then the proportion of the solute atoms that can be accepted will remain very low. This is because the crystal lattice is distorted by the different sizes of atoms. In the case of steel, carbon atoms, being much smaller than the iron atoms, form in the interstices between the iron atoms. The proportion of carbon that can be accommodated for most applications is usually much less than 1%. The distortion of the lattice helps to increase properties such as strength and hardness. Ductility and electrical conductivity will decrease for the same reason.

Thermal equilibrium diagrams (phase diagrams)

In order to understand how well different elements mix, a map is produced showing different structures in zones. Taking a particular mixture of elements A and B, melting and mixing the elements fully, the mixture is cooled very slowly to ambient temperature. The temperature will fall gradually although at certain points the temperature will remain constant for a period of time (Figure 3.2.10). These points represent a change of state. These points of isothermal arrest are then plotted against composition, to form lines on the map.

They are called thermal equilibrium diagrams when generated under conditions where there is no exchange of heat with the environment during the process. In practice, data is generated under conditions where there is minimal heat loss to the environment. The resulting map is called a phase diagram. Such diagrams are often used to explain the changes occurring in alloys which are being heated or cooled. There may be considerable variations in the behaviour of alloys being heated or cooled unless the rates of heating or cooling are very slow; much slower than would normally be the case.

An idealized diagram for two metals, which are completely soluble in each other, is shown in Figure 3.3.2.

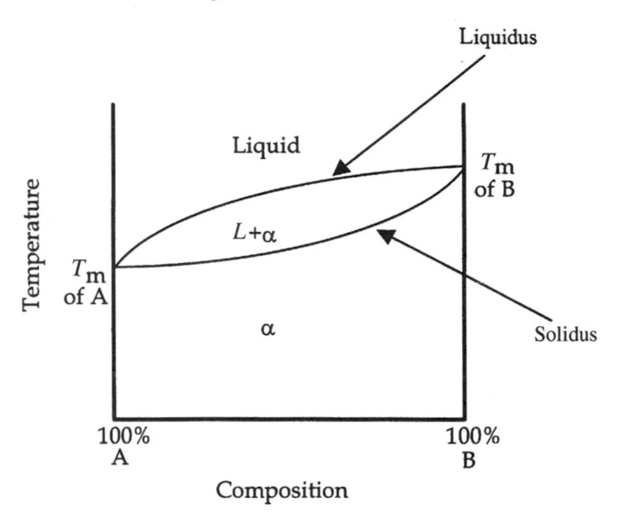

Figure 3.3.2 *Thermal equilibrium diagram for two metals that are completely soluble in each other*

In this case, the melting temperatures of the two pure metals are shown at the appropriate compositions (100% A and 100% B). For any other composition, the melting point is no longer at one temperature. The liquid cools first through the upper boundary line, called the *liquidus*. At this point, solid begins to form around the element B. Notice that it is now liquid at a lower temperature than would be the case in the pure B. The rest of the material remains liquid, L. By the time the lower boundary line, called the *solidus*, is reached, all the material has solidified into a form labelled α. Hence the phase between the liquidus and the solidus is labelled ($L + \alpha$).

Solid solutions are traditionally given Greek symbols; α, β, γ, etc. Each of these phases has a variable composition. For example, the α phase in Figure 3.3.2 can vary from 100% metal A to 100% metal B. The diagram, however, shows what phases are present at any temperature for any alloy in the system.

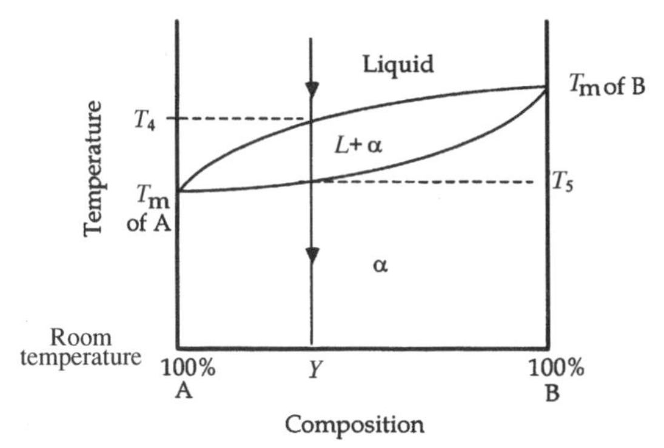

Figure 3.3.3 *Cooling of alloy containing Y% of B*

We will now consider an alloy in the system containing Y% of B. The proportion of A will therefore be (100–Y)%. This is illustrated in Figure 3.3.3.

At temperatures above the liquidus (T_4), the alloy will be a liquid solution of A with Y% B. Below the solidus (T_5), it will be a solid solution of A and B with an overall composition of Y% B. Now consider the cooling of the alloy from the liquid state to room temperature.

The alloy will remain a liquid until temperature T_4 is reached. At this point, it starts to solidify forming α. T_4 is below T_m for metal B, but above T_m for metal A. Therefore there is a tendency for metal B to solidify. The first solid metal to form will have more than Y% of B. (It will not be completely B.) It will be rich in metal B. As cooling continues between T_4 and T_5 the solid formed will be rich in B while the liquid will become increasingly rich in A. Solidification will be complete at T_5. Each crystal will grow in a spiky, 'tree-like' form called a dendrite. As you will see in Chapter 5, this peculiar shape of the crystal is due to the fact that as the melt solidifies, it gives off its excess energy (called latent heat). Consequently, the area around the newly solidified crystal warms up sufficiently to stop further crystallization. Crystal growth then moves to another point on the surface. The crystal that is formed is rich in B at the centre and rich in A between the dendrite arms as you can see in Figure 3.3.4.

The dendritic pattern can be seen when the section of metal is viewed under the microscope because of differences in composition. This process by which differences develop in the composition is called coring. The slower the rate of cooling, the less coring there will be. Coring is harmful as the properties of the alloy will vary across the grains and could lead to serious weaknesses in structure in some parts. In fact, some engineering failures are caused by coring.

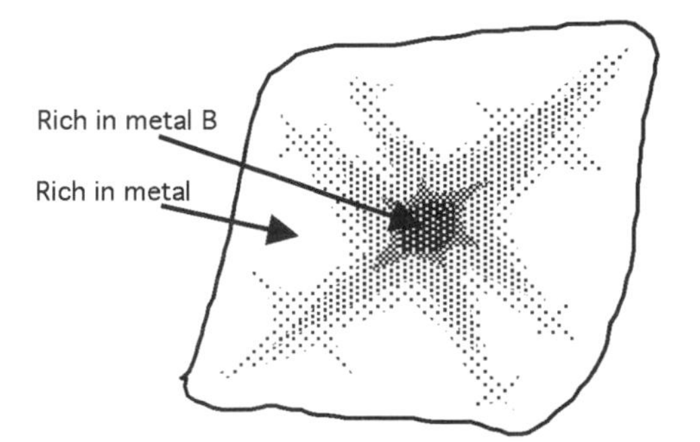

Figure 3.3.4 *Crystal growth showing differences in composition through the section*

Metals that are partially soluble in the solid state and form a eutectic

As this is the most common type of equilibrium diagram for metals, as illustrated in Figure 3.3.5, we will consider several scenarios in some detail. First, we need to define some further terms. As you can see from the diagram, there are three compositions at which the liquid changes from liquid to solid at a single temperature. These are the 100% A and 100% compositions as well as the composition labelled *E*. This composition is called the eutectic composition as the metal behaves just like a pure metal.

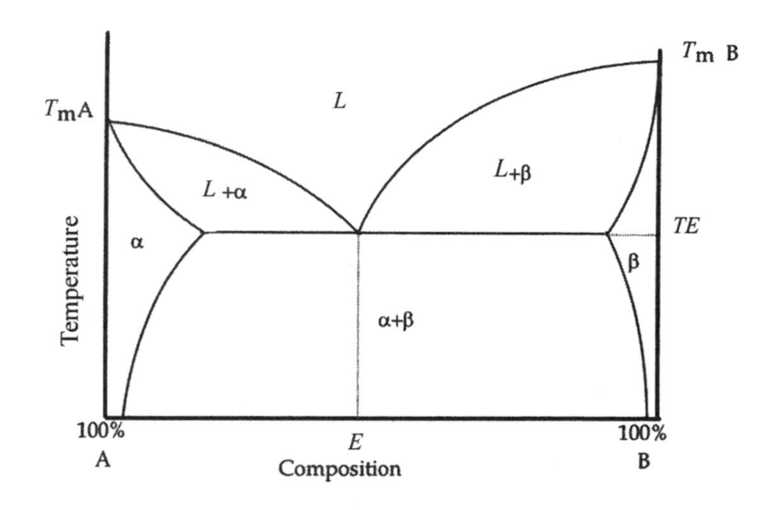

Figure 3.3.5 *Thermal equilibrium diagram for two metals that are partially soluble in each other. α is a solid solution of B in A, β is a solid solution of A in B, α and β are insoluble in each other. E = Eutectic composition, TE = eutectic temperature*

Consider now the cooling of the melt with eutectic composition. The liquid will be a solution with atoms of metal A and metal B uniformly mixed. There will be no change until the liquid cools to a temperature of *TE*. The thermal equilibrium diagram shows that just above *TE* the alloy is completely liquid. Just below *TE* it is completely solid. The resultant solid will consist of a fine mixture of α and β as illustrated in Figure 3.3.6.

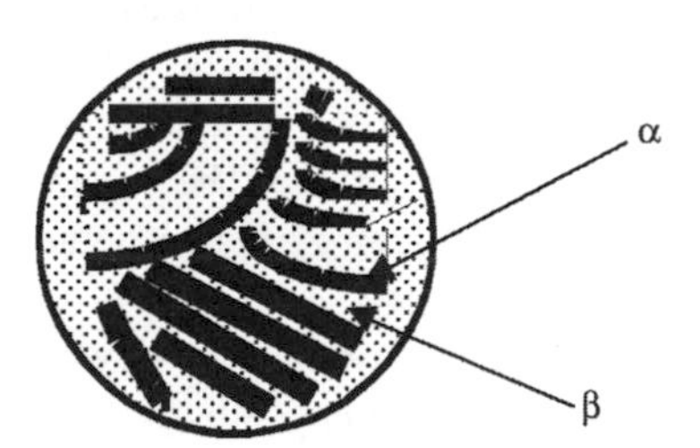

Figure 3.3.6 *Two phases formed for a eutectic composition*

Consider now an alloy of composition X% B (Figure 3.3.7).

Solidification starts when the liquidus is reached at a temperature T_1. On cooling through the *L*+β region, solid dendrites of β form. Coring

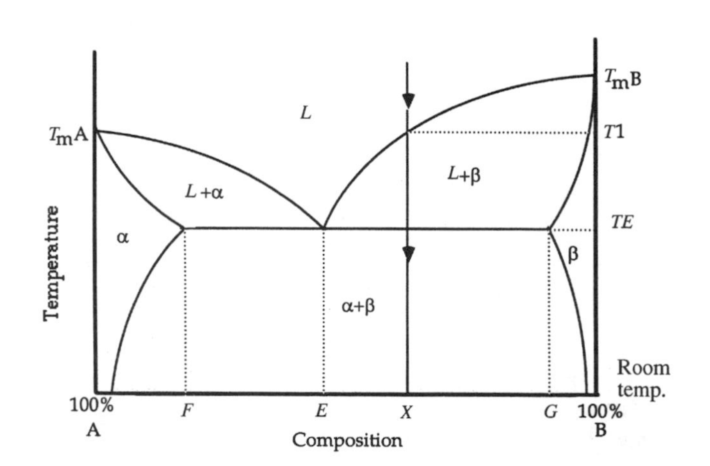

Figure 3.3.7 *Cooling of alloy containing X% of B*

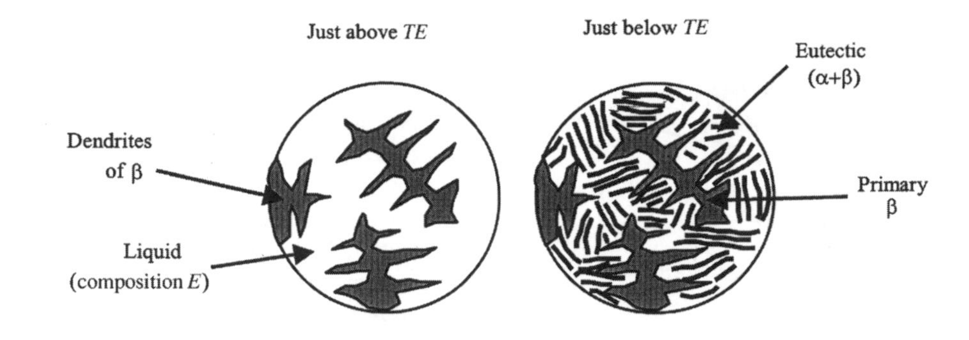

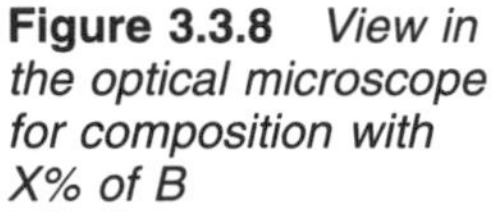

Figure 3.3.8 *View in the optical microscope for composition with X% of B*

occurs as described earlier and so the dendrites are rich in metal B. The liquid therefore becomes richer in metal A until at temperature *TE* the composition of the liquid is *E*. At *TE* the eutectic liquid solidifies to give a eutectic solid. The final structure will consist of grains of β with the eutectic composition seen earlier in Figure 3.3.7 making up the rest of the material. This is illustrated in Figure 3.3.8.

There is little change as the alloy cools from *TE* to room temperature. All alloys with compositions between *E* and *G* will solidify in the same way to give a structure consisting of primary β + eutectic. The closer the composition is to *E* the less primary β there will be and the more eutectic. Similarly, all alloys with compositions between *E* and *F* will solidify to give primary α + eutectic.

The third scenario we will consider is one with *Y%* of B (Figure 3.3.9).

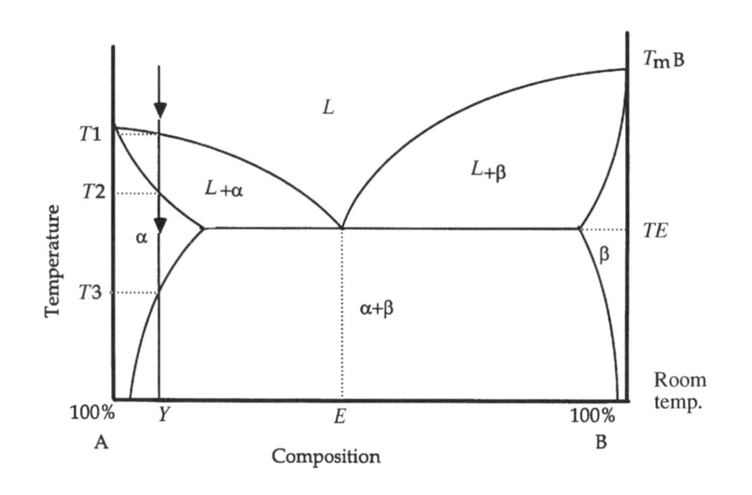

Figure 3.3.9 *Cooling of alloy containing Y% of B*

In this case, solidification starts at *T*1. Dendrites of α will form as the alloy cools through the *L*+α region. The alloy will be completely solid at *T*2, and the structure will consist of α crystals only. A view under the optical microscope is illustrated in Figure 3.3.10. The only visible features will be the grain boundaries.

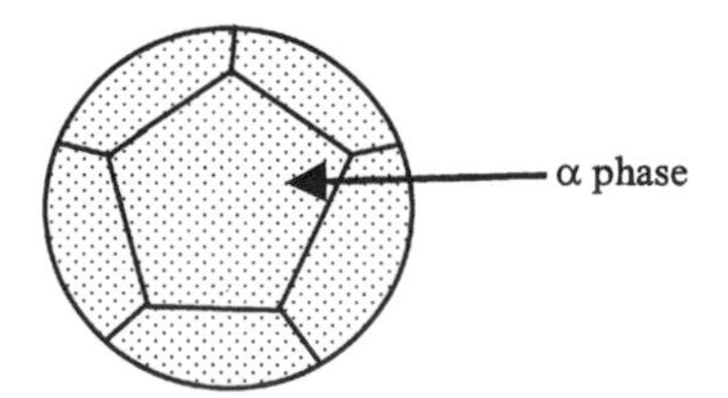

Figure 3.3.10 *View in the optical microscope at T2*

There will be no change until temperature *T*3 is reached when the *solvus* line is crossed. This line represents a change of phase within a solid. It indicates that the α phase can no longer take more *Y%* B and so tries to reject excess B atoms. The β phase will therefore form or from the α phase. It is difficult for a solid to change its form and so β will

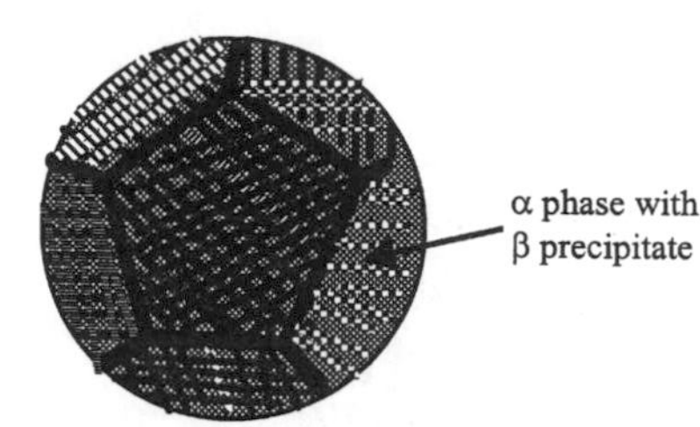

Figure 3.3.11 *View in the optical microscope at T3*

precipitate where there is most room for it – at the grain boundaries of the α and along some crystal planes (Figure 3.3.11). The precipitation involves the rearrangement of the atoms into the β form.

The final structure consists of the α phase with precipitated β particles concentrated at grain boundaries and other sites.

Consider now another form of structural change that can form in partial solutions in metals. This is the peritectic type of transformation for composition *P*, illustrated in Figure 3.3.12.

TP is the temperature at which the peritectic reaction takes place. This arises when there is a large difference between the melting temperatures of the two constituents. For example, in brass, copper has a T_m of 1083°C and zinc has a T_m of 420°C. When an alloy of composition *P* cools, solidification starts at temperature T_1. On cooling through the *L*+α region dendrites of α form. The dendrites grow until, at T_P the remaining liquid reacts with the α to give a new solid, β. All the material will be in the form of β crystals and will remain so until the temperature T_2 is reached. If the cooling rate is low enough, some α may be precipitated in the β.

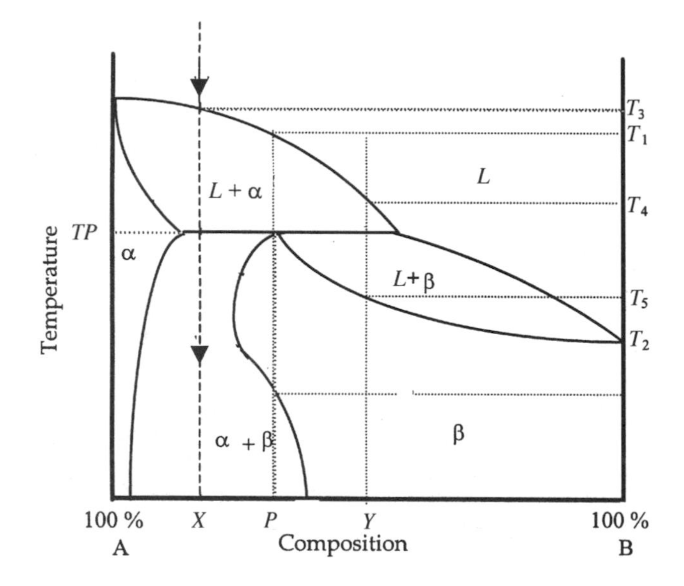

Figure 3.3.12 *Cooling of composition with P% of B*

When an alloy of composition *X* cools, solidification starts at T_3. α dendrites form on cooling through the *L*+α region. At T_P there is a higher ratio of α:liquid than there was with the alloy of composition *P*. At T_P the peritectic reaction takes place.

$L + \alpha \rightarrow \beta$

In this case there is an excess of α and so all the liquid reacts with some of the α to give β. As illustrated in Figure 3.3.13, the regions of α decrease in size and the remainder take up the β form.

When an alloy of composition *Y* cools, solidification starts at T_4. Dendrites of α form from the liquid. When the temperature T_P is reached, the peritectic reaction takes place as before. There is, however, an excess of liquid in this case. All the liquid reacts with some of the α

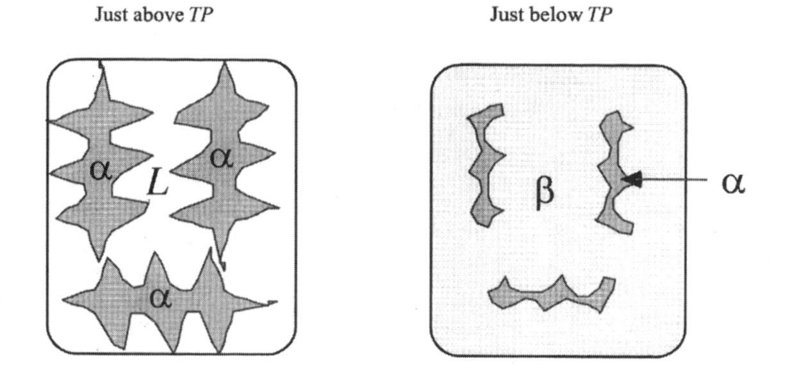

Figure 3.3.13 *View in the optical microscope for composition with P% of B*

to give β. After the peritectic reaction is complete the structure will consist of solid β in liquid (L + β). The solid β continues to form until solidification is complete at the temperature T_5. The structure will then consist of crystals of β. There will be no further change to room temperature.

Polymer alloys

In the case of polymers, the mixing can be achieved at three levels. At the stage where the polymer is created (polymerization), two or more different basic monomer units can be combined to give a polymer with the best of the properties of each. In the case of thermoplastic polymers, different monomers may be joined together to create a copolymer. The possible arrangement along the chain is shown in Figure 3.3.14.

Thermoplastic elastomers, which are highly elastic like a rubber at ambient temperature but can be moulded like a thermoplastic, are usually copolymers. Varying the proportions of each monomer will

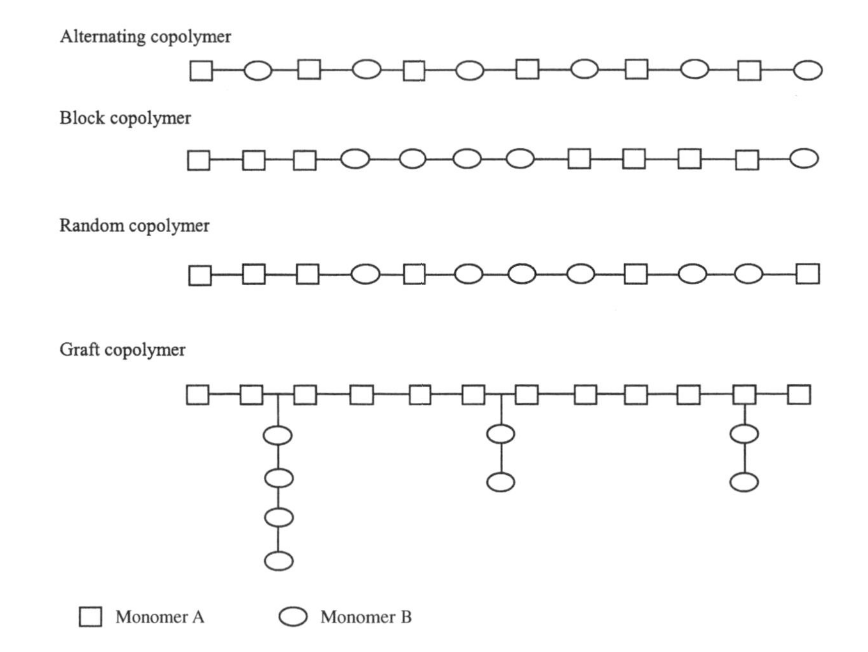

Figure 3.3.14 *Possible arrangements of two different monomers in a copolymer*

modify the properties of copolymers. One example of a copolymer is acrylonitrile butadiene styrene (ABS).

At a second level, two or more polymers may be combined in a compounder where the materials are brought to a melt and mixed, to create a new polymer called a blend. Any polymer processor can experiment by mixing polymers in the hope of producing a blend with better qualities. Unlike metals where the constituents of each metal are small, regular atoms, in polymers, large chains produce complicated structures even without mixing.

Semicrystalline polymers produce partly regular or crystalline structures. Blending will tend to disrupt this regularity. Blending is most valuable when the toughness (see 'Toughness and impact strength', page 126) needs to be improved. For example, polystyrene is brittle at room temperature and everyone who has ever owned a plastic (polystyrene) ruler knows that if it is flexed by too much, it will snap! By adding a thermoplastic elastomer to the melt, a high impact polystyrene (HIPS) is produced. When this material is prepared and viewed using an optical microscope, it looks like a piece of salami (Figure 3.3.15). Although the rubber and the polystyrene are incompatible, they are completely intermixed with sections of one wrapping the other. HIPS is very tough although not as strong as polystyrene. It can take knocks and so objects such as toys and TV cabinets were made from HIPS. In recent years, HIPS has given way to ABS, the equivalent copolymer. Truly compatible polymer mixes are rare.

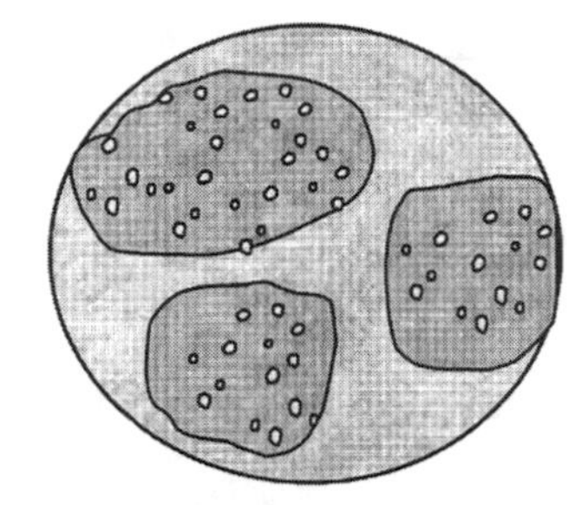

Figure 3.3.15 *View in the optical microscope of the high impact polystyrene blend*

At the third level and most commonly, additions are made to any polymer, copolymer or blend to give the materials in most products. These additives help to:

- Enhance the processability of the material using stabilizers and lubricants.
- Modify the properties of the material using plasticizers, fillers and fire retardants, for example.
- Reduce the cost of the product using fillers.
- Improve the appearance of the product using colourants.
- Improve the life of the product using stabilizers.

Defects in materials

Real materials have much lower strengths than their theoretical values. The strength is reduced because of defects. The defects may be:

- *Gross*: For example, a poor design feature such as a keyway in an axle with a very small radius at the corners would act as a stress concentrator (Figure 3.3.16). Therefore, any stress applied to the part will be increased by a factor depending on the severity of the stress concentration and the component could fail prematurely.
- *Microscopic*: Small defects left behind during processing can also act as stress concentrators. For example, casting defects such as porosity and grain boundary defects.
- *Crystalline*: Sub-microscopic defects that form naturally within the material such as dislocations. We will look at the latter in the next section.

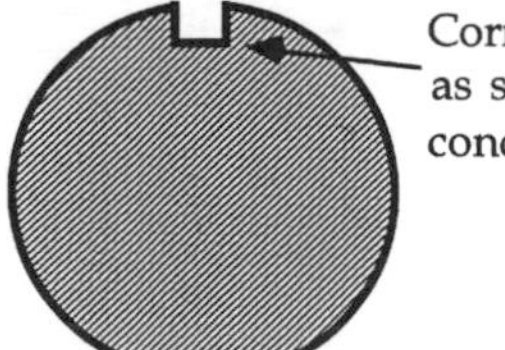

Figure 3.3.16 *Effect of small radii in components under load*

Point defects

Within the crystal, there may be irregularities in the pattern of arrangement. This is because atoms carry kinetic energy in the form of vibration in the solid. The following types of point defects can develop as illustrated in Figure 3.3.17:

- *Vacancies*: These are positions without atoms. As large numbers of atoms are organized into a solid, it is inevitable that some gaps will appear. It is also possible for individual atoms to be displaced from one position to another by diffusion. The vacancy is called a *Schottky defect*.
- *Interstitial defects*: Displaced atoms can occasionally slot themselves into the gap between the regular position of atoms. Such atoms form *self-interstitial* defects. They cause a great distortion of the structure. They usually occur in a pair with a vacancy and as a pair is referred to as a *Frenkel defect*. It is more likely that smaller atoms can diffuse into the interstitial space. These interstitial atoms could be of gases such as oxygen and nitrogen which are plentiful in the atmosphere as well as carbon (see why this is the case by referring to the periodic table).
- *Substitutional defects*: Atoms of foreign elements may replace the atoms of the host metal in their normal positions, substituting for them. They could either be impurities that come from the environment in which the metal is formed and so are usually unwanted or they could be desirable additions made to improve the properties of the metal.

Such irregularities not only exist in metal structures but also in ceramics and the crystalline segments of semicrystalline polymers. In all cases, they can act as stress concentrators since the crystal lattice is distorted around the defects and there is strain energy associated with them. As the kinetic energy of atoms is increased, by increasing temperature, for example, it follows that there will be more movement of atoms within the crystal and so more defects form.

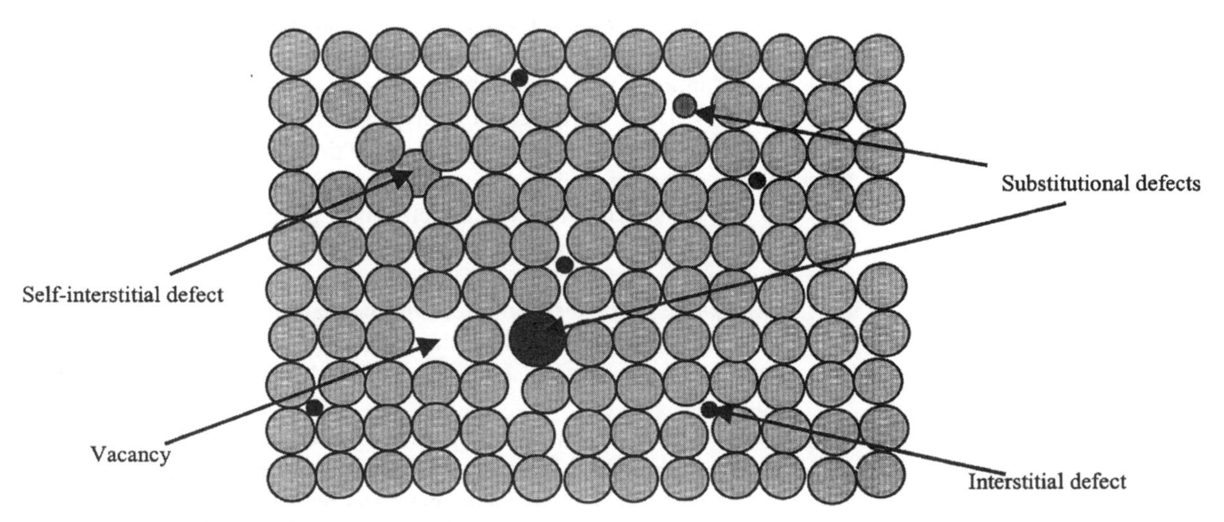

Figure 3.3.17 *Point defects in crystalline solids*

Line defects

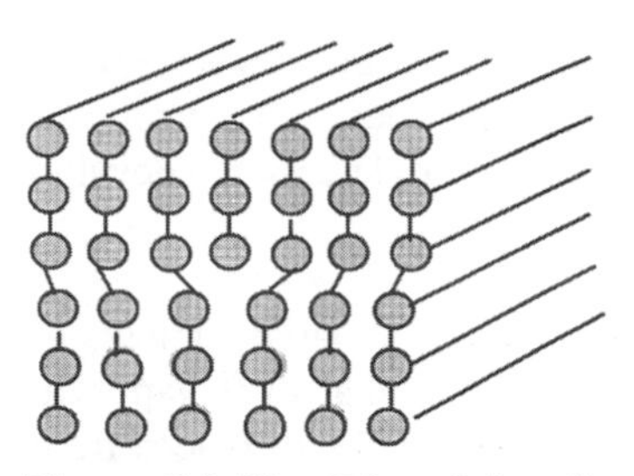

Figure 3.3.18 *Edge dislocation in a crystal*

An *edge dislocation* can be considered as an extra half plane of atoms in the crystal. By convention they can be considered as either positive (⊥) or negative (⊤) dislocations (Figure 3.3.18). The extra plane of atoms increases the cross-section dimension of the crystal by one plane although a view along the direction of the arrow in Figure 3.3.18 will not show any significant displacement. Another kind of dislocation is the *screw dislocation*. In this case, a view along the arrow will show a misalignment of the planes by shear. This will look like rotation (or screw-like) movement of the plane of atoms.

Activity 3.3.1

You can simulate the packing of atoms in a metal using tennis balls, marbles, beads or ball bearings laid out so that they are mostly touching. Using marbles of three different sizes (with at least 50 of the middle size and a few of the other sizes), you can show substitutional and interstitial solid solutions, defects and grain boundaries. Try to explain under what conditions, as alloys solidify, some are completely soluble while others are partially soluble or completely insoluble. How would the arrangements change if the marbles were replaced by strings of beads?

3.4 Properties of materials

In the previous section, we saw how the needs of materials in particular environments are identified in terms of the properties of a material. Some properties are given in Table 3.1.

It must be noted that whenever properties are given in tables, they may have been obtained from different sources and therefore are always subject to some variability. They can only be treated as typical for that class of material. For example, the tensile strength of a material is dependent on the precise composition and the nature of the manufacturing process and heat treatments that the material is subjected to. In the case of mild steel the value may vary from about 400 MN/m^2 to greater than 550 MN/m^2. In the case of polypropylene, it depends on the degree of crystallinity in the material. The melting point of a material will change according to the additives or alloying elements added. Properties such as density, modulus of elasticity and electrical resistivity will only change by a small amount, providing the levels of additions are low. They are more readily relied upon. However, factors such as the temperature of the test can affect all materials and especially polymers at ambient temperature. Even the moisture content in the atmosphere can affect properties of some polymers and natural materials such as woods.

We will now relate these properties back to the fundamental structure of the atom and the forces that control how they combine. This is important since good choice of materials requires an understanding of how to go about looking for the materials. The properties of a material are dependent to a large extent on the freedom of the electron (discussed in Chapter 4, Section 4.2) to move about an atom and between atoms. As we have seen the different bonding mechanisms cause different degrees of control on the electron. In the case of metallic bonding, the electron has the greatest degree of freedom as the free electrons form a mobile 'sea'. Hence metals have a high degree of electrical conductivity. When an electrical potential is applied across a piece of metal such as copper, electrons will drift towards the positive end of the battery or other source of potential. Their movement is impeded by the presence of other electrons, which cause repulsion, as well as the positive ion cores, which attract. The electrons also possess kinetic energy and so move randomly throughout the structure. As the temperature increases, the kinetic energy increases. This increases the random motion of the electrons and so the resistance to flow in the direction of the applied potential increases. Hence good conductors become less effective at higher temperatures. In a semiconductor, the effect of increasing temperature is to release more electrons from the positive ion core into the valence band. Hence the conductivity increases.

Chemical reactions especially at the surface of a material are dependent on the freedom of electrons. The steel surface of a ship at sea experiences corrosion because positive iron ions can dissolve in water where they react with negative ions while the electrons left behind take part in other reactions to form hydrogen gas (see Chapter 5). In the case of a polymer, moisture will enter the material but will not generally react since the covalent bond will lock the atoms more tightly into the polymer structure.

Thermal properties depend more on how well the atomic structure in a material is ordered, since thermal energy is stored as kinetic energy within every constituent of the atoms. If one end of a bar of metal is heated relative to the other, the additional energy is transferred into vibration of the atoms around their cores and into kinetic energy of the free electrons. At every atom, you can imagine vibrations along three axes (x-, y-, and z-axes) as well as rotation about the three axes. The vibrational energy is transferred rapidly across the metal as in a slinky spring, because there is a high degree of order in the structure. In the case of a ceramic formed by ionic bonding, again there is a high degree of order and so energy is transferred quickly, making them good thermal conductors. Structures such as silica glass have a much more open structure and so energy is transmitted less readily. This makes them poorer conductors although anyone holding onto a glass container containing a hot drink will eventually notice that the heat has been transferred to the outside of the container, although at a much slower rate! Polymers have the most open structures and so have very low thermal conductivity. Providing they do not melt or burn, a polymer can make an excellent insulator. A good example in the home is the modern polypropylene kettle body. The body can be made in one piece with an integrated handle that can be picked up without scalding oneself. Even in metals the conductivity is dependent on the regularity of the structure. Grain boundaries, cavities, impurity particles, and different phases can all give rise to significant differences between published values of any property. It is in the skill of the manufacturer of the solid to optimize properties by good choice of process and good quality control so that the properties do not change significantly from one batch of material to another.

Activity 3.4.1

Returning to Table 3.1.1 compare the range of values of each property for metals, ceramics and polymers. What features are common to each class of material?

Electrical conduction

All materials are potentially able to conduct electricity, since electricity is the flow of electrons in an ordered fashion. In the case of electrical conductors such as copper, a relatively high current can flow for a given electrical potential because the resistance to flow is relatively small. In semiconductors such as silicon or germanium, this flow may not happen unless the right conditions exist, releasing electrons from the host atoms. In the case of insulators such as silica, alumina or PVC, flow does not happen except at very high potential difference when the material breaks down. The different behaviour of conductors, semiconductors and insulators is dependent on the availability of 'free electrons' and a path along which the electrons can flow. In the case of liquids, the flow is dependent on the free movement of positive or negative ions.

In a solid, electricity flows when electrons are able to escape from an atom and skip across to a neighbouring atom and then onto another and so on in a given direction. These free electrons are abundant in good conductors such as copper. When an electron leaves an atom, the atom becomes positively charged and another takes its place to maintain a neutral charge in the atom. Normally at room temperature, the free electrons will move between atoms randomly. However, when an electric potential is applied, there will be a tendency to move in one direction.

Band structure in solids

Electrons in an atom are arranged strictly according to rules, in shells and subshells as we have seen already. The number of electrons in any subshell is limited to two, each with opposite spin according to the Pauli exclusion principle. When atoms come together to form a solid, the electron orbitals of adjacent atoms overlap and their energy states (defined by shell and subshell) are altered slightly so that no more than two electrons (of opposite spin) are in any one state. This is illustrated in Figure 3.4.1. In this example, if the material is in the form of a gas, the electrons of each atom would be a long way away from each other. They exist with energy levels of E_1 and E_2. When the atoms are locked into place as a solid, E_1 and E_2 each becomes a 'band of energy levels'.

With the creation of a band of energy levels, the difference in energy between shells becomes smaller and hence, the energy required to move from a filled, bound valence shell to a free shell also becomes smaller. The highest energy level occupied by electrons in a metal is called the Fermi energy, E_F. When an electrical potential is applied, electrons

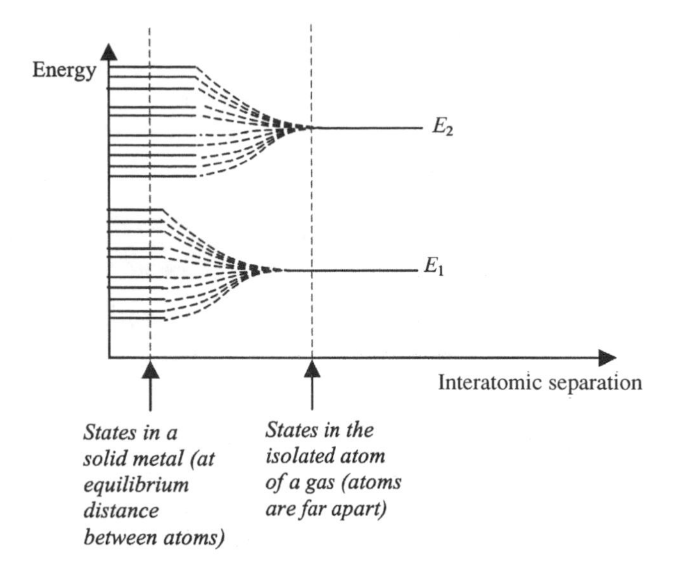

Figure 3.4.1 *Band of energy levels in a metal*

are able to move relatively easily and thus contribute to electrical conduction. In the case of a covalently bonded material such as diamond (all carbon) or polyethylene (a hydrocarbon), all the *s* and *p* subshells are filled and so none are available close enough to allow electrons to overcome the energy gap. The band gap is said to be too large.

The rate of flow of electrons in a material is characterized by Ohm's law according to which,

$$V = IR$$

where V is the applied electrical potential across a material, I is the electrical current and R is the electrical resistance. This resistance R is related to the dimensions of the conducting material and also to the nature of the energy levels and therefore freedom for electron movement in the material. We can define this resistance due to the freedom of electrons alone in terms of the *electrical resistivity*, ρ, of the material. Hence,

$$\rho = \frac{RA}{l}$$

where A is the cross-sectional area of the specimen and is its length as illustrated in Figure 3.4.2. With the dimensions now separated out, the electrical resistivity becomes a material property.

As you can see from Table 3.1.1, copper, silver, gold and aluminium have the lowest value of ρ and so it is not surprising that they are all

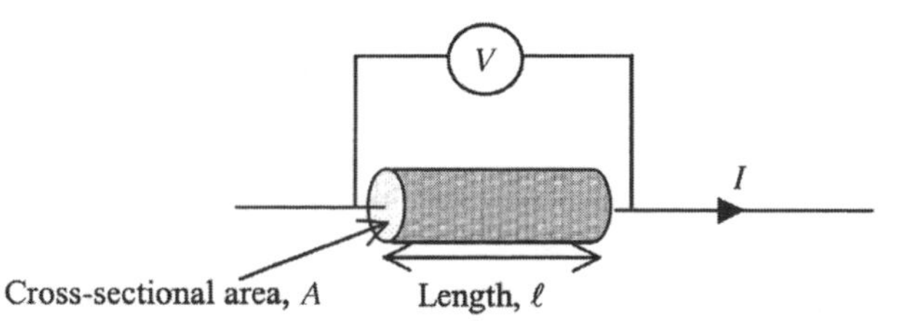

Figure 3.4.2 *Measurement of the electrical resistivity of a conductor*

used as electrical conductors. Copper is widely used to carry current in electrical devices. Gold is preferred in printed circuit boards where only very small quantities are required since it is more stable in the environment although more expensive (see Chapter 5 on corrosion).

Activity 3.4.2

The electrical cable of a hairdryer consists of several copper strands in a sleeve. If the total length of the cable is 5 m what is the resistance of the cable given that there are 20 strands of copper wire in the cable, each of diameter = 0.24 mm. Using the value for ρ in Table 3.1.1, calculate its resistance. What would the resistance be if mild steel strands of the same diameter were used?

Answer: For copper, $R = 1.88\ \Omega$; for mild steel, $R = 17.68\ \Omega$.

Semiconductors

In some covalently bonded materials such as silicon, some electrons can be pushed into the higher unoccupied state by supplying energy in the form of either heat (raising the temperature), radiation (light) or an electrical field. Each of these electrons leaves behind a vacant site called a vacancy or hole. If there are sufficient numbers of electrons and holes, the material may become conducting. The energy gap between the filled valence band and the free, conduction band must be small enough for this to happen. Elements in group 14 can conduct electricity by freeing electrons in this way. The element tin has such a small energy gap that it is regarded as a metal anyway, whereas, the elements silicon and germanium can become conductors at high temperature. These latter elements are important semiconductors. Other common semiconductors include the element selenium as well as the compounds gallium arsenide, zinc selenide, and lead telluride. At low temperatures, pure semiconductors behave like insulators.

Doping

The conductivity of semiconductors can also be improved by adding impurities. In this case, the material is 'doped' by replacing one of the host atoms with an atom of another element that has a different number of valence electrons, so that either an extra electron or a hole is generated in the structure. Consider the illustration of a doped silicon (Si) crystal (Figure 3.4.3). Each silicon atom has four valence electrons (represented by dots). In the covalently bonded structure, each of these electrons combines with an electron from the adjacent atoms to form a pair. Hence each atom has eight valence electrons and therefore a filled outer shell. If an element of phosphorus (P) with five valence electrons replaces a silicon atom, a free electron is created in the lattice. This will be available for conduction. Phosphorus can be taken into the structure as it is adjacent to silicon in the periodic table and so is approximately the same size. There is negligible distortion of the crystal lattice.

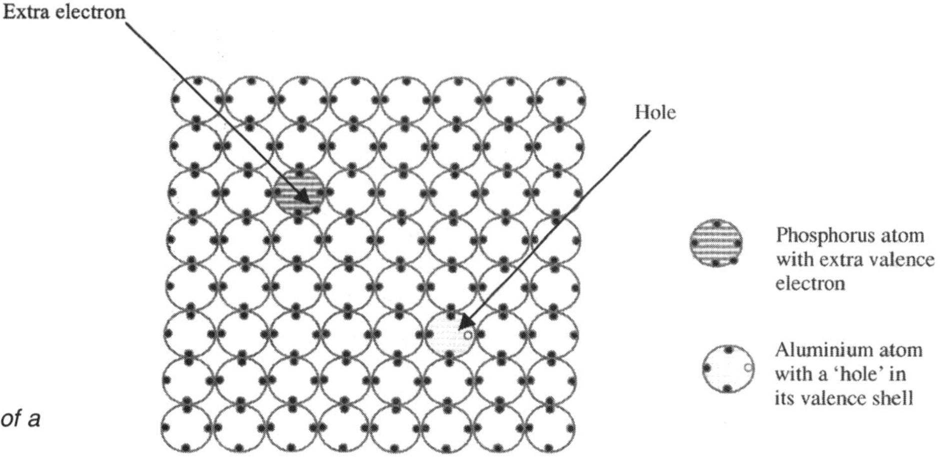

Figure 3.4.3 *Doping of a silicon crystal*

Because of the extra electron provided by each phosphorus atom, this material is termed n-type (negative type) silicon. In p-type (positive type) silicon, atoms with three valence electrons such as aluminium can be added. This leads to a deficiency of electrons, or the creation of holes. Providing that the concentration of either electrons or holes is large enough, the material will behave as an electrical conductor.

Elastic behaviour

Review – stress and strain

When a load is applied to a test piece of solid material, it will resist that load up to its limit of capability. If the load is shared across the cross-section of the material, we can define stress as the force per unit area of surface, at right angles to it and on which the force acts (Figure 3.4.4). Hence, if a load F is applied normal to a surface of cross-section A then,

Stress, $\sigma = F/A$

The stress is now independent of the geometry of the test piece. The load can cause either tension (pulling force) or compression (pushing force) in the material along the direction of the load. This is also illustrated in Figure 3.4.4 where the direction of the applied load is indicated by arrows.

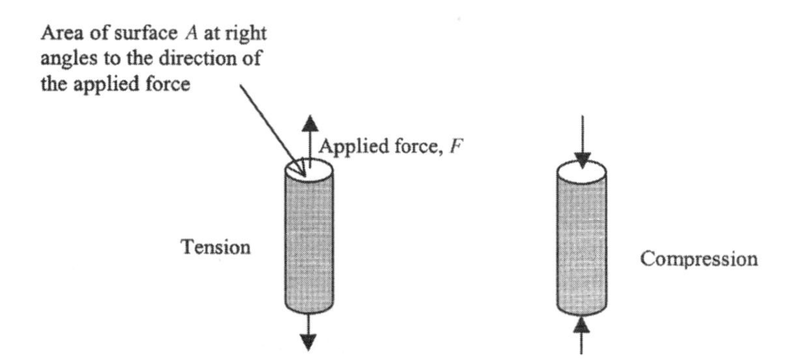

Figure 3.4.4 *Application of a force on a test piece*

The stress is dependent on the applied force and the cross-sectional area only. It does not depend on either the materials from which the test piece is made, or the length of the piece being tested. However, under the effect of the stress, the material will deform and the amount of deformation will depend on the material used. The deformation is measured along the direction of applied stress. Again, in order to make this independent of the size of the test piece, the term strain is defined as a ratio of the deformation, δ, along the direction of load to the length, l, of the test piece in that direction, i.e.,

Strain, $\varepsilon = \delta/l$

This measure is a ratio of the change in length over the original length. Quite often, the strain is given as a percentage of the original length. This is done by just multiplying the above ratio by 100. That is to say, if a bar of length 50 mm extends by 1 mm due to a tensile load,

Strain, $\varepsilon = 1/50 = 0.02$

or,

$$\varepsilon = (1/50) * 100 = 2\%$$

Constant strain-rate test

The properties of a material either in tension or compression are usually obtained from a test where a defined shape of material is subjected to load under constant strain rate. In a tensile test, the specimen is clamped between two platens on the test rig as illustrated in Figure 3.4.5 and one platen is allowed to move at constant speed relative to the second. As the specimen is pulled, it resists and a load cell on the test rig monitors this resistive load. In the case of a compression test, the specimen may be cylindrical or rectangular with flat ends that are parallel to the platen surfaces. In addition to the strain rate, the temperature must also be kept constant.

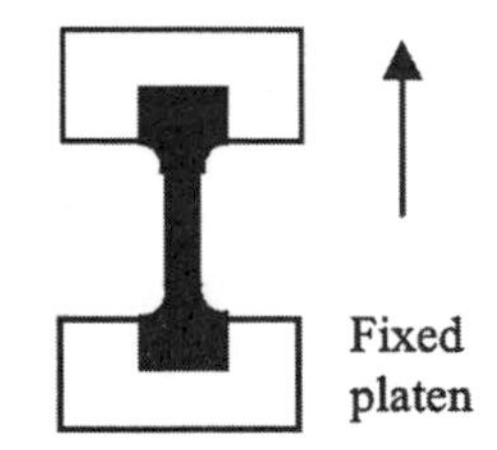

Figure 3.4.5 *Arrangement for a standard tensile test*

The tensile test for polymers is easily performed to defined standards such as BS2782, ASTM D638 and ASTM D790. When tests are conducted to standards, the shape and size of the specimen are specified. The test piece for tensile tests may have a cross-section that is rectangular, square or circular. In all cases, the cross-sectional dimensions at the grips are much greater than the dimensions of the middle section that is free to deform. There is also a shoulder section where the dimensions change progressively to the dimensions of the free section (Figure 3.4.6).

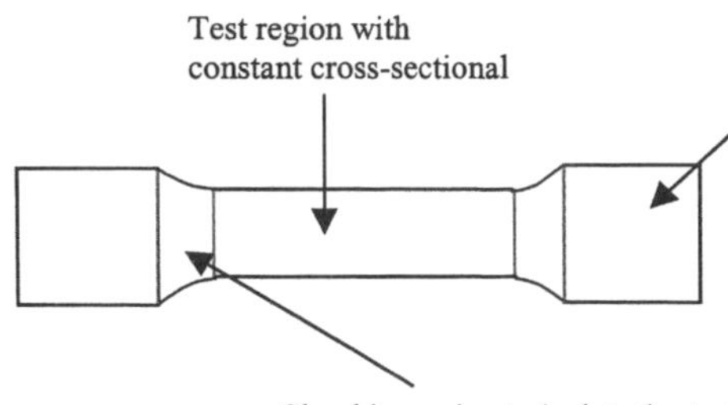

Figure 3.4.6 *Test piece for a tensile test*

Young's modulus

If the stress is small enough not to cause permanent deformation in the test piece, then the extension of the test piece will depend on the modulus of elasticity, E (also called Young's modulus). This is the rate of change in length with applied load, along the direction of the applied load. It is given by the gradient of the initial straight line section of the stress/strain graph as illustrated in Figure 3.4.7,

$$E = \sigma/\varepsilon$$

(The particular graph shown in Figure 3.4.7 illustrates the behaviour of steel. We will be looking at this in detail later, but for the present, we will concentrate on the early part of the graph only.)

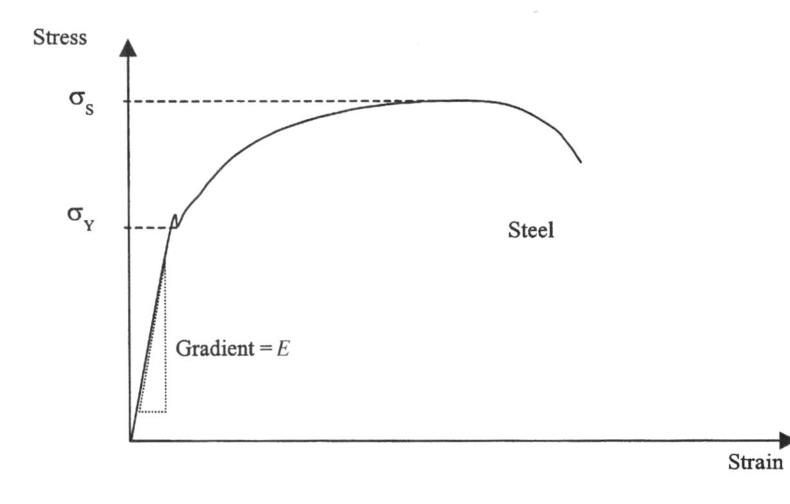

Figure 3.4.7 *The nominal stress–strain behaviour of a steel obtained from a tensile test*

The Young's modulus is a measure of the resistance to elastic deformation in a material. On an atomic level, it is therefore a measure of the force needed to pull the atoms or molecules apart. In general, this is also referred to as the stiffness or rigidity of the material. The Young's modulus is commonly quoted in textbooks. Since it is a measure of the atomic or molecular force, the temperature of the test will influence its value as well as the presence of weaknesses or flaws in the material.

The stress/strain curve is a straight line for many materials in the elastic region, i.e.

$\sigma \propto \varepsilon$

This relationship is called Hooke's law.

Since the Young's modulus is a measure of the rate at which the material deforms under load, a low modulus could lead to excessive deformation. This could mean the malfunction of a structure such as an aircraft wing, a bicycle frame or an umbrella. Imagine an aeroplane whose wings were not sufficiently stiff as illustrated in Figure 3.4.8. Excess flexing would alarm passengers but also lead to a loss of lift.

Figure 3.4.8 *Effect of high deflection in bending*

Equally, if the material is too stiff, the structure would not be able to respond to deformation passed on from another part of the structure and could fail in a brittle manner. Again, in the case of the aircraft wing, vibrations in the structure could cause the wing to flap and lead to levels of deformation where a material of high stiffness would fail. This is how a pane of glass when tapped with a hammer fails.

In an ideal material, without any flaws, the Young's modulus should give an indication of the stiffness of the bonds between atoms or molecules. In practice, the stiffness is reduced by the presence of flaws of various types. To measure the modulus accurately the extension must be measured accurately in the elastic region. This is done using a sensitive instrument called an extensometer.

Shear deformation

So far we have considered the effect of a load acting along the centroidal axis of a symmetrical test piece. The load is said be an axial load. In general, however, an applied load can also cause shear stresses. This will always be the case when the load does not act along the axis of symmetry of the test piece. Consider the example in Figure 3.4.9 where a solid piece of material, A, is shown being pushed downwards by one load acting through the block B against two blocks, C and D, which will provide upward resistive forces.

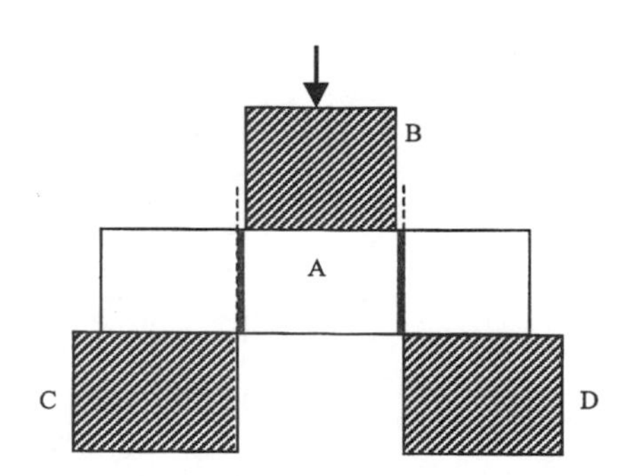

Figure 3.4.9 *Illustration of deformation by shear*

If the structure is in equilibrium, that is to say, it does not move under the loads, then the sum of the upward forces will equal the downward force. As the forces are acting inwards onto the block, there will be a compressive stress in the block. In practice, this will vary across the block as the load is unevenly applied. Also, since the upward loads are not in line with the downward load, there will be a shear force. An example of this situation may be a stamping machine or a hole punch used for paper. When the gap between block B and blocks C and D is very small, we have true shear. This is where one plane of the material in our test piece slides past an adjacent one without any stretching or contraction of the bonds between them.

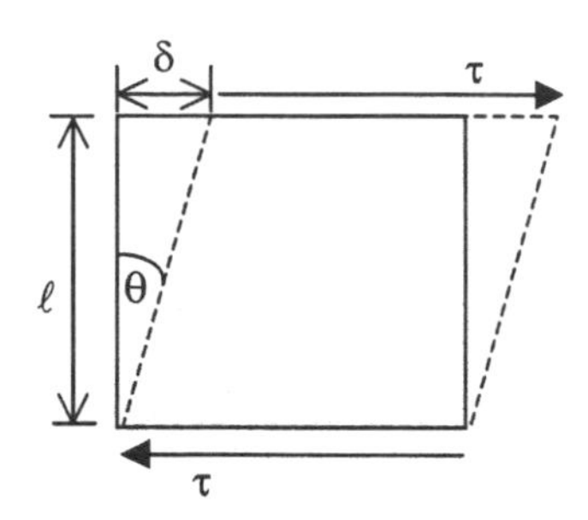

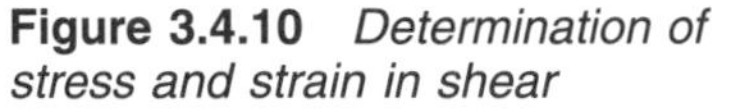
Figure 3.4.10 *Determination of stress and strain in shear*

In this mode, stress acts within a plane and at right angles to the direction of its influence as illustrated in Figure 3.4.10. Shear stress, τ, is defined as the ratio of applied force to the area parallel to which it acts.

The shear strain, γ, is given by the ratio of displacement, δ, to the distance, l, between planes over which the shear stress acts. Hence,

$$\gamma = \delta/l$$

If the stress is small enough not to cause permanent deformation in the test piece, then a shear modulus of Elasticity, G, can be defined such that,

$$G = \tau/\gamma$$

In practice shear stresses are generated when shafts and tubes are subjected to torsion and thick bars subjected to bending as shown in Figure 3.4.11.

The yield stress, σ_Y, is defined as the stress beyond which there will be a permanent change in the material and therefore the dimensions of the test piece as a result of the applied stress. This permanent change is termed plastic deformation. Up to the yield point (Figure 3.4.12), the deformation (from O to A) is said to be elastic. It is recovered fully as soon as the load is removed.

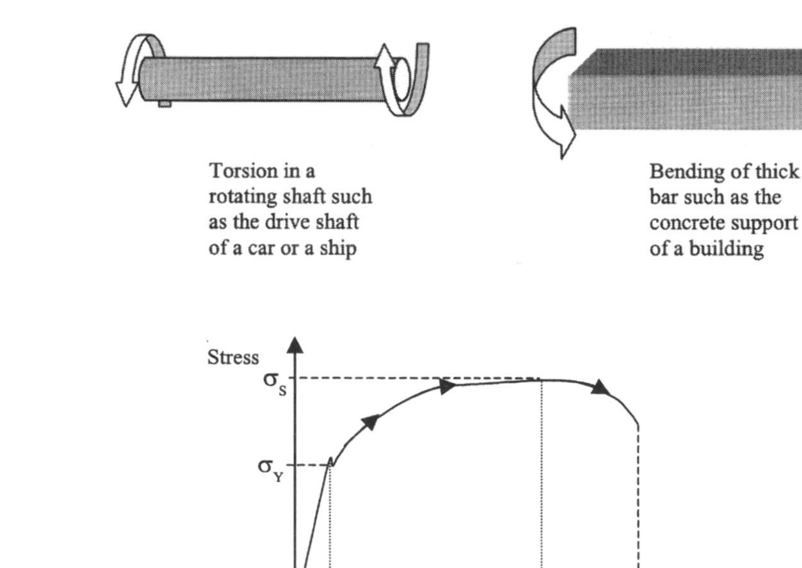

Figure 3.4.11 *Practical examples of shear*

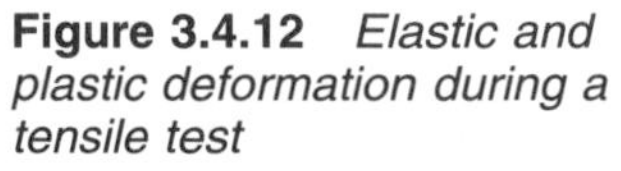
Figure 3.4.12 *Elastic and plastic deformation during a tensile test*

Plastic deformation and failure

Yield point and tensile strength

In steel, this is easily identified since the stress in the material drops slightly before increasing again. σ_Y is important for the designer since this is the limiting stress for any material in a component designed to carry load. If, for example, the material in the frame of a bicycle is subjected to a stress greater than its yield stress, then the frame will warp permanently and the rider would notice the loss of balance. Although the yield stress is quoted in textbooks, the value is very dependent on factors such as the test temperature and the speed with which the material is pulled. Polymers are especially affected by these two factors. The value for most materials is also affected by any previous heat treatment as will be explained later. Small variations in composition and the presence of flaws in the material also make this value difficult to predict. Up to the yield point, the material behaves in an elastic manner. That is to say, when any applied stress is removed, the material will not show any affect of that load.

In most materials there is the added complication that there is no clear boundary between the elastic and plastic regions. In this case an approximate value, called the proof stress, is defined and quoted. The proof stress, σ_P, is obtained by drawing a line parallel to the initial slope but displaced by a fixed value of strain (Figure 3.4.13). For example, if a designer believes that he or she can allow the material to deform by a strain of 0.1% without affecting the performance of the component, then the 0.1% proof stress is used, this being the point where the parallel line from 0.1% strain meets the actual curve.

Proof stress can be determined at 0.1, 0.2 or 0.5% strain. The designer will choose a value according to how critical the onset of plastic deformation would be for the particular application. Of course, if the customer demands a high level of accuracy, the cost of the product is likely to increase significantly. The strain used should be quoted when the value of proof stress is given.

Beyond the yield point, when plastic (permanent) deformation begins, the stress that the material can take increases further. This should not happen in the ideal metal solid, as this is an indication that planes of atoms have begun to slide past each other to cause large-scale change of shape. In practice, the impurities and other defects in the metal that we discussed earlier (in Section 4.9) distort the metal

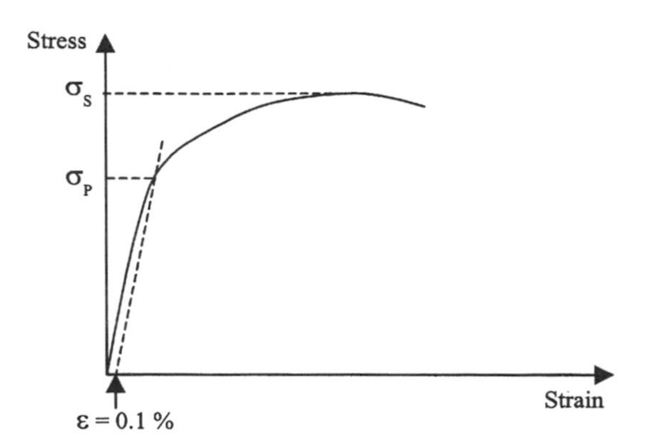

Figure 3.4.13 *Determination of proof stress*

structure and therefore increase the resistance to sliding. This phenomenon is called work hardening. The rate of work hardening is an important parameter affecting the ease with which the material can be fabricated by cold working. It is also important in structures designed for safety such as crash barriers and car bumpers. If a car is involved in a crash, for example, the crumpling action of the bumper and other parts of the structure absorb much of the energy and so reduce the chance of injury to the passengers. The material gets stronger as it crumples and so gives increased resistance to further crumpling.

The maximum stress that a material can take before failing is often the easiest parameter to obtain. This is commonly referred to as the strength of the material. Labelled σ_S, this is also called the ultimate tensile stress (or UTS). It is the ratio of the maximum load a material can take per unit of original cross-sectional area. No account is taken for any change in cross-sectional area during the test. Hence, although the curve shows a decrease in stress with increasing strain after the σ_S value (note the arrows in Figure 3.4.12), it is more likely that the stress remains constant or increases slightly. The load, however, decreases because the cross-sectional area of the material at its weakest point reduces as the material 'necks' as illustrated in Figure 3.4.14. Between points A and B, the test piece extends uniformly and so every part of it will be under the same stress and strain. The material becomes longer and thinner *uniformly* by plastic deformation. From B to C, the stress will change as the cross-sectional area decreases and the increasing strain will be localized to the neck region. Eventually at C, the centre of the necked region will fail.

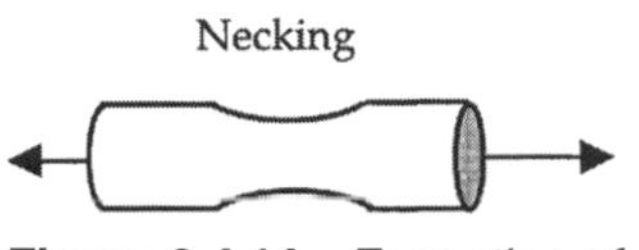

Figure 3.4.14 *Formation of a neck beyond the point of maximum load in a tensile test*

Ductility

Another important concept is ductility. This is a measure of the *plastic strain* and therefore permanent strain. If we look at the stress–strain behaviour of steel again, we can see that most of the deformation is plastic but there is some elastic strain that is recovered. The work done to cause plastic deformation is used up in changing the internal structure of the material. This can sometimes be felt as a slight heating up of the test piece. The ductility is measured directly from the fractured specimen after tensile testing. There are two ways of recording ductility. The first involves measuring the total strain to failure or,

$$\%\ \text{Elongation} = \frac{\text{final length} - \text{original length}}{\text{original length}} \times 100\%$$

Alternatively, the equivalent change in cross-sectional area can be measured giving,

$$\%\ \text{Reduction of area} = \frac{\text{original area} - \text{final area}}{\text{original area}} \times 100\%$$

Ductility is important because solid materials are shaped by plastically deforming them. For example, car body panels are made by pressing steel sheet into shape. This process is called *cold working*. Steel has an elongation of approximately 35% whereas aluminium alloy of similar strength has an elongation of approximately 10%. Therefore, it would be more difficult to make such complex car body shapes from aluminium alloys. Furthermore, as we saw earlier, if a car is involved in a crash, the ductility will allow crumpling and energy absorption as the metal distorts. When combined with the work-hardening effect, steel proves to be an extremely tough structural material.

When a piece of steel or another material fails, there is some elastic recovery (Figure 3.4.15). We often notice this because of the loud bang at failure. This is caused by the sudden release of the elastic energy that is remaining in the material. Energy is released at the speed of sound and like an earthquake, causes a lot of damage within the material structure and has the potential to cause serious harm to anyone nearby. For this reason, components designed to carry shock loads must be inspected regularly because progressive damage may not be visible externally.

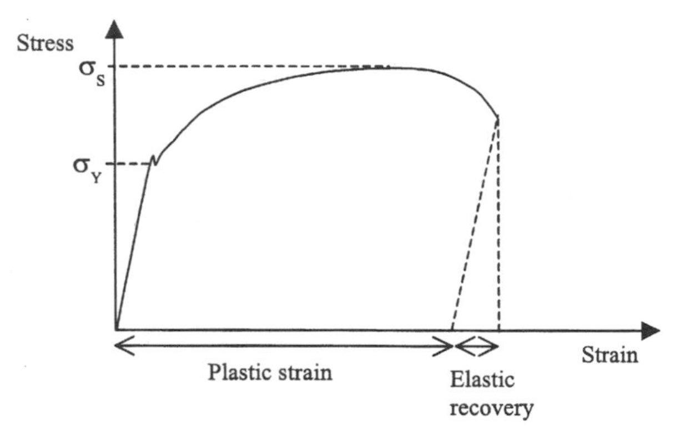

Figure 3.4.15 *Elastic recovery at failure*

Although our discussion has concentrated on metals, there are similarities between all materials, as we will see now. Most metals show behaviour similar to that shown in Figure 3.4.13. In this case there is no clear sign of a yield point but there is work hardening and also necking. The degree to which the material work hardens will depend on the metal, and particularly on mechanisms for strengthening the metal. Necking may not be seen with the more brittle metals of the close packed hexagonal type.

Activity 3.4.3

Take a piece of Blu-Tack and gently deform it by pulling. What can you say about its elasticity? Does the material return to its original shape when you stop pulling? Take a piece of metal plate such as the top of a discarded 'baked beans' tin or an old plastic credit card and bend it? How far can you bend it before it is permanently distorted? Then repeatedly flex it back and forth about its original plane so that the object is permanently distorted. Does the material always show ductility? Try to explain the changes taking place.

Brittle behaviour

Brittle materials behave as illustrated in Figure 3.4.16(a). Ceramics, such as silicon carbide and glass at ambient temperatures will show almost linear change in strain with stress. The final failure is elastic without any plastic deformation. When a test piece has failed, the failed pieces should all fit together with minimal distortion. This is classed as brittle behaviour. Amorphous polymers at well below their glass transition temperature, T_G, such as polystyrene at ambient temperature also behave in a brittle manner. Such polymers are said to be glassy. Even semicrystalline thermoplastics and rubbers become brittle at well below T_G for the material. Thermosetting polymers such as polyester or epoxy will also give little plastic deformation before they fail. Ceramics and thermosetting polymers are similar in that they consist of three-dimensional networks of atoms with relatively large spaces between atoms when compared with metals. All the bonds are strong and directional and so there is little freedom for plastic deformation. Amorphous thermoplastics and glasses, however, are essentially 'frozen' viscous solids like ice. They can deform if the load is applied at a very low rate. For example, glass panels in vertical Victorian windows have been known to show a thickening in the lower half as glass has flowed due to gravity over decades. This change in deformation with time, described later as creep, is a property of viscous solids.

Amorphous polymers well above their T_G but well below their apparent melting temperature, T_M, will exhibit behaviour of the type shown in Figure 3.4.16(b). This is described as 'rubbery' behaviour. The rate of loading is particularly important since as the rate is increased, the material will apparently become stronger (the maximum stress it can take will increase) and the elongation to failure will decrease. This

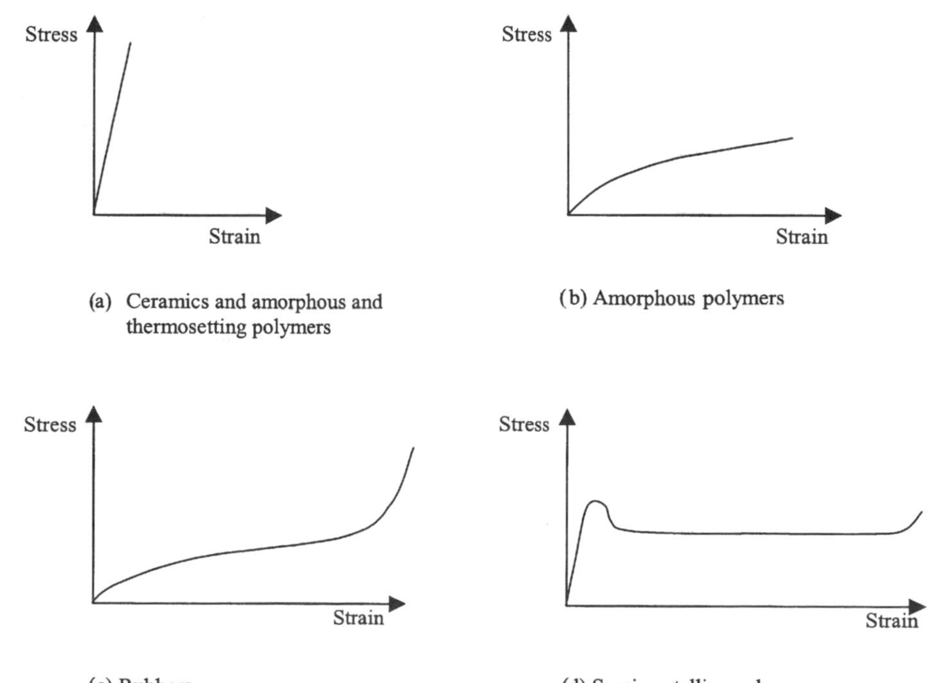

Figure 3.4.16 *Typical tensile test curves for different materials*

shape is similar to that of rubber in the early part of the curve (Figure 3.4.16(c)). This is not surprising because the amorphous thermoplastics are very similar in structure to rubbers (also known as elastomers). They differ only in that in the case of rubbers, the long polymer chains are joined together at a few points, called crosslinks. These crosslinks are responsible for increasing the rubber stiffness, as it is extended, and also for returning the material to its original shape even after an elongation of more than 500%. If the number of crosslinks is increased, the stiffness of the rubber will increase until eventually the material becomes a thermosetting plastic.

Figure 3.4.16(d) describes the behaviour of a semicrystalline thermoplastic at well above the T_G of the material. In this case, the initial extension is similar to a rubber or amorphous thermoplastic but at a critical point the crystalline segments in the material begin unravelling. This is responsible for the large permanent elongation seen in materials such as HDPE, polypropylene and polyamide 6,6.

When plastic deformation of metals takes place planes of atoms slide over each other in the crystal. This occurs on the planes, which are most densely packed with atoms. They are called *slip planes*. FCC metals are very ductile because they have many close packed planes.

Activity 3.4.5

To pull most materials to failure can often be difficult and dangerous. For example, if you were brave enough to pull a rubber band until it snapped, it could hurt but you will also hear it snap and if you held the broken pieces together, they should be the same size as the original rubber band. This is an example of brittle failure. (*So why is rubber said to be elastic?*) The material is elastic up to the point of failure and when it fails, it returns all the elastic energy in the form of sound and the painful whiplash you experience! A safer option would be to take a piece of writing chalk and to bend it until it snaps. Again, if you can find all the pieces, they should form the original shape. Brittle materials can fail explosively and therefore do a lot of damage!

Dislocation mechanism of slip

Figure 3.4.17 shows the atoms above and below a slip plane. There is an extra column of atoms above the slip plane (Figure 3.4.17(a)). This represents a positive edge dislocation centred at atom 4. When a shear stress acts along the slip plane, atoms 1–7 move slightly to the right relative to atoms a–f. Atom 4 is closer to atom d than atom 5. The bond between 5 and d is broken and 4 bonds to d (Figure 3.4.17(b)). The centre of the dislocation is now at atom 5. This process continues. Atom 5 bonds to e and 6 bonds to f. This process continues, possibly until the edge of the crystal is reached. This is illustrated in Figure 3.4.17(c).

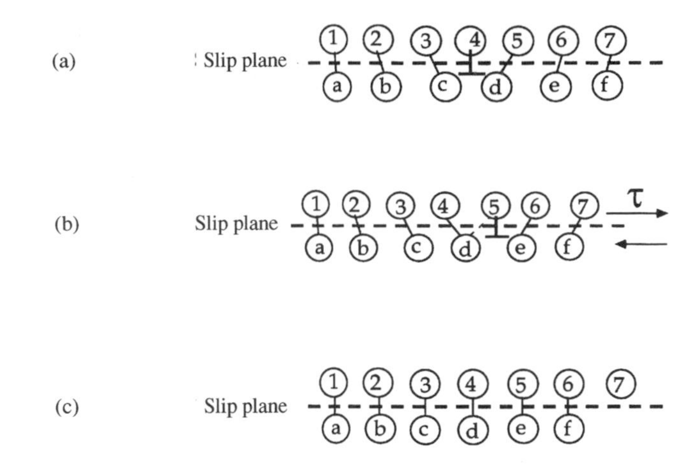

Figure 3.4.17 *Progress of a slip plane during deformation*

We can view plastic deformation as the movement of dislocations through a crystal. Hence, to make a metal stronger it is necessary to obstruct the movement of dislocations through the crystal.

Obstructions to dislocations can take the following forms:

- Grain boundaries
- Impurity atoms
- Solute atoms (solute hardening)
- Second phases
- Other dislocations

Dislocations are formed naturally but when they interact with grain boundaries, internal defects and surface irregularities, while under stress, certain sites can become sources of dislocations. Under the influence of a high shear stress, dislocation sources are capable of producing dislocations that can move through the crystal.

Deformation of solid polymers

Unlike metals, solid polymers exhibit viscoelastic behaviour when subjected to a load. As the name implies, this behaviour is a combination of viscous flow and elastic displacement. All fluids are viscous to one degree or another. We can visualize the significance of viscous behaviour by imagining the flow of engine oil or treacle. When heated, the oil or treacle flows more freely. That is to say, its viscosity decreases. If fluid in a container is stirred, all the work done is absorbed and the fluid may heat up slightly. Hence the viscous element in the polymer absorbs and dissipates. The elastic element is the equivalent of a spring. If the spring is displaced, the energy is stored, since as soon as the force acting on the spring is released, it returns the energy by returning to its original stable state. The deformation of the polymer will therefore show both energy dissipating characteristics and energy absorbing characteristics. It is also dependent on the following factors:

- The rate at which the load or deformation is applied.
- The time for which the load is applied.
- The previous loading history.
- The temperature of the test.

Polymer behaviour can be classified into a few different types since the mechanisms by which they deform are similar and independent of the nature of the material. They are controlled by:

- Strength of primary molecular bonds.
- Strength of secondary molecular bonds.
- Chain geometry.
- The presence of crosslinks or crystallinity.
- The energy available for movement of chains by rotation and translation.

The graphs in Figure 3.4.18 show typical results of a tensile test on a semicrystalline thermoplastic at different temperatures.

Curve (a) – At T_1 ($<< T_G$), there is almost linear extension with increasing load up to failure. Failure occurs in a brittle manner. Such behaviour is seen with amorphous and semicrystalline thermoplastics and thermosetting polymers.

Curve (b) – At T_2 ($\sim T_G$), a yield point may be observed. Necking begins as the load reaches a peak and falls. Failure occurs in a ductile manner. This is true for semicrystalline polymers.

Curve (c) – At T_3 ($>> T_G$), necking is followed by cold drawing. This is only possible with semicrystalline thermoplastics.

Curve (d) – As temperature is increased further ($T_4 > T_3$), necking is no longer observed. This rubber-like behaviour is seen with both semicrystalline and amorphous thermoplastics.

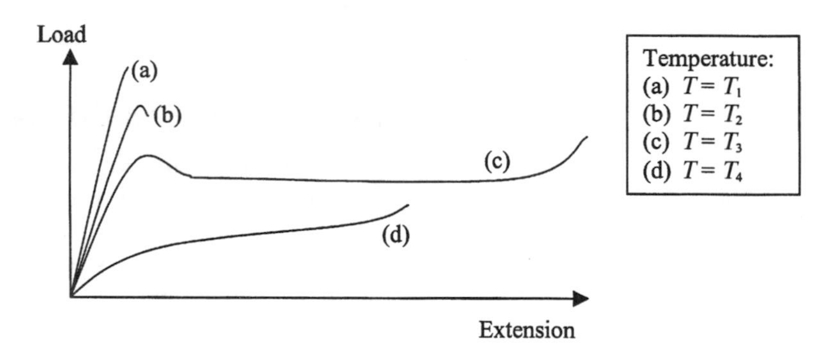

Figure 3.4.18 *Tensile test behaviour of a semicrystalline thermoplastic at different temperatures*

We will use curve (*c*) in Figure 3.4.19 to explain the behaviour of thermoplastics as it contains all the features seen in curves (a), (b) and (d). This represents the behaviour of a thermoplastic at a temperature $T >> T_G$.

As the specimen is forced to extend at a constant velocity, the load increases initially in an elastic manner. This point cannot be determined easily but has been identified with point A in the graph in Figure 3.4.19. The load eventually reaches a maximum at point B. Until this point is reached, the specimen would have deformed uniformly. However, at point B, the deformation becomes localized. The load drops quickly to point C as the cross-sectional area at a point along the specimen reduces.

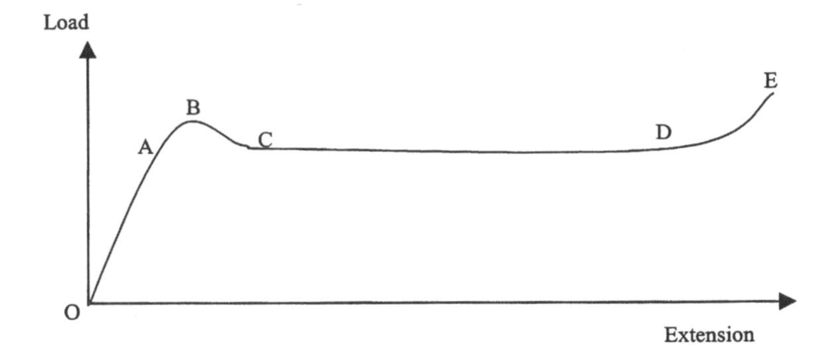

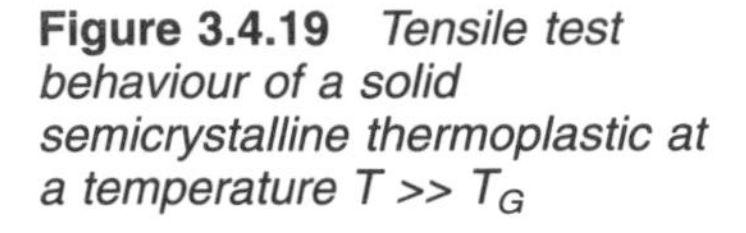

Figure 3.4.19 *Tensile test behaviour of a solid semicrystalline thermoplastic at a temperature $T >> T_G$*

This is described as necking. Between points C and D, three regions are clearly defined as illustrated in Figure 3.4.20. The deformed region (iii) remains stable while the remaining material undergoes necking. Necking is triggered as the shoulder region (ii) moves through region (i) with little permanent deformation.

The stress calculated from the load at C is regarded as the *yield stress* of the original material. Although the load drops, the actual stress in the material remains approximately constant as the specimen extends between C and D while the deformation is concentrated in the necking region, (ii). This process by which the cross-section is reduced is called *cold drawing*. Eventually, when all the material has undergone cold drawing, the load increases (from D to E) until the material fails.

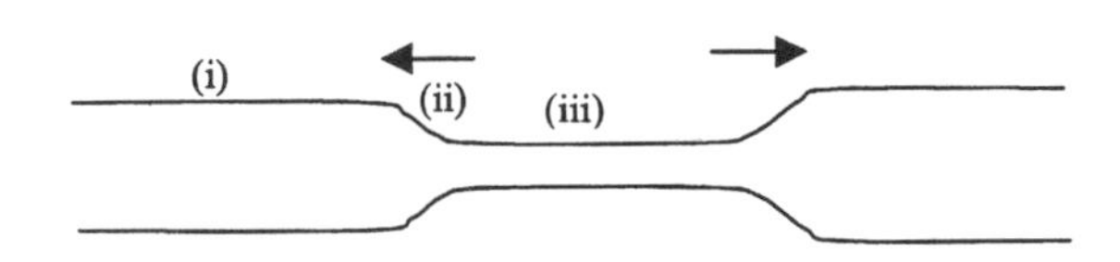

Figure 3.4.20 *Process of cold drawing starting from a neck*

The behaviour described for curve (c) is generally seen for semicrystalline polymers when the test temperature is significantly greater than the glass transition temperature T_G. In the case of high density polyethylene ($T_G = -10°C$) and polypropylene ($T_G = -27°C$), tests at ambient temperature will produce this result. It is also important to maintain the strain rate at a low level. On increasing the strain rate, the behaviour observed becomes more like curve (b). At very high strain rates, brittle failure as indicated by curve (a) will be seen.

Activity 3.4.6

You may all have seen the material called 'silly putty'. It behaves like Blu-Tack when it is moulded in the hand, but if it is rolled up into a ball, it can also bounce as though completely elastic. Now roll a piece into the shape of a cylinder. Then pull the two ends very gently so that the piece is in tension. Describe the behaviour. Now pull on the ends by jerking as fast as possible. Is the behaviour any different? Explain what is happening.

Hardness

A very useful and relatively simple measurement of material properties is the *hardness*. It is difficult to define in a scientific sense and the value depends on the method by which it was measured. One very simple measurement method gives the *Mohs' scale of hardness*. It gives a rating to materials according to their resistance to scratching. Materials are compared by scratching each against one with a known Mohs' value. The one that is more resistant to scratching is given the higher hardness value. Table 3.4.1 shows the relative position of the standard materials in the scale:

Table 3.4.1 The Mohs' hardness scale

Mohs' number	*Material*
1	talc
2	gypsum
3	calcite
4	fluorite
5	apatite
6	orthoclase felspar
7	quartz
8	topaz
9	corundum
10	diamond

This method is valuable for comparing minerals and can be used to identify specific mineral types. With engineering materials a more accurate method is a measure of the resistance to penetration when a loaded indentor is applied to the surface. There are several standard tests using different indentors with different loads. Tests are carried out particularly in industry where a simple method is needed to compare materials on a regular basis.

For example, a manufacturer of pins or nails would typically manufacturer at least several thousand pins every hour. The manufacturing operations for converting a piece of thin cylindrical rod into a pin with a head and a tapered end are complex and need a good knowledge of the behaviour of the material. If the material is too brittle, the material will break up during the process. If the material is too ductile, the pin may lose its shape during formation. Therefore checks have to be kept on the quality of the material bought from suppliers. This is done by carrying out a hardness test on a few pieces periodically. Small changes in hardness value could indicate that either the heat treatment of the material or composition may be unacceptable. The particular batch of material may then be rejected and returned to the supplier before the products are made.

The precise details of the test method are specified, for example, in British Standards (BS). The methods involve the measurement of the size of the indent left behind. The common methods are as follows:

(1) Brinell test – a hardened steel ball is forced into a surface under a precise load and maintained for a fixed period of time according to BS 240 (1986). When the load is removed, the diameter of the crater left behind is measured using an optical microscope with a calibrated scale. The Brinell hardness number, H_B, is defined as,

$$H_B = \frac{\text{applied load (kg)}}{\text{surface area of the indent (mm}^2\text{)}}$$

In order to test a large range of materials, the ratio of applied force, F, to the square of the diameter, D, of the ball is specified for different materials.

(2) Vickers test – a diameter pyramid is used as the indentor in the Vickers test according to BS 427 (1961). The Vickers hardness, H_V, is also given by,

$$H_V = \frac{\text{applied load (kg)}}{\text{surface area of the indent (mm}^2\text{)}}$$

The pyramid shape of the indentor leaves a square impression that is again measured using the calibrated scale of an optical microscope. In this case, the applied load can be varied depending on the hardness of the material.

(3) Rockwell test – various indentors ranging from steel balls to diamond cones can be used to test a range of materials according to BS 891 (1962). The Rockwell hardness is a measure of the depth of the indent produced on a surface. This is obtained as a reading on a dial. This method is quick and so is popular in industry. The method can also cover a greater range of materials than the other methods.

As we have seen, hardness measurements are relative. The value measured does not indicate one property only. In the case of metals, a permanent indent is left behind. Therefore, it is a measure of the work done in causing plastic deformation. In some metals such as steel, it could be a measure of the yield stress. At the other extreme, when a hardness test is conducted on a rubber, the deformation is wholly elastic and therefore temporary. It is therefore a measure of the modulus of elasticity. The results from the different tests cannot be converted to each other universally, although conversion tables are available for specific types of material, e.g. BS 860.

Hardness can be related to the wear resistance of metals and ceramics but is not in itself a fundamental property. Its main use is in quality control in manufacturing processes where it could indicate that other, more important, properties are within specification. The tests are quick and easy to carry out and require little preparation. Tests do not damage the object apart from leaving a small mark on the surface.

Since hardness tests are carried out on one very small area of the surface, a true representation requires several measurements to be taken. It is also worth remembering that surface properties could be considerably different from internal properties in components that have been specially treated.

Toughness and impact strength

Materials are classified as being tough, ductile or brittle according to their behaviour when loaded to destruction. A tensile test could be performed on steel and a curve such as illustrated in Figure 3.4.15 obtained. As previously mentioned, stress is just force per unit area and strain is the extension per unit length of specimen. Hence, the graph shown gives a measure of the work done (= force × extension) per unit volume (= length of specimen × cross-sectional area). That is to say, the area under the curve is equal to the work done in destroying the specimen per unit volume of material. This is just the energy consumed. We can use this idea to define a *tough* material as one that requires a large input of energy to cause it to fail. Hence the material with the highest toughness is not necessarily the strongest material, where the maximum stress is highest, nor is it the one that is the most ductile. Compare the three materials illustrated in Fig 3.4.21.

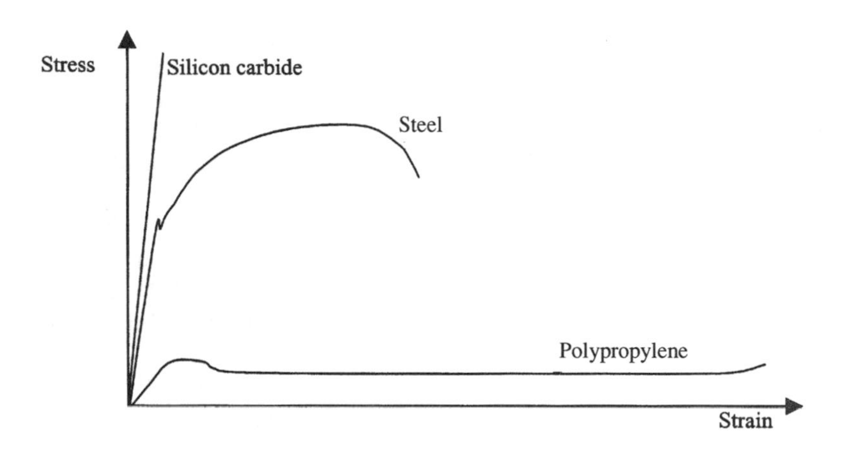

Figure 3.4.21 *Tensile test behaviour of a range of materials*

The strongest material is silicon carbide since it can take the highest stress before failing, typically 600 MN/m^2. However, it has a very low strain to failure which may be 0.3%. The most ductile material would be polypropylene since it will extend the greatest before failure. It may have a strain to failure of 600% but a maximum stress of 25 MN/m^2. The steel, however, may have a tensile strength of 500 MN/m^2 and a strain to failure of 150%. Hence, the area under the curve for steel is the highest, making steel the toughest of the three materials. The high performance ceramic silicon carbide is the least tough material.

The silicon carbide is the only *brittle* material in the example in Figure 3.4.21. It will fail as soon as it reaches the limit that it can extend elastically. A good way of looking at this is to take the broken pieces produced when a specimen has failed and try to put them together so that they form their original shape. A truly brittle material is one that can be put together to form the original shape exactly like a broken vase. Most materials will have some degree of ductility that will lead to a distortion of the shape of the component.

Toughness and brittleness are not opposites (unlike hardness and softness). In most engineering metals they appear to be opposites, i.e.

tough metals are usually not brittle and brittle metals are usually not tough. That is not true for all materials; pure lead, for example, is neither tough nor brittle. Wood, on the other hand, is both tough and almost brittle. Modern composites are also tough and usually nearly brittle. We must be careful not to refer to such materials as brittle since most materials will deform if only slightly. They are better referred to as 'non-ductile' for that reason.

Toughness is a most important engineering property, whereas brittleness is very undesirable. Whether the designer is called upon to design the forks on a bicycle, the body of a kettle or the hull of a boat, the maximum stresses exerted should only cause elastic deformation. However, in all cases, he or she will need to consider what might happen if there is an overload that takes the material beyond its elastic limit. For example, it has been shown that the maximum load on the knee joint of a person jumping off a height of one metre onto a solid surface could be three times as high as the maximum load during walking. If the rider of a bicycle accidentally encounters a small crater in the road, the load on the bicycle frame could again be much higher than under normal road conditions and under extreme conditions could exceed the elastic limit of the material. Modern racing bikes are made from materials such as carbon-fibre reinforced epoxy. This combination is very strong and stiff, but as riders test the bikes to their limit, the bikes can fail catastrophically and break apart because the material combination is non-ductile. Compare this with the ordinary bicycle made from steel tubes. It may lack the performance capability of a sleek racer, but under shock load it will protect the rider. At worst, it will bend out of shape because of the ductility of the material. In fact in most cases, the designer will prefer to use a material with some ductility even though the component is designed to deform elastically only. This is because of the need for safe operation. If the bike fails catastrophically, without warning, the rider could be seriously hurt and the manufacturer could face a compensation claim.

In practice the results of a tensile test are not very useful as a measure of toughness. The tensile test is carried out relatively slowly, whereas most applications requiring toughness happen under shock loads that happen very quickly. We have considered the shock load on a bicycle. Other knocks also lead to sharp increases in stress for a very short time. The most obvious example of a component designed to withstand shock loads is a car bumper. Hence tests need to be done at a rapid rate. The material behaviour can be very different at high rates of loading. Generally speaking the higher the rate of loading, the more non-ductile the behaviour of the material. Even steel, which is normally tough, may fail by brittle fracture at temperatures below 0°C. This could not be simulated easily by a tensile test.

The impact test is designed to measure the toughness of a material. In this test, a loaded object is dropped at speed onto the material being tested and the energy to fracture is determined. The common methods for testing are:

- The drop weight test – a given mass is dropped from a certain height onto a disc of the test material suspended around its rim. The minimum load required to just fracture the disc is used to determine the fracture energy.

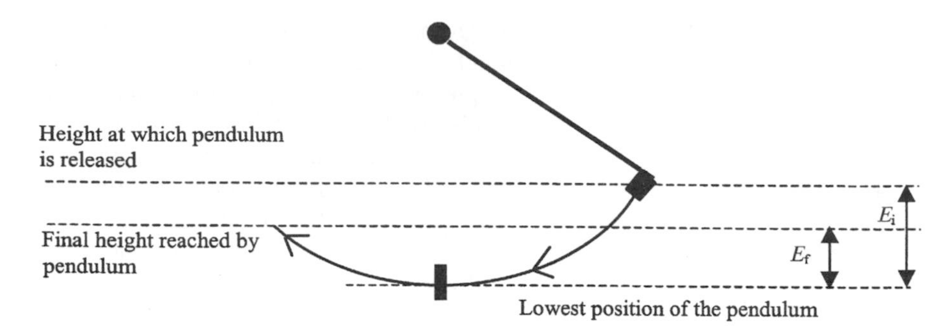

Figure 3.4.22 *Pendulum impact test for the determination of toughness*

- Charpy test – a bar of square cross-section and given dimensions is supported horizontally at its ends and is struck at the centre of its span as shown in Figure 3.4.22. The energy to cause failure is simply given by the energy lost by a pendulum as a wedge-shaped impactor at its head, strikes and fractures the test piece. This energy lost is determined by the difference between its initial height and its final height. This is equal to $(E_i - E_f)$.
- Izod test – this is very similar to the Charpy test except that the test piece is gripped vertically and the free half of the test piece is separated by the impact from the gripped part.

Notch impact tests are simple and can be carried out quickly. The results, however, are at best relative, giving only an indication of the material properties. In practice, many pieces would need to be broken as the failure is dependent on microscopic weaknesses within the material which, by their nature, are randomly produced. By cutting a notch into the centre of the test piece, the results of the Charpy and Izod tests can be made more repeatable since the notch will then act as the dominant flaw within the test piece. The dimensions of test pieces and test details are specified in standards such as BS 131.

Activity 3.4.7

When cutting into a surface by drilling, we would use different drill bits for metals, wood and masonry. Explain why this is the case in terms of the properties of the material being cut and the material in the drill bit. Hint: both the hardness and impact behaviour of each material is important in the selection.

Creep

Under the effect of a load, an elastic material will produce an instantaneous strain response. However, under some circumstances, the strain may continue to increase over a period of time. This is referred to as *creep*. A typical graph for metals is shown in Figure 3.4.23.

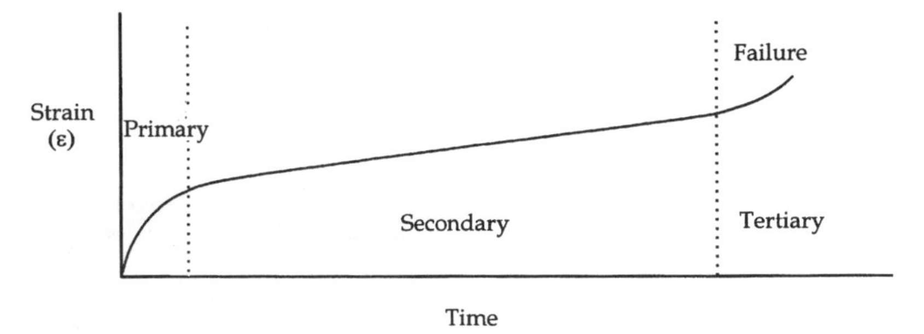

Figure 3.4.23 *Creep behaviour of a metal*

All materials can exhibit creep when operating at a temperature greater than about $0.4\,T_m$, where T_m is the *absolute temperature* at which the material melts (measured in kelvin).

Review – absolute temperature

The absolute temperature is a true measure of the level of energy within a material. Hence the thermodynamic analysis of a material is always best evaluated in absolute temperature. The unit for absolute temperature is the kelvin (or simply K). The lowest possible temperature is zero kelvin. Our usual reference is the melting point of ice at normal atmospheric pressure, which is equal to 0°C, which is the same as 273 K. The boiling point of water at normal atmospheric pressure is equal to 100°C, which is the same as 373 K.

For example in a steel, with a melting temperature of 1500°C (=1773 K), creep must be considered when the operating temperature is greater than,

$$0.4 \times 1773\,\text{K} = 709\,\text{K} = 436°\text{C}$$

Hence this is a problem for high temperature use, e.g. gas turbine engine blades. Many plastics creep at normal ambient temperatures, since their melting temperatures range from as low as 115°C (= 388 K) to just over 300°C (= 573 K) for the high performance polymers such as polyetheretherketone (PEEK). Creep is often the limiting factor in design with plastics. Ceramics are more resistant to creep than other materials since they are stable solids at temperatures in excess of 1500°C. Creep tests are carried out on specimens similar to those used for tensile testing. A load is applied and extension is measured periodically. The temperature is accurately controlled. Tests may last from a few hours to several years and so equipment must be robust and reliable. The results are plotted as creep curves (Fig 3.4.23).

A typical creep curve will show three regions. The primary region lasting a short time shows the fastest rise in strain. In metals, this is treated as being equivalent to the normal elastic deformation of the material. The secondary region takes the longest time of all. The rate at which strain increases with time, or the *creep rate* can be defined in this region and is used for predicting the long-term strain in a metal. Tertiary creep is runaway strain leading to failure. The onset of tertiary creep is difficult to predict, and so components need to be carefully monitored as they approach the end of their life.

Activity 3.4.8

Pile up some fine sand on a flat surface. Tap the surface a few times and see what happens to the shape of the pile. Now take a piece of silly putty, roll it into a ball and leave it on a flat horizontal surface for about 20 minutes. Explain how and why the shape has changed. Try to find a link between the behaviour of the two materials. Hint: in both cases, energy has to be given to the material to cause a change.

Fatigue

Fatigue is the failure of a material under *fluctuating* or *oscillating* loads. The failure stress is much less than the value at which the material would yield under static loading. Testing can be simple, but very sophisticated experiments may be used. A simple laboratory test involves rotating a specimen with a mass suspended from one end (Figure 3.4.24).

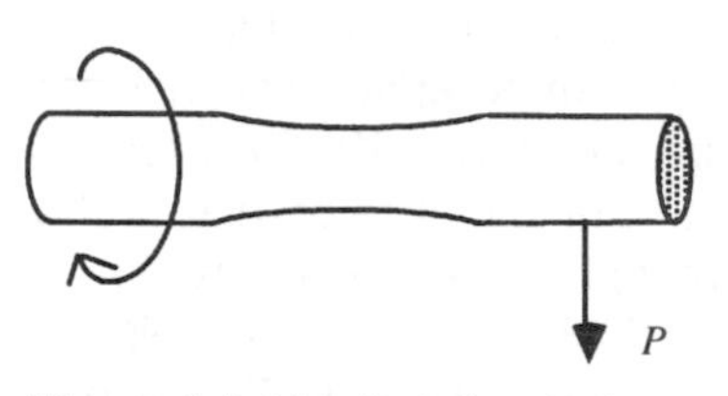

Figure 3.4.24 *Rotating test piece under load at one end to determine fatigue characteristics*

Many tests are carried out with different loads and the results are plotted as graphs of stress (S) against number of cycles to cause failure (N), the so-called *S/N curves*. The results for steel (Figure 3.4.25) show an important limit.

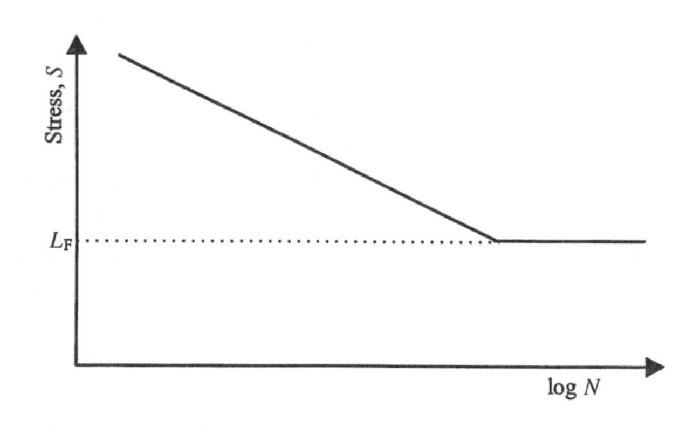

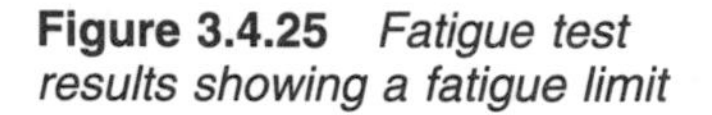

Figure 3.4.25 *Fatigue test results showing a fatigue limit*

L_F is the stress called *fatigue limit* below which failure will not occur no matter how many cycles there are. This is a useful design parameter since fatigue failure can be completely avoided if the stress is below this value. Most materials, however, do not have a fatigue limit. Their behaviour is as illustrated in Figure 3.4.26.

Such materials will fail at any stress given a sufficiently large number of cycles. The *endurance limit*, L_E, is then given for design purposes. This is the stress below which failure will not occur for a specified, usually large, number of stress cycles. During the life of many high

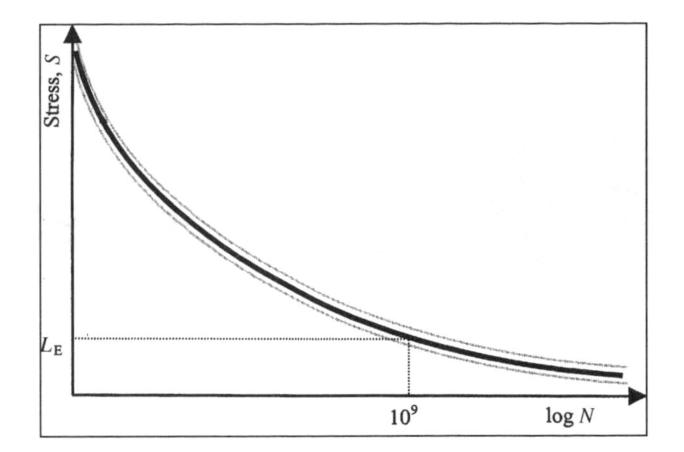

Figure 3.4.26 *Fatigue test results showing an endurance limit and a band to indicate the uncertainty of results*

performance components, they may experience between 10^8 and 10^{10} cycles of loading. Fatigue behaviour is very sensitive to slight defects in surfaces that act as initiators for cracks to begin growing. This gives considerable scatter in the experimental results. Thus, *S/N* curves are often plotted as probability bands of failure rather than as lines.

Many pencil cases are made as a single piece consisting of the base and cover connected by a narrow strip that acts as a hinge and a press-fit closing mechanism. The case is opened and shut many times during its life and therefore the material experiences fatigue. Polypropylene is used in this application as it has a very high endurance limit.

Further reading

Callister, W. D. *Materials Science and Engineering*. John Wiley & Sons, 1999

Meyers, M. A., Chawla, K. K. *Mechanical Behaviour of Materials*. Prentice Hall, 1998

John, V. B. *Testing of Materials*, Macmillan, 1992

4 Processes

Summary

This chapter deals with the processes available to designers. It is important that a designer appreciates how the processes are carried out as their design often sets and constricts the processes available to manufacturing.

The chapter starts with taking an overview briefly describing many of the processes available. The chapter then discusses in three sections the main different means of producing a product's features.

Section 4.2 looks at the processes involving removing some of the starting material. Section 4.3 looks at the processes which conserving material by changing the starting shape. Section 4.4 looks at the process of joining together other prepared components.

Objectives

By the end of this chapter, you should be able to:

- appreciate that there is a large range of processes available for changing the shape and characteristics of starting materials to produce components and finished products (Section 4.1):
- understand the mechanism and forces involved in the material removal operations and their effect on the finished component (Section 4.2);
- understand the material conservation processes such as casting and forming (Section 4.3);
- understand the methods of forming permanent joints between two components (Section 4.4).

4.1 Selecting a manufacturing process

A manufacturing process takes raw, or semi-finished material in one form and converts it into another form. There are normally a choice of several materials, starting shapes and associated processes that can be functionally satisfactory to make the new form. Criteria such as resultant strength, weight and other properties need to be considered as well as the process characteristics such as time, quality and cost.

Processes come under one of four main categories:

- Material removal, e.g. cutting or machining.
- Material shape changing, e.g. casting, forging or bending.
- Material joining, e.g. welding, coating or assembly.
- Material characteristic changing, e.g. heat treatment.

A fairly comprehensive list of process is given under 'Manufacturing processes – a brief overview', below. The main processes in the first three categories are discussed in some detail within this chapter in the following sections. The material characteristic changes are discussed in Chapters 3, 5 and 6 which deal with materials.

A model of any individual process would include the starting and finishing material characteristics, the geometry of the material and the tool and the flow of energy. The process is the interaction between those that change the starting material during the operation. The geometric change is effected by the relative movement of the tool and the workpiece and/or the shape of the tool itself. The final materials will include any waste produced as a byproduct. The prime energy process considered is normally from the tool on the workpiece, but may be involved in changing thc statc of the material. Energy is lost at each stage and often produces an energy waste that has to be specifically countered.

Manufacturing processes – a brief overview

Non-shaping processes

These do not alter the basic shape of the process but either add a coating or change the surface characteristics and texture.

Chemical-based actions

- Chemical vapour deposition (CVD): A high temperature wear and corrosion resistant surface coating technique.
- Electroless plating: Alternative to electroplating for nickel–phosphorus wear and corrosion resistant coating.
- Electroplating: Used to coat a wide variety of materials with another metal which has better surface wear or corrosion properties.
- Anodizing: Used for producing a thick oxide coating to give protective or decorative finishes to aluminium.
- Phosphating: Applied as a protection to steel parts before painting.
- Toyota diffusion process: Mainly used on tool steels to increase life by applying a layer of alloy carbides.

Thermal-based actions

- Physical vapour deposition: Used on metals and ceramics tools to give a decorative finish or wear and corrosion resistance.
- Thermal spraying (hardfacing): Used for repair of components and to provide wear, heat and corrosion resistance in a variety of materials.

- Laser surface treatment: Used on cast iron and steels to give localized heat treatment or to bond a metal to a variety of materials.
- Plasma nitriding/carburizing: Low temperature surface hardening by increasing carbon content or forming nitrates on the surface.
- Plasma arc spraying: Used for repair of components and to provide wear, heat and corrosion resistance in a variety of materials.
- Carbonitriding/carburizing: Thermochemical process to increase surface layer of carbon or nitrides to give hard wearing.
- Injection flame hardening: Mainly used on ferrous metals, especially cylindrical components and gear teeth to change surface hardness.
- Ion implantation: Used for tool steel and carbides to give wear and fatigue resistance.
- Nitrotec process (oxygen enhanced): Used on mild steel and low alloy steel pressings to increase strength, and wear, fatigue and corrosion resistance.

Mechanical-based actions

- Shot peening: Used with a variety of metals to improve surface finish and characteristics through work hardening.
- Shot blasting: Used to clean surface prior to coating.

Shaping

These processes change the shape of the starting material.

Material removal processes

These processes 'cut' away material to leave the finished component.

Chemical actions

- Electrochemical machining (ECM): Used to produce complex shapes in difficult to machine materials.
- Electrochemical grinding (ECG): Ideal for thin and hard metals.
- Photochemical machining: Selective etching away of materials.

Thermal actions

- Electrical discharge machining (EDM): Produces shape by removing material by high frequency sparks in a dielectric fluid. Used in a variety of materials, mainly tools.
- Electrical discharge wire cut (EDWC): Similar to EDM, but for cutting a continuous path.
- Laser cutting: Good for precision cutting of low melt point material.
- Plasma arc cutting: Material removed by a high velocity stream of ionized gas.
- Flame cutting: Material removed by oxidizing in a stream of high purity oxygen.

Mechanical actions

- Single point cutting: Material removed by passing a single cutting edged tool over workpiece. Can be rotational or linear. Very common process.
- Multi-point cutting (linear): Material removed by a series of 'teeth' – sawing and broaching.

- Multi-point cutting (rotational): Material removed by a series of defined teeth – milling and drilling.
- Grinding: Material removed by a series of small grit embedded in a matrix.
- Centreless grinding: Specialist grinding of a cylindrical workpiece without using holding devices.
- Creep feed grinding: Special high material removal grinding process.
- Water jet cutting: Material removed by high velocity water jet. Ideal for non-metallic material and soft metallic materials.
- Abrasive jet cutting: A wide variety of materials can be processed. Similar to water jet cutting but with an abrasive material added. Media can also be air.
- Ultrasonic machining: Slow process for highly brittle material such as glass and ceramics.

Mass saving process

These processes attempt to conserve the starting material by altering the outer shape.

Solid deforming action

- Blow moulding: Used to make thin wall plastic and glass bottles and containers.
- Hot forging (open die): Metal is shaped between flat or simple contoured dies – depends on skill of operator for shape. Temperature is over 65% of melt point of component material to reduce stress on dies.
- Hot forging (closed die): Metal is shaped between finish contour dies.
- Hot/cold rolling: Shape produced by multi-passes through shaped rollers.
- Hot/cold extrusion: Material is forced through a shaped hole which gives the cross-section profile.
- Upset forging: Cross-section of part of the component is increased.
- Isothermal precision forging: High cost forging process normally limited to air-space components of titanium alloy.
- Rotary forging (GFM): Similar to hot forging but multi-hammers employed.
- Orbital forging: Useful in producing large flanged items as reduced area of contact.
- Axiforce process: In effect a two-stage forging process.
- Cold forging: Used in forging and extruding ductile materials. Temperature is less than 25% of the melt point.
- Warm forging: Used in mid-strength materials to give some stress relief on dies. Temperature is 25%–65% of the melt point.
- Stretch forming: Sheet material deformed by stretching over a block.
- Deep drawing: Sheet metal drawn to a deep recess.
- Fluid and rubber die forming: Sheet forming where one half of die is replaced by a more flexible pressure.
- Metal spinning/flow turning: Forms a sheet into a hollow shape.
- Vacuum forming: Used for thermo-forming plastic sheet.
- Conform process: Continuous extrusion of softer metals.

Liquid starting material processes

- Injection moulding: Can produce complex plastic mouldings.
- Sand casting: Simplest material moulding system – uses sand to contain molten metal in desired shape.
- Metal mould casting: Similar to sand casting but moulds made out of metal allow their reuse.
- 'V' process (vacuum sealed moulding): Alternative to sand casting where a thin plastic film is vacuum formed over the sand surface.
- Low pressure sand casting: Similar to die casting but using zircon sand moulds used for aluminium alloys.
- Investment casting: Ceramic mould produced from wax pattern.
- Resin shell casting: Resin shell mould produced on metal pattern.
- Ceramic mould casting: Refractory moulds used for high melting point alloys. Plastic for lower melt points.
- CLA/Hitchiner process: Similar to investment casting, but uses vacuum fill rather than gravity.
- Gravity die casting: Uses metal moulds for highly fluid alloys.
- High pressure die casting: Metal moulds but restricted to lower melting point alloys.
- Low pressure die casting: Mainly aluminium application for low quantity runs.
- Vertical high pressure die casting: Variation on high pressure with vertical filling to overcome some problems.
- Acurad process: Variation on high pressure using a secondary plunger to overcome some problems.
- Pore-free die casting: Uses an active gas to reduce air inclusion during mould fill.
- Hyperforge process: Combined casting and forging operation.
- Squeeze form/casting: Uses a liquid charge into closed dies.
- Compression moulding: Simple forming of thermoset plastics from preheated charge.
- Transfer moulding: Similar to compression moulding but has combined heating stage.
- Monomer casting: Used for many plastics. Charge is liquid which polymerizes in mould.
- Full mould casting. Similar to investment casting but uses foam for expendable pattern and sand for mould.
- Centrifugal casting: Produces a long hollow cylinder through filling a rotating mould.
- Osprey process: Spray deposits molten metal into a mould. Blank is further processed.
- Continuous casting: A specialized process that allows continuous casting of fixed cross-sections. Used for steel and non-ferrous metals.
- Injected metal assembly: Lead or zinc alloy injected into a joint to form assembly.

Slurry starting material processes

- Contact moulding: Used for laying down glass reinforced thermosetting plastics.
- Slip casting: Used to produce hollow shapes in clay, concrete, ceramics and metals.

- Powder cold press and sinter: Production of powder components by pressing then heating in controlled atmosphere.
- Powder forging (hot press): As cold press with addition of forging after heat treatment.
- Hot isometric pressing (HIP): Heat treatment of cast and sintered components in hot pressurized inert gas.
- Pressureless sintering: Sinter product produced by compaction and heating.
- Cold isostatic pressing: Compaction of powder by uniform gas/fluid hydrostatic pressure.
- Extrusion of powders: Powder compressed by extrusion ram pressure.

Joining processes

Complete fusion of materials

- Friction welding: Joining of dissimilar metals, plasticized by high friction heat produced by spinning.
- Electron beam welding: Vacuum or gas protected – high very localized heat with deep penetration.
- Laser beam welding: Similar to electronic beam using laser for heat source.
- Manual metal arc (MMA) welding: Flux coated electrode rod used. Electricity used to melt small portion of base material and filler rod.
- Gas metal arc welding – TIG/MIG/MAG: Inert gas used to shield molten materials.

Bonding of materials

- Soldering: Low temperature process where joining material is below the melt point of the base materials. Seldom joining to base material.
- Brazing: Low temperature process where joining material is below the melt point of the base materials. Some joining to base material.
- Adhesives: Commonly used to join dissimilar materials. Normally does not react with base material.
- Mechanical fasteners: Wide range of disassemble and non-disassemble types available.

4.2 The material removal processes

This section covers the main material removal processes. It uses the 'cutting' action of a single point tool to describe the basic effects involved, and the speeds and power needed to economically produce a component. It then describes briefly a range of the other material removal techniques available.

Material removal involves starting with a piece of material which is larger in all dimensions than the finished component. Sections of unwanted material are then removed, i.e. cut away, to leave a finished, or semi-finished, component.

Material removal does not change the basic properties but it will cut across the natural grain of the material which may produce stress points. There may be localized changes at the surface nearest to the removal process. The material removed is usually heavily deformed with changed properties and is often contaminated. It has a lower value than the starting material does.

Components can be worked on by manual processes such as filing and chiselling, but we shall only examine the main industrial processes in this text. We use this process to remove large volumes of material quickly, and to achieve high surface finish and dimensional accuracy. We can use it as:

- The initial process (e.g. cutting a long bar into shorter lengths for further processing).
- A finishing process following another process to improve surface finish on tolerances (e.g. after forging or sand casting).
- A complete conversion process for a component (e.g. turn a bolt complete from a bar).

The cutting operation

Cutting processes are classified by the number of cutting faces involved. We shall use a single point tool to discuss the detail of the operation – multi-point tooling has the same action at each of the points in use.

In single point tools, only one tool face is used for the operation. Motion can be applied to the workpiece, the tool or both, depending on the weight of the workpiece. The heavier and bulkier the workpiece, the more likely it is to be kept stationary.

The processes using a single point cutter are:

Rotational. Here the workpiece or the tool moves in a circular path. These are the most common cutting processes of external turning and internal boring as shown in Figure 4.2.1. The machines used range from

(a) Straight turning

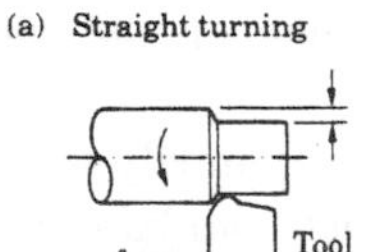

(b) Taper turning

(c) Profiling

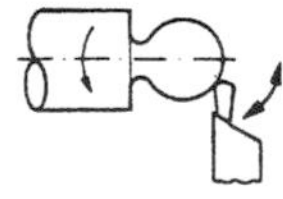
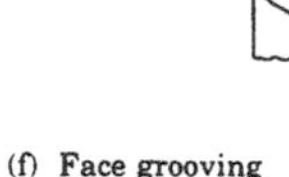

(d) Turning and external grooving

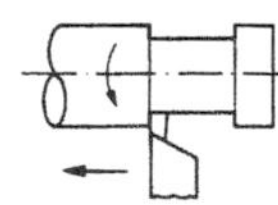

(e) Facing

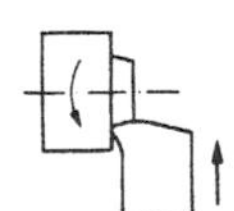
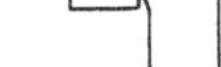

(f) Face grooving

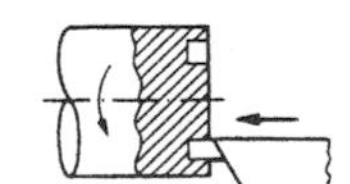

(g) Form tool

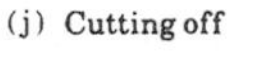

(h) Boring and internal grooving

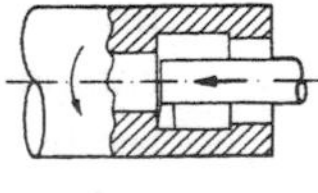

(i) Drilling

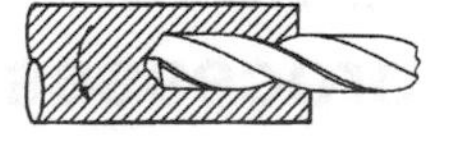

(j) Cutting off

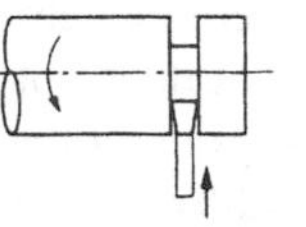

(k) Threading

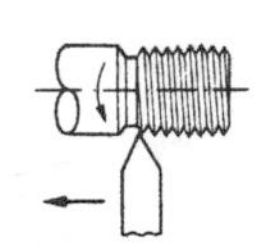

(l) Knurling

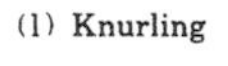
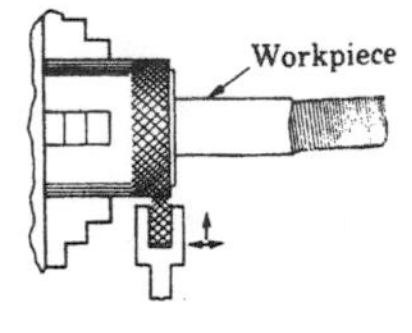

Figure 4.2.1 *Typical turning operations. (From Doyle and Each, c. 1985* Manufacturing Processes for Engineering Materials. *Reproduced with permission of Pearson Education Inc., Upper Saddle River, NJ07458)*

manually controlled units through a series of semi-automatic units to the computer controlled units of today. Trepanning for cutting large holes also comes under this classification but more commonly uses two tools for balancing forces.

Linear. Here the workpiece or the tool moves in a straight line. Processes are shaping, where the tool moves and planing, where the workpiece moves. These machines were largely used as an alternative to milling (see 'Multi-point tools', page 155). The shaping process has largely fallen into disuse, but the planer is still in use for large sections.

We shall initially have a look at the detailed effects of what is happening during the cutting operation and demonstrate through an example the calculations involved in working out an economic operational time when using a single point tool.

The basic operation

The operation is really not a cutting action as you would achieve with a knife. It is actually one of shear (see Figure 4.2.2). We are not going into a full description of the forces involved in the cutting action within this textbook, but some detail is required to appreciate what is happening. In the cutting action, there are three component forces as shown in Figure 4.2.3. We are not going to go deeply into a full explanation of these forces involved, but we do need to appreciate at the design stage that they occur as the component must resist them.

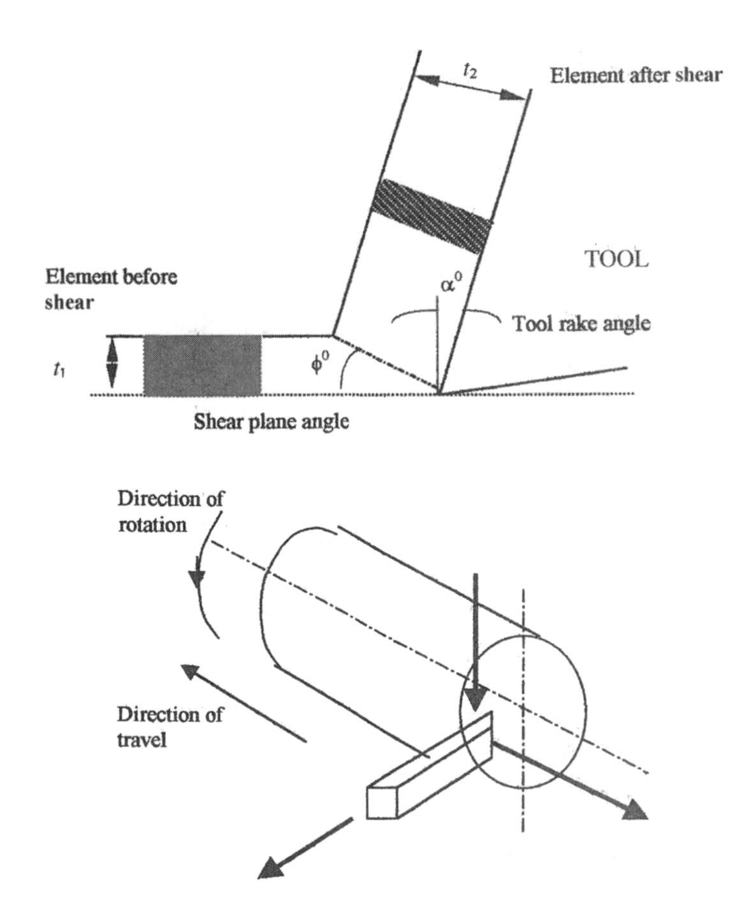

Figure 4.2.2 *Action at the tool face in material removal*

Figure 4.2.3 *Forces involved at cutting point are shown in thick lines*

The tool moves though the material and subjects the material immediately before it to a compressive pressure. That pressure builds up until the compressed material is eventually sheared from the remainder – the only unrestricted path for it is along the tool face as in Figure 4.2.1. This sheared material is termed the chip (swarf is used as a collective name for the chip produced). Note that the chip thickness, t_c, is more than the depth of cut, t_d, this is due to the way the sheared material is reformed as it moves away from the workpiece.

Activity 4.2.1

Take a knife, hold it next to an edge of a pad of butter with the blade at a right angle to the pad and scrape it across the butter. What happens to the butter at the blade edge?

Try the same action but this time vary the tilt of the knife. Is the butter coming off the same thickness with the different angle?

Much of the work done produces heat in the sheared metal. Most of this heat is carried away with the chip, dissipating to the tool, the coolant used and the surroundings. Some, however, does enter the workpiece affecting the immediate material – this can give rise to changes in that surface material's characteristics due to a form of heat treatment. The strains involved in the shearing process also can change this immediate material causing surface imperfections and work hardening.

The cutting is often carried out with coolant, but sometimes the material or the tool performs better without any cooling. Coolant can be neat mineral oil, synthetic oil or water-based emulsion with additives to counter corrosion and bacterial growth. Sometimes air, or pure oxygen is used instead of a fluid.

Functions of a coolant are:

- To prevent thermal distortion in the workpiece and the tool.
- Increase tool life by reducing the operating temperature.
- Reduce friction between the chip and the tool.
- Improve the surface finish

The coolant can be a danger to health, especially if allowed to become airborne or if left untreated in the machine which allows bacterial growth to take place. Some fluids can cause skin problems, such as dermatitis or cancer, so it is important that operators do not come into extended contact with them.

Cutting speed

The speed at which the tool moves through the base material is dependent on the combination of the machinability of the material being cut and the tool being used. The cutting tools are subject to a variety of conditions:

- High pressure from the cutting forces.
- High heat produced raises the operating temperature and gives thermal shock.

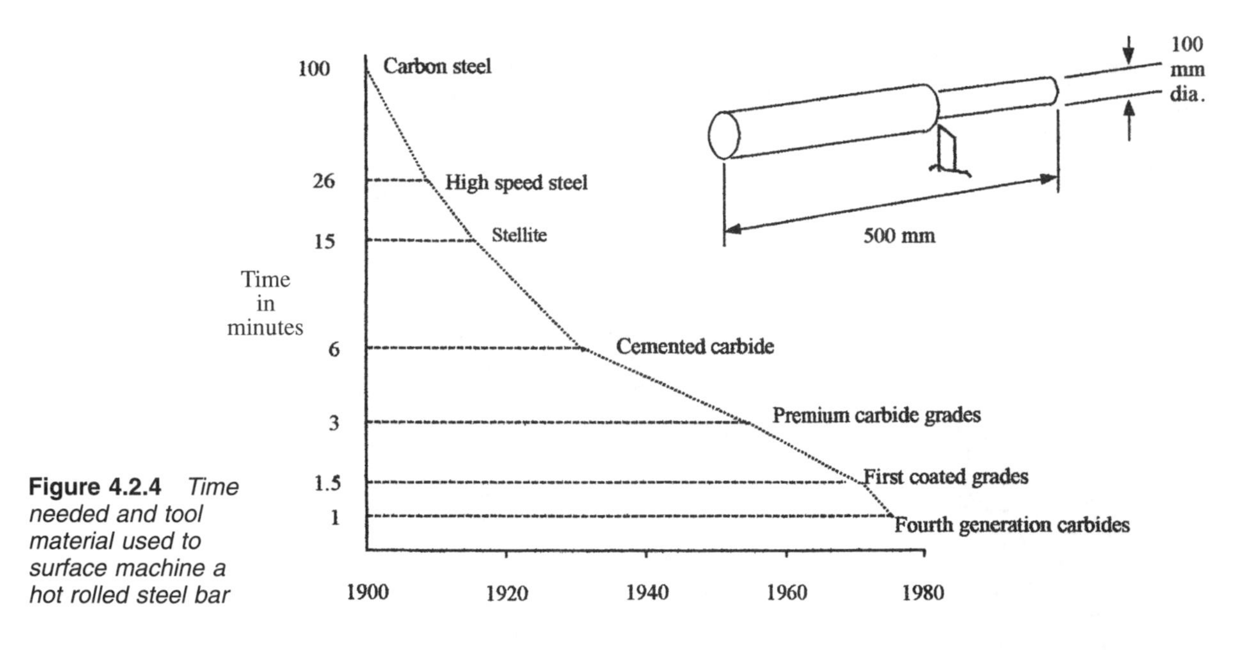

Figure 4.2.4 *Time needed and tool material used to surface machine a hot rolled steel bar*

- Shock loads from initial contact, interrupted cutting and variations in the material structure.
- Abrasive wear from the sliding action of the chip and contact with workpiece.
- Attrition wear from the chip 'welding' to the tool and then being broken away.
- In addition the body of the tool will be at a cantilever and subjected to a bending moment.

These conditions mean tool materials have to be a blend of different characteristics of hardness and toughness. The selection of the tool material is critical to ensure that it lasts an economic time while retaining its shape and sharpness.

The developments in cutting tool materials in the twentieth century have led to an increase in cutting speeds resulting in substantial drop in machining time as shown in Figure 4.2.4. The improving tooling has also contributed to the requirement for higher power cutting units with increased rigidity.

Cutting tool manufacture itself is a specialized subject. If you are required to select a cutting tooling, it is best to contact the tool suppliers. They will give you information relating to the material being cut and the different tools with the preferred operating conditions to select as shown in the extract in Figure 4.2.5.

The basic formula for calculating tool life is Taylor's:

$$V \times T^n = C$$

where

V = surface speed in metres/min
T = tool life in minutes
n = constant determined by insert grade and material
C = ideal surface speed in metres/min which will give 1 minute life – determined by insert grade, material, feed, depth of cut, coolant used, etc.

GENERAL TURNING Cutting data

First choice geometry and grade recommendations

A

Application	T-MAX P	CoroTurn
Extra finishing	CNMG 12..	
Finishing	CNMG 12..	CCMT 09..
Medium	CNMG 12..	CCMT 09..
Light roughing	CNMG 16..	CCMT 09..
Roughing	CNM. 16..	CCMT 09..
Heavy roughing	CNM. 19..	

ISO	CMC Coromant Material Classification	HB Hardness Brinell	MATERIAL	APPLICATION	Negative inserts T-MAX P: Geometry	Grade	Insert type	Nose radius	Rec. a_p mm	Rec. f_n mm/r	Rec. v_c m/min	Positive inserts CoroTurn (T-MAX U) (Single sided): Geometry	Grade	Nose radius
[illegible]	[illegible]	[illegible]	Non-alloy carbon steel	*Extra finishing* *Finishing* *Medium* *Light roughing* *Roughing* *Heavy roughing*	QF PF PM PR PR HR	4015 4015 4025 4025 4025 4025	G G G G M M	04 08 08 12 12 16	0,5 0,4 3,0 4,0 5,0 10.0	0,12 0,2 0,3 0,4 0,5 0,8	430 395 325 290 260 205	PF (UF) PM (UM) PR (UR)	4015 4025 4025	04 08 08
	[illegible]	[illegible]	Low-alloy steel	*Extra finishing* *Finishing* *Medium* *Light roughing* *Roughing* *Heavy roughing*	QF PF PM PR PR HR	4015 4015 4025 4025 4025 4025	G G G G M M	04 08 08 12 12 16	0,5 0,4 3,0 4,0 5,0 10,0	0,12 0,2 0,3 0,4 0,5 0,8	465 425 330 290 265 210	PF (UF) PM (UM) PR (UR)	4015 4025 4025	04 08 08
	[illegible]	[illegible]	High-alloy steel, annealed	*Extra finishing* *Finishing* *Medium* *Light roughing* *Roughing* *Heavy roughing*	QF PF PM PR PR HR	4015 4015 4025 4025 4025 4025	G G G G M M	04 08 08 12 12 16	0,5 0,4 3,0 4,0 5,0 10,0	0,12 0,2 0,3 0,4 0,5 0,8	340 295 220 195 180 145	PF (UF) PM (UM) PR (UR)	4015 4025 4025	04 08 08
	06.2	[illegible]	Steel castings, low-alloy	*Extra finishing* *Finishing* *Medium* *Light roughing* *Roughing* *Heavy roughing*	QF PF PM PR PR HR	4015 4015 4025 4025 4025 4025	G G G G M M	04 08 08 12 12 16	0,5 0,4 3,0 4,0 5,0 10,0	0,12 0,2 0,3 0,4 0,5 0,8	220 200 150 135 120 95	PF (UF) PM (UM) PR (UR)	4015 4025 4025	04 08 08
M	05.21	180	Stainless steel, austenitic, bars/forged	*Finishing* *Medium* *Light roughing* *Roughing* *Heavy roughing*	MF MM MR QR HR	2015 2025 2025 4035 4035	G G G M M	08 12 12 16 16	0,4 3,0 3,0 5,0 10,0	0,2 0,3 0,35 0,5 0,8	250 180 165 135 95	MF (UF) MM (UM) MR (UR)	2015 (1025) 2025 2025	04 08 12
	05.52	180	Stainless steel, austenitic/ferritic, bars/forged (Duplex)	*Finishing* *Medium* *Light roughing* *Roughing* *Heavy roughing*	MF MM MR QR HR	2025 2035 2035 235 235	G G G M M	08 12 12 16 16	0,4 3,0 3,0 5,0 10,0	0,2 0,3 0,35 0,5 0,8	255 235 205 120 92	MF (UF) MM (UM) MR (UR)	2015 (1025) 2035 (235) 2035 (235)	04 08 12
M-S *Super alloys*	23.22	R_m 1050	Titanium alloys	*Finishing* *Medium* *Roughing*	-23 -23 QM	H10A H13A H13A	G G G	08 08 12	1,0 2,0 4,0	0,2 0,3 0,4	65 53 49	UM(G) KM (UM) KR (UR)	H10A H13A H13A	04 08 08
	20.22	350	Heat resistant alloys Ni-base	*Finishing* *Medium* *Roughing*	-23 -23 QM	H10A H13A H13A	G G G	08 08 12	3,0 3,0 4,0	0,2 0,35 0,4	40 20 20	UM(G) KM (UM) KR (UR)	H10A H13A H13A	04 08 08
K	[illegible]	[illegible]	Grey cast iron, high tensile	*Finishing* *Medium* *Roughing*	KF KM KR	3005 3015 3015	G G A	08 12 16	0,5 3,0 4,0	0,2 0,35 0,55	250 210 180	KF (UF) KM (UM) KR (UR)	3005 3005/3025 3015/3025	04 08 12
	[illegible]	[illegible]	Nodular cast iron, Perlitic	*Finishing* *Medium* *Roughing*	-KF KM KR	3005 3005 3005	G G A	08 12 16	0,5 3,0 4,0	0,2 0,4 0,55	270 220 190	KF (UF) KM (UM) KR (UR)	3005 3005/3025 3015/3025	04 08 12
[illegible]	[illegible]	HRC 60	Hard steel	*Finishing* *Medium* *Roughing*	.NMA .NGA	7020 670	A A	08 12	0,2 0,2	0,1 0,15	150 100	.CMW	CB20	08
[illegible]	30.22	90	Aluminium alloys	*Finishing* *Medium* *Roughing*	-23	H13A	G	12	3,0	0,35	2000	.CMW AL AL	CD10 CD1810 H10	04 08 12

SANDVIK Coromant

Figure 4.2.5 *Cutting data (reproduced courtesy of Sandvik Coromant)*

Cutting data **GENERAL TURNING** v_c

Negative and positive inserts
CERAMICS, CBN AND PCD

Rec. a_p mm	Rec. f_n mm/r	Rec. v_c m/min	Geometry	Grade	Insert type	Nose radius	Chamfer	Rec. a_p mm	Rec. f_n mm/r	Rec. v_c m/min
0,4	0,11	485								
0,8	0,2	370								
2,0	0,25	345								
0,4	0,11	520								
0,8	0,2	375								
2,0	0,25	350								
0,4	0,11	380								
0,8	0,2	250								
2,0	0,25	235								
0,4	0,11	245								
0,8	0,2	175								
2,0	0,25	160								
0,4	0,11	255								
0,8	0,2	205								
2,0	0,3	180								
0,4	0,11	310								
0,8	0,2	285								
2,0	0,3	235								
0,5	0,15	70								
0,8	0,2	60								
2,0	0,25	55								
0,5	0,15	30	**.NGN**	670	N	12	T01020	1,0	0,12	432
0,8	0,2	15	**.NGN**	670	N	12	T01020	2,0	0,2	351
2,0	0,25	15	**.NGN**	670	N	12	T01020	3,0	0,3	271
0,4	0,11	265	**.NGA**	650	A	12	T01020	1,5	0,3	550
0,8	0,2	250	**.NGA**	6090	A	16	T02520	3,0	0,4	500
2,0	0,3	230	**.NGA**	690	A	16	T02520	3,0	0,4	500
0,4	0,11	300	**.NGA**	650	A	12	T01020	1,5	0,3	400
0,8	0,2	270	**.NGA**	1690	A	16	T01020	3,0	0,4	350
2,0	0,3	240	**.NGA**	690	A	16	T02520	3,0	0,4	350
0,2	0,1	150	**.NGA**	7020	A	08	S01020	0,2	0,1	150
			.NGN	670	N	12	T01020	0,2	0,15	100
			.CMW	CB20	W	08	S01020	0,2	0,1	150
0,5	0,1	2000	**.CMW**	CD10	W	04	F	0,5	0,1	2000
1,5	0,3	2000								
1,5	0,3	2000								

Insert geometry and grade recommendations

1. Find the nearest material type — CMC code number (Coromant Material Classification). See also the Material Cross Reference list on pages J 4 – 8.

2. Establish the working area:

Finishing
Operations at light depths of cut (D.O.C.) and low feeds

Medium
Medium to light roughing operations.
Wide range of D.O.C.and feed rate combinations.

Roughing
Operations for maximum stock removal and/or severe conditions. High (D.O.C.) and feed rate combinations.

Extra finishing:	f = 0,05 - 0,15 mm/r a_p = 0,25 - 2,0 mm
Finishing:	f = 0,1 - 0,3 mm/r a_p = 0,5 - 2,0 mm
Medium:	f = 0,2 - 0,5 mm/r a_p = 1,5 - 4,0 mm
Light roughing:	f = 0,4 - 1,0 mm/r a_p = 3 - 10 mm
Roughing:	f = 0,5 - 1,5 mm/r a_p = 6 - 15 mm
Heavy roughing:	f = ≥ 0,7 mm/r a_p = 8 - 20 mm

3. Find the insert most suitable for the toolholder that has been selected. If the operation is affected by intermittent machining, vibrations or limited machine power, this should be taken into consideration in the final choice of insert.

Note!
If difficulty should arise in finding a suitable insert for the toolholder selected, it may be a better solution to choose an insert first and then go back and choose a suitable toolholder, or maybe even a different toolholder system.

Insert type

The fourth sign in the ordering code = insert type. See code key on page A 6.

CNM**G**

- G = Double sided insert with chip breakers
- M = Single sided insert with chip breakers
- A = Flat insert with hole
- N = Flat insert without hole
- W = Flat insert with hole, screw clamping

a_p mm = Recommended starting value for cutting depth in mm

f_n mm/r = Recommended feed value in mm/ revolution

v_c m/min = Recommended starting value in m/min for cutting speed

Figure 4.2.5 *Continued*

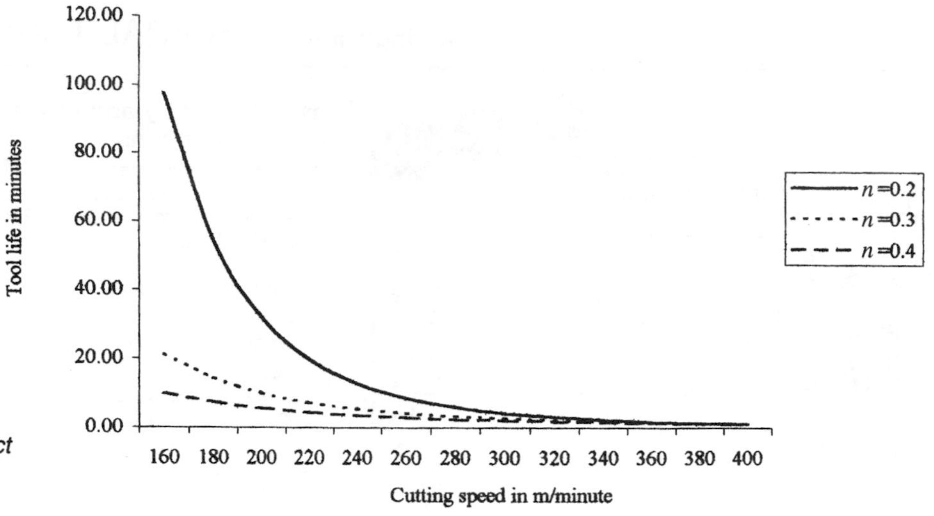

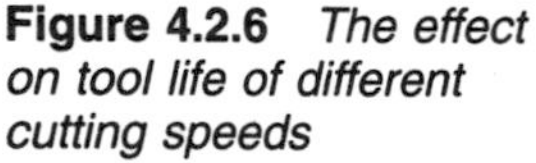

Figure 4.2.6 *The effect on tool life of different cutting speeds*

The significance of the above formula is that within a range, tool life is exponentially dependent on the cutting speed. From it we can see that an attempt to reduce the operation time by cutting at a high surface speed can result in a significant reduction in the tool life. If we examine a range of cutting speeds, we can see the change in the tool life. Figure 4.2.6 shows the effect on tool life of different tool materials in relation to cutting speed.

The data shown in a cutting tool manufacturer's handbook will tell you the expected tool life that different tools will have. In the extract in Figure 4.2.5, the expected life is 15 minutes. However, as the case study (page 150) shows tool life is only one factor in determining cutting conditions to give optimum operations.

Power required

It is possible to theoretically calculate the power used from the forces involved, but in practice this is made difficult due to the inconsistency in the structure of the material being cut. A standard value is therefore used to give an indicative specific cutting energy (SCE) required to remove different material as in Table 4.2.1.

Table 4.2.1 Range of specific cutting energies

Material	*Range of Brinell Hardness*	*Specific Cutting Energy* ($J\,mm^{-3}$)
Carbon steel	*150–300*	*2.0–3.0*
Stainless steel	*150–200*	*3.2–3.6*
Hardened alloy steel	*250–500*	*3.2–3.9*
Extra hard steel	*550–620*	*5.8–6.2*
Low tensile grey CI	*150–200*	*1.0–1.3*
Malleable CI	*120–250*	*1.7–2.2*
Chilled CI	*400–500*	*4.2–4.5*

The material's Brinell hardness value can be used to give an approximate SCE based on the ratio of $1\,J\,mm^{-3}$ per 100 Brinell value. This approximate relationship is modified depending on the characteristics of the material, but gives a reasonable first guide.

The actual value is further modified by:

- Changes in the tool face (rake angle α in Figure 4.2.2): Increasing the angle reduces the force but also reduces the strength of the tool point. Decreasing the angle increases the force, but also increases the strength of the tool point. A reasonable adjustment in SCE is 1.5% per degree change.
- The feed rate: Perhaps surprisingly the higher the feed rate, the lower the cutting force. This is due to a combination of factors, which are not covered here, the main one being an increase in the wedging effect of the tool.

As shown in Table 4.2.1, the power required is related to the volume of the material being removed in relation to time. To find the material removal rate (MRR) we take the cross-sectional area of the cut (feed, f, multiplied by depth of cut, d) and multiply it by the cutting speed, V, i.e. MMR = $V \times f \times d$. Be careful here that all these are in the same unit of measurement – say mm. To ensure we maximize the power available we then use the following formula to check the power needed:

Power required = MRR × SCE for the material

If we know the power available and the material, we can work out the possible material removal rate by transposition as shown in the example on page 148.

Note that not all of the power going into the cutting machine is available at the tool point – there are losses in the machine mechanism due to gears, friction, etc.

Surface finish and texture

Modern machine tools can accurately position tooling and component. With low wearing tools there is a pressure to attempt to achieve maximum metal removal rates even with the finishing cuts. To remove large quantities of metal, roughing cuts are used with corresponding poor surface finish. Finishing cuts to give a better surface finish require low feed rates, shallow depths of cut and/or large diameters at the tool point (which requires extra power). In machining terms there is no such thing as a perfect surface. Quality in surface finish and accuracy involve extra time or an increase in power due to changes in tool shape – these increase cost.

The machining symbols give an indication of the specified finish and the lay required by the function. The actual surface produced can be split into five components (see Figure 4.2.7):

- Tool point shape and feed used give the basic underlying form.
- Roughness: Small-scale irregularities caused by the cutting action.
- Waviness: Cyclic characteristics of the machine tool – workpiece movements.

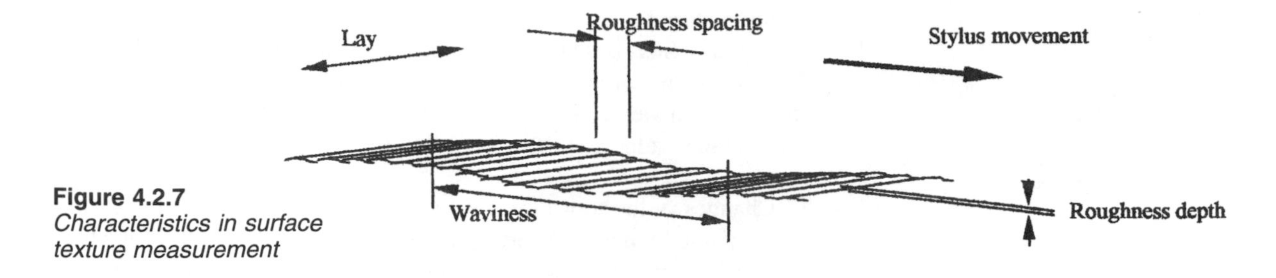

Figure 4.2.7
Characteristics in surface texture measurement

- Form: Gross errors resulting from items such as inaccurate tool setting, distortion of the workpiece, inadequate support, slide-way variation and wear.
- Lay: Surface marking caused by direction of tool movement over the surface.

In addition the 'surface integrity', the difference between the base material and the material of the surface, and the sub-surface, may suffer weaknesses due to work hardening, small cracks or residual stresses.

Measuring surface roughness

Surface roughness can be measured by different processes. A common one is through drawing a sharp-pointed diamond stylus over the surface across the lay direction. To eliminate errors of form and waviness only a small set length is sampled as in Figure 4.2.7.

Once the trace is produced, it can be measured and the surface roughness calculated. This can be expressed in different ways, the R_a method uses the sum of the areas above and below a mean line divided by the sampling length (see Figure 4.2.8):

$$\text{Surface roughness, } R_a = \frac{(\text{sum of areas } r) + (\text{sum of areas } s)}{L}\ \mu\text{m}$$

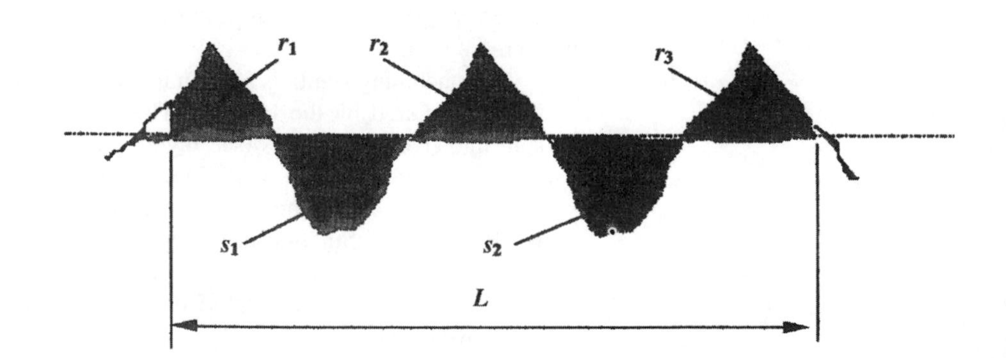

Figure 4.2.8
RMS measurement

Unwanted extras – burrs and swarf

The final aspects we are going to consider are the unwanted byproducts of the cutting process.

The burr is not a separate item that is produced. It is an unwanted feature left at the end of a cut in the remaining material. It is formed at the end of a cut by the compressive force bending over the last piece of material rather than shearing it off (see Figure 4.2.9). This burr is sharp and is potentially dangerous to anyone handling the component. It can also cause damage to another component or prevent the component entering a fixture or assembly process. It must therefore normally be removed. This can be through a final operation on the main machine, or as a separate process later.

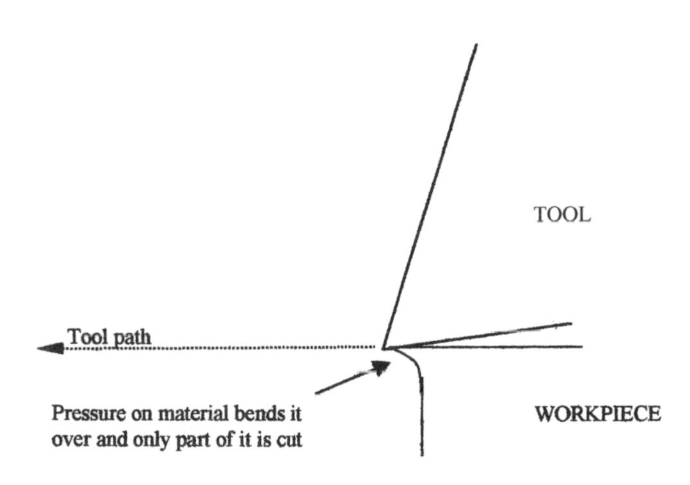

Figure 4.2.9 *Formation of burr at end of cut*

Swarf is the name given to all the chip that comes off. The chip, under ideal cutting conditions will flow off in a continuous ribbon-like stream, with its characteristics largely that of the parent material. Unfortunately the forces involved in the cutting process mean that the chip has been subjected to both physical and thermal distortion. It will therefore probably have different characteristics to its parent.

In addition because it has a relatively large surface area, that surface will probably be heavily contaminated by oxidation and other chemical reactions degrading the material. If a coolant has been used then that, or its residue, will be coating the surface and further devaluing the swarf collected.

Under ideal cutting conditions the shape of the chip (i.e. a very hot, long and thin ribbon of sharp edged material) is potentially very dangerous and machines require guarding to prevent operators and passers-by being injured. The long ribbon of swarf is not in an ideal form to collect and store. It requires to be broken up as it leaves the tool face or it will tangle on the tool or the workpiece causing damage.

Example 4.2.1

As you can see from the following example, the whole process of selecting operating conditions is a long, iterative one.

If we have a requirement to cut a carbon steel bar (see Figure 4.2.10) on a machine with a 10 kW motor, what feed, speed and depth of cut would we select? The SCE for carbon steel is $2.5\,\text{J mm}^{-3}$. Net power available is 8 kW after allowing for losses. This is the equivalent of $8000\ \text{J sec}^{-1}$.

$$\therefore \text{MRR} = \frac{8\,000\ \text{J sec}^{-1}}{2.5\,\text{J mm}^{-3}}$$

$$= 3200\,\text{mm}^{-3}\ \text{sec}^{-1}$$

We then have to decide on the tool, the surface cutting speed, the depth of cut and the feed, while taking into account the desired tool life. If we consult the toolmaker's data (see Figure 4.2.5), we find that using a PR4025(G) insert will give a tool life of 15 mins when roughing carbon steel at a surface speed of $260\,\text{m min}^{-1}$. As the starting size of the material is 100 mm (i.e. 0.314 m circumference) this equates to a rotational speed of 828 revolutions per minute.

We now have to examine both the component and the machine to select the actual cutting speed, the depth of cut and the feed. The nearest available speed is 850 revolutions per minute, which we select. This means one revolution takes 0.07 seconds. This gives an initial actual surface speed of $4450\,\text{mm sec}^{-1}$ for this component. This is slightly higher than the recommended surface speed, but as the diameter being cut reduces in size the smaller circumference means that a lower speed is actually used for most of the cuts. Some machines do have the ability to programme in a change of rotational speed as the diameter reduces, but in this case we shall assume a constant speed. Dividing the surface speed into the MRR means a cross-sectional area of $0.66\,\text{mm}^2$ can be cut.

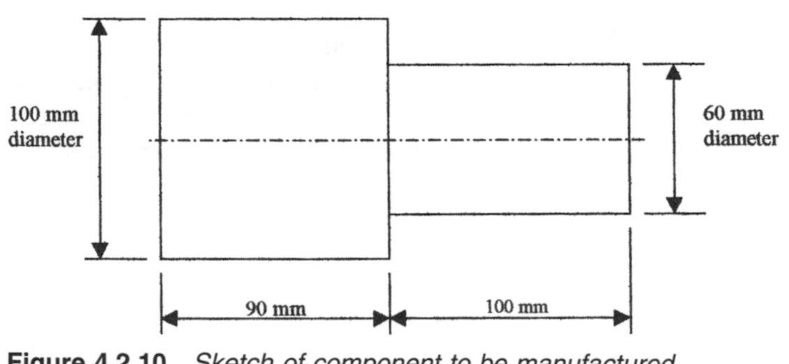

Figure 4.2.10 *Sketch of component to be manufactured. Material carbon steel*

We now have to select the actual feed and speed. These will determine:

- How long one cut takes, i.e. the time taken for one traverse of the workpiece
- The number of times the tool has to be reset, i.e. brought back to the starting point at the end of each cut.
- The desired surface finish.

If we assume a roughing cut, then the surface finish does not matter, we should therefore be concentrating on minimizing the combined time of cutting and tool resetting. We already commented that a heavy feed reduces the power required, therefore we will select a feed of 0.3 mm rev^{-1}. This means a depth of cut of 2.2 mm to achieve the cross-section area of 0.66 mm^2. The feed and depth of cut are well within the limits laid down by the tool manufacturer.

It will take 333 revolutions to traverse the 100 mm cut at the feed of 0.3 mm rev^{-1}. Therefore at 0. 07 sec per revolution one cut takes 23.3 sec. The total depth of material to be removed is 20 mm, of which we will leave approximately 1 mm for the finishing cut. Therefore number of cuts is $20/2.2 \approx 9$. Total cutting time is 9×23.3 seconds = 210 seconds = 3 mins 30 seconds.

To this needs to be added a time allowance for the movement involved to reset the tool for the start of each cut. This will vary from 10 seconds on a manual machine to under 1 second when under NC control. This adds extra time to the operation.

From the example we can see how important it is to have high power available at the tool point and to make full use of it by selecting the correct cutting conditions. This enables a minimum of resetting.

Note that the tool material selected gave only 15 mins operation under these conditions. This means that the tool would need replacing after only four components. In this case the insert will be turned to a new point. It was the requirement for constant tool changing which led to the introduction of inserts in tooling, rather than having to change the whole tool. The full benefits from using inserts are:

- Speedy replacement of cutting point.
- Accurate repositioning of the tool point on replacement.
- Allowing a tough shank to be used to cushion a hard but brittle insert.
- Reducing need to resharpen tools.
- Reducing the expensive material used at the cutting point.

Case study

Decision making on CNC units at Connelly & Co., Iron Foundry, Bulawayo, Zimbabwe

Decision making in choosing operating parameters with CNC units is a complex one due to the way the various costs interact – many of these costs and interactions can only be established with experimentation over the materials and machines.

Normally, the most important factor to be considered in CNC working is to ensure that the actual cutting process is in use to a maximum. The main areas where savings can be made are:

- Load/unload operation.
- Setting-up costs.
- Repositioning of tool between cuts.
- Cutting tool usage – insert cost and tool changing cost.
- Actual cutting cost – power, labour, deterioration (i.e. maintenance).
- Material costs.

Load/unload operation

Unloading. It is important that the completed workpiece be removed efficiently and quickly and transferred to a point very close to the chuck/fixture. While the new item is being machined, the old one can have ancillary work done, such as cleaning and deburring, and be transferred to another spot.

Loading. Again the new item should be brought very close to the chuck/fixture while the previous one is being machined. As much as possible of the clamping operation should be carried out off the machine so the loading operation is at a minimum.

Setting-up costs

Again as much as possible should be done off the machine so that the chuck/fixture is quickly located and clamped in a designated position.

The same applies to all tooling – it must be fully preset off the machine with offsets noted so that when it is located on the machine the tool tip's position is accurately known and the offsets keyed in.

The setting costs can be reduced through minimizing changes required between batches. This is achieved by grouping items into families using the same chuck/fixtures and tooling configurations.

Tool path

Although one advantage of CNC units is that tools can be quickly repositioned between cuts, this should not blind the programmer to minimizing tool repositioning. Therefore the tool cutting should be at a maximum. This is achieved by heavy roughing cuts and ensuring that multiple paths used follow criteria of maximum length.

Tool costs

Calculating tool costs is a complex operation – some information can be obtained from the manufacturers, but this often has to be modified through experimentation on the materials being cut by the different grades of inserts.

Taylor's basic formula for tool life is

$$V \times T^n = C$$

where:

V = actual surface speed in metres/min
T = tool face life in minutes
n = constant determined by insert grade and material
C = the surface speed in metres/min which will give 1 minute life – determined by insert grade, material feed, depth of cut, coolant used, etc.

The significance of the above formula is that within a range, tool life is exponentially dependent on the cutting speed. An Excel worksheet has been used to produce a graph (see Figure 4.2.6) showing tool lives under different V and n factors for a standard value of C. It could be that operating under slower cutting speeds can reduce overall costs (see later) where tooling costs are relatively high compared to marginal operating costs.

One problem is that the workpiece material is not homogeneous and can include hard skin layers, inclusions and blowholes which can all reduce tool life. The cutting conditions affect tool life through wear, chipping, localized adherence and the frictional heat produced, and significant material variations can cause tool fracture.

Material, item design and raw material quality (pre-CNC)

The main influence on the CNC (computer numerical control) machining costs are item design and excess material. Therefore the main reduction in CNC costs will arise through improvements in quality from the foundry. As has been seen, the foundry is capable of high quality work, but sometimes quality problems slip through and affect the CNC operations:

- Hard material, not rectified by heat treatment.
- Different grade of material, perhaps at the end of a melt during changeover.
- Blowholes.
- Inclusions.
- Rough surfaces.
- Sand scabbing (not seen).
- Misalignment of patterns or core.

These can cause:

- Reduced tool life.
- Extra cuts.
- Slower feeds and speeds.
- Extra power input.
- Breaking down of set-ups to send material for rectification causing extra set-ups.
- More scrap.

Excess material basically means extra cuts, including repositioning tools for these and more of the tool's life used up to produce an item. It is therefore essential that quality improves in the foundry to both reduce material input, machining cost and waste production. Note that similar problems can arise where the raw material comes from other than foundry sources – e.g. plate cutting and pre-machining.

Selected cutting speed

Obviously the selected cutting speed has a major impact both in the cutting operation and the tool life as in Taylor's formula above. Also see later.

Overall costs

If we take:

$C1$ = marginal cutting costs/min – machine operator, power, deterioration, etc.
$C2$ = set-up costs – programming, tool pre-setting, chuck/fixture preparation.
$C3$ = tooling costs – cost per insert edge or cost per reground edge.
$C4$ = tool changing costs/min – machine operator or setter.
$C5$ = load/unload costs/min – machine operator.

With:

t = time tool removing metal.
t_c = time to change tool.
t_s = time to set up job.
t_h = time to load/unload.
T = tool life.
N = batch size.

We have to minimize, bearing in mind delivery requirements:

$$\text{Cost per item} = (C1 \times t) + (C2/N) + (C3 \times t/T) + (C4 \times t/T) + (C5 \times t_h)$$

Unfortunately often when we optimize one cost component we tend to worsen the effect of another as in Table 4.2.2.

Table 4.2.2 Effect of changes in setting

Change	*Reduce*	*Increase*
Increase cutting speed	$C1 \times t$ and lead time	$C3 \times t/T$ and $C4 \times t/T$
Improve presetting	$C5 \times t_h$	$C2/N$

With large batches, it is worthwhile investing in experiments to determine the optimum cutting speed. With small batches two different approaches can be taken:

Minimum cost or maximum throughput

To minimize cost we set cutting speed,

$$V_{min} = C/[(1/n-1) \times (t_c \times C2/C1)]^n$$

where

C = constant from Taylor's tool-life formula
n = constant from Taylor's tool-life formula
Tool life, $T = (1/n - 1) \times (t_c \times C2/C1)$

To maximize throughput, we may wish to ignore costs such as tooling and setting:

Cutting speed, $V_{max} = C/[(1/n - 1) \times t_c]^n$

To maximize profit, we select a cutting speed between V_{min} and V_{max}. The graphs in Figure 4.2.11 show the effect on cost and throughput as the cutting speed changes.

Design pointers

In addition to the above details on the process, it is worthwhile emphasizing some points to bear in mind while designing:

- Starting size: Because of the large amount of expensive waste that can be produced, use the minimum starting size. It may be more economical to produce the starting material in a process such as forging or casting (see Section 4.3) to reduce this waste.
- Material: To maximize machinability, it is sometimes useful to start with a material in an easy to machine condition and then modify it afterwards by surface treatment or heat treatment for service characteristics.
- Form/Features: Remember a tool has to access the material to remove it, therefore a clear tool path is required. Swarf may also be trapped inside a complex internal shape.
- Work-holding: Because of the high forces involved, the material has to be securely held during the cutting process – make allowances for this as the holding force may damage the surface.

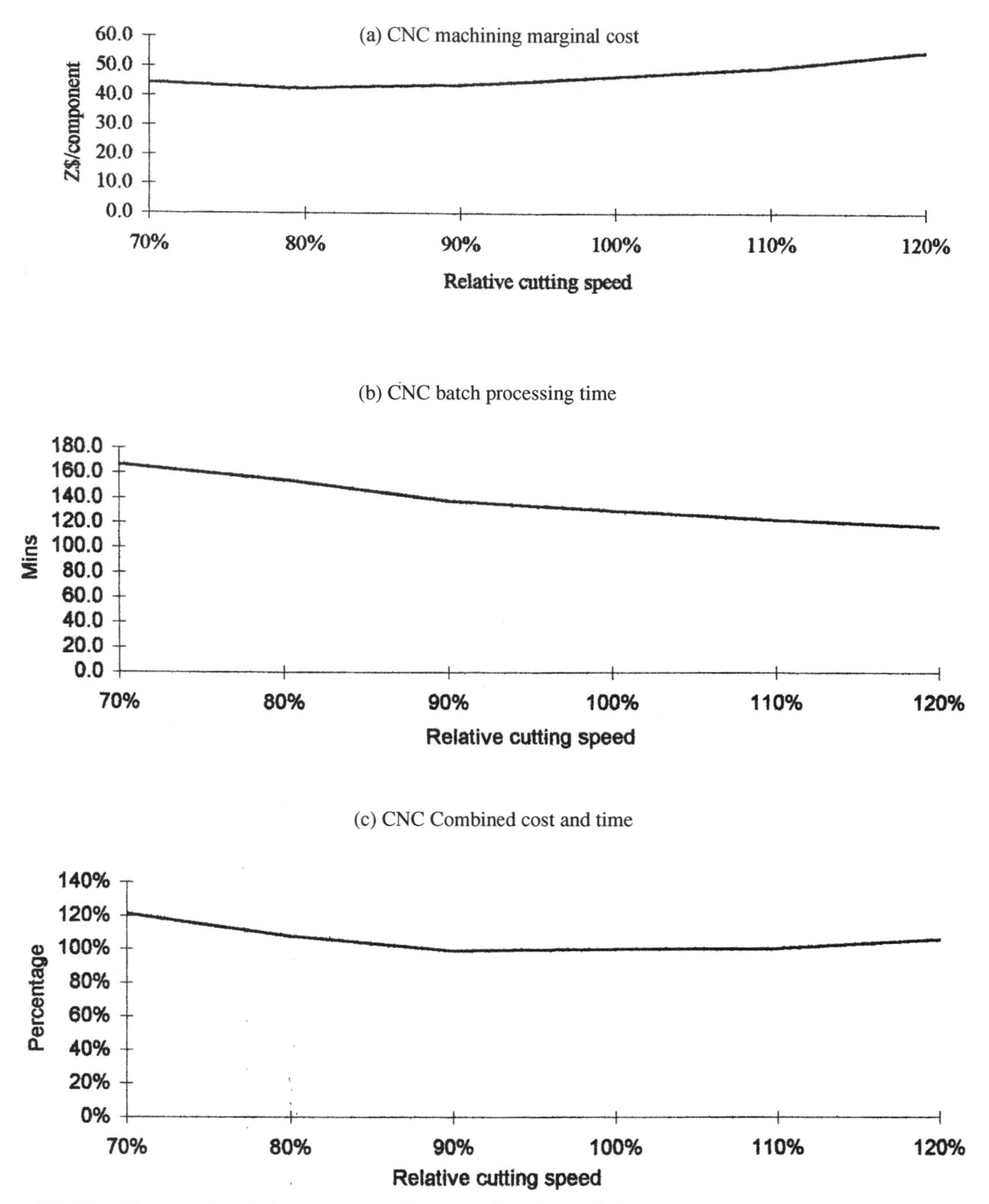

Figure 4.2.11 *Changes in minimum cost and processing time relative to standard cutting speed. (From Doyle and Each, c. 1985* Manufacturing Processes for Engineering Materials. *Reproduced with permission of Pearson Education Inc., Upper Saddle River, NJ07458)*

- Distortion: The high forces and elevated temperatures will subject the component to forces which may distort it. Avoid long, slender components as they will bend away from the forces.
- Sharp internal corners: This requires a very sharp point on the cutting tool. A sharp point is very weak and therefore will break or wear quickly.
- Coolant marks: Often the coolant may react with the parent material and cause unsightly marks or rusting. A finish treatment may be required.

Multi-point tools

There are a multitude of different machines and tooling arrangements in this classification. The condition at each of the points is similar to that in single point cutting.

Rotational

The tool comprises a series of what are basically single point tools arranged round a circle. Processes include milling, drilling and trepanning.

Milling

Used for producing flat or straight profiled surfaces. Can use a variety of tool profiles cutting on side, face or end of tool as Figure 4.2.12. Milling forms the basic cutting action on modern CNC (computer numerical control) machining centres.

As with single point tooling, consult the tool manufacturer for operating parameters. The surface speed is the peripheral speed of the tool and feed is normally given as mm per tooth. Some machine feed will have to set as mm per minute which will mean a calculation from the feed per tooth data.

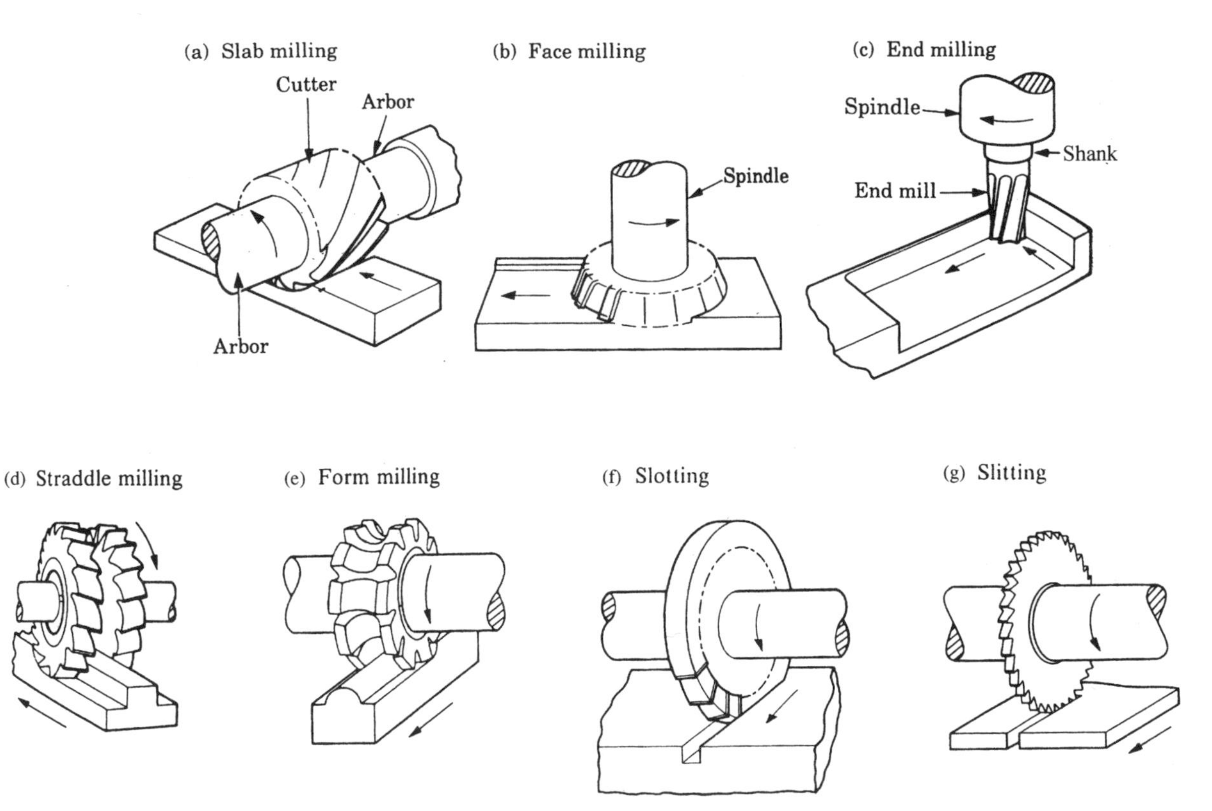

Figure 4.2.12 *Typical milling operations. (From Doyle and Each, c. 1985* Manufacturing Processes for Engineering Materials. *Reproduced with permission of Pearson Education Inc., Upper Saddle River, NJ07458)*

Maths in action

To find feed in mm per second knowing:

- Feed per tooth, f_t mm (taken from tool manufacturer's catalogue)
- Cutter diameter, D mm
- Number of teeth in cutter, N (normally related to cutter diameter)
- The cutting speed, V mm sec^{-1} (taken from tool manufacturer's catalogue)

Revs per minute, $R = (60 \times V)/\pi D$ revs per sec

Feed setting $= f_t \times R \times N$ mm per minute.

Note: ensure that f_t, D and V are all in the same units.

As we have seen already on page 144, power depends on the material removal rate. Therefore in milling to find the metal removal rate, we take the depth (mm) of cut, feed in mm per minute and width (mm) of cut (which may be the full width of the cutter diameter).

In milling we can either use down (also known as climb) milling or conventional (up) milling (see Figure 4.2.12).

Down milling takes a heavy cut first on the top of the component. One component of the cutting force acts downwards and adds to the clamping force. It should only be used where there is anti-backlash facility on the feed mechanism as the initial bite can dig in and attempt to move the workpiece. It produces a shock on the component and tool tip and tool life will be shorter. Down milling should not be used where there is a hard surface, such as scale.

Up milling takes a lighter cut initially on the end of the component. It is gentler on the tool tip, but it tends to pull the component up out of the fixture. The latter is a problem on thin components.

Trepanning is a combination of milling and single point cutting as normally two single point cutters are arranged to cut round a circular path. It is used to cut holes in sheet materials and wood and plastics.

Drilling and reaming

Drilling is similar to end milling, but the profile of the tip (see Figure 4.2.13) means a large axial thrust force acts through the shank. The SCE is some 50% higher than in single point cutting. Consult drill suppliers for recommended speeds and feeds.

Very accurate tolerances on hole diameter are difficult because of the uneven forces on the tool point causing slight wandering off the centre point. If close tolerances are required, normally boring or reaming will be needed as a second operation. Accurate centre positioning is also difficult unless guiding the drill in a jig, which increases the tooling cost

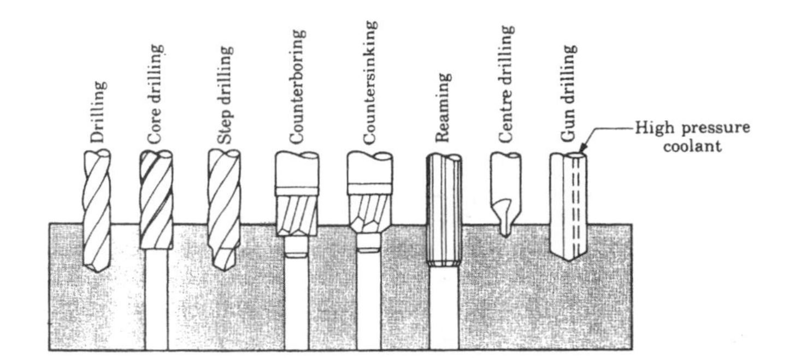

Figure 4.2.13 *Typical drilling/reaming operations. (From Doyle and Each, c. 1985* Manufacturing Processes for Engineering Materials. *Reproduced with permission of Pearson Education Inc., Upper Saddle River, NJ07458)*

for a job. The starting surface for a drilled hole must be perpendicular to the hole axis to prevent the point being diverted away from its position.

Reaming is carried out by a tool similar to a drill except for the flute pattern and the tip profile (see Figure 4.2.13). It is made to close tolerances and removes very little material so that it is subjected to little wear. High diametric tolerances and a good surface finish can be produced through reaming – but it depends upon having a drilled hole first, which limits the positional tolerance to that of drilling as it can only follow that hole.

Deep holes require special care as the chip can only escape back along the flute of the drill. This increases the forces involved often leading to drill breakage. Because of this, special drills are designed for deep hole drilling.

Drilling will leave a burr on the inside of the hole as the drill comes through the material – this normally will need to be removed.

Linear

The tool comprises a series of single point tools arranged in a straight line. Processes include sawing and broaching.

Sawing

A saw blade is a thin straight metal blade with a series of teeth as in Figure 4.2.14. Each tooth is a single point tool removing a small amount of material. The teeth are slightly offset to reduce binding and rubbing

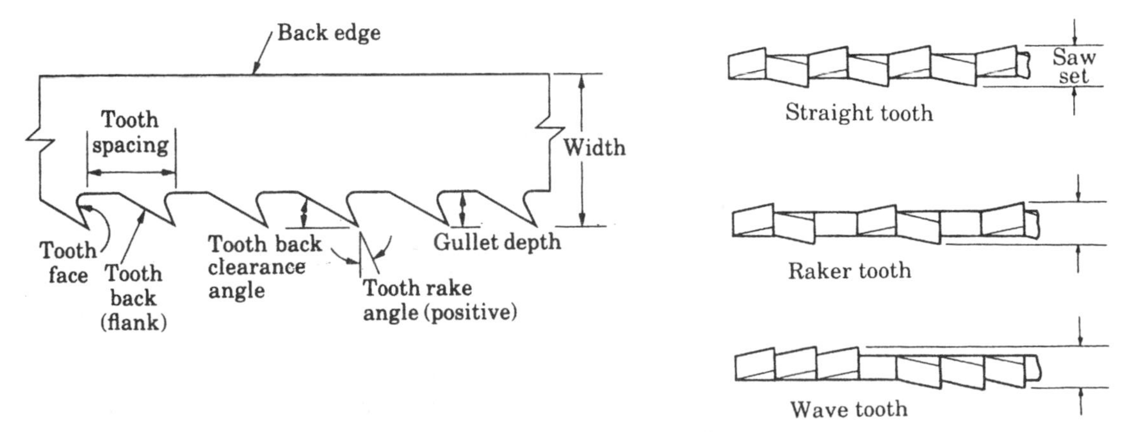

Figure 4.2.14 *Saw blade profiles. (From Doyle and Each, c. 1985* Manufacturing Processes for Engineering Materials. *Reproduced with permission of Pearson Education Inc., Upper Saddle River, NJ07458)*

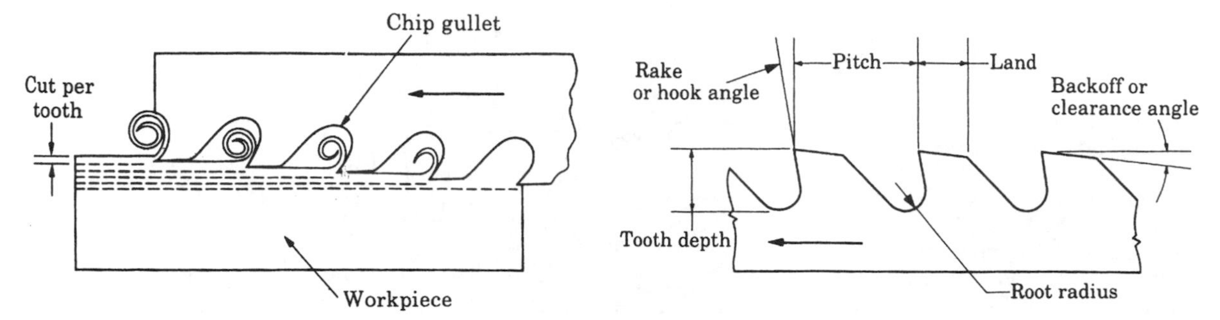

Figure 4.2.15 *Broach profiles. (From Doyle and Each, c. 1985* Manufacturing Processes for Engineering Materials. *Reproduced with permission of Pearson Education Inc., Upper Saddle River, NJ07458)*

during cutting. The number of teeth is dependent on material and its thickness as several teeth should be in contact at any one time.

Manufacturers have a range of blades available. Circular sets can be used to cut a pocket in a variety of materials such as wood.

The thin width of the saw blade means that only a little material is removed during the sawing operation. However, it is difficult to cut to an accurate finish because this thinness means the blade is then flexible.

Variations in sawing are:

- Power hacksaws: Backwards and forwards action with cutting only in one direction. The return movement normally does not cut and can be carried out at a quicker speed.
- Bandsaw: Blade is continuous and therefore always cutting.
- The so-called circular saw is not really a saw. It is a slitting milling cutter.

Broaching

A broach resembles a saw blade, but instead of having a flat thin profile, the cross-section is contoured. By drawing this over a surface we can impart this profile onto its shape. They can be used externally, or internally, and can impart a good finish with close dimensional control.

The teeth are progressive (see Figure 4.2.15) each one cutting a little deeper into the parent material. This enables the cross-section to start out as one shape and progress towards another – as in making a round hole into a square one.

Abrasives and other processes

Grinding

Grinding is basically the action of many small cutting faces. Grinding does not normally remove much material although there are special grinding processes which can do so. It does produce a good surface finish with good dimension tolerances possible.

The cutting faces are formed by a multitude of small hard grits that are supported in a glassy matrix which is formed into the wheel shape. These wheels can grind either on the side, the face or both (see Figure 4.2.16).

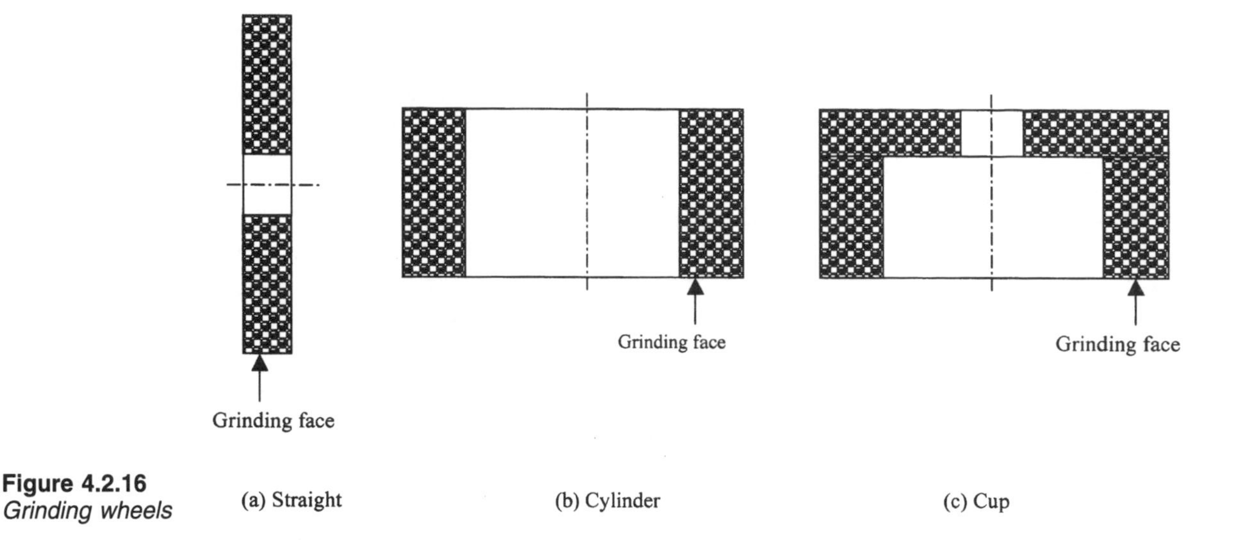

Figure 4.2.16
Grinding wheels

If one of the grits wears, the cutting force on that grit increases. This causes the grit either to break, forming a new cutting face, or to get pulled out altogether from the supporting matrix. The profile of the tool is therefore continually changing in size and contour which means it needs to be regularly inspected and reformed.

The drawbacks of the grinding process are:

- Each grit can only remove a very small amount of material.
- The grits are not optimally oriented causing arbitrary rake angles meaning:
  - Some chip formation is as in conventional cutting, but often at sub-optimal angles.
  - There is some rubbing between the matrix and the workpiece where no grit exists.
  - Ploughing can occur where the grit does not cut at all but plastically deforms the material forming a groove and raised sides.
- Extra energy is required – most of which is translated into heat, which needs removal by flood coolant to prevent workpiece distortion.

The wheel structure is normally porous, often significantly so. This porosity can carry cutting coolant to the cutting action, but weakens the strength of the supporting matrix. This is partially good as it allows blunt grits to escape easily, but we do not want any correctly formed grit to be also pulled out.

The wheel is 'dressed' by knocking out some grits exposing new grits for cutting. The voids between grits can both contain removed material and carry coolant direct to the material. If the bond is too strong and holds in the grits the removed material will fill these voids causing an increase in non-productive rubbing and reducing actual cutting. The wheel wear is caused by:

- Attrition: Creates wear flats on the grits reducing cutting action and increasing power needed and results in extra frictional heat.
- Fracture: Either in matrix, or in grit itself. Essentially a self-sharpening process but causes a change in the wheel form.

The matrices which are used are vitrified, resinoid, rubber and metal. We designate wheels on the basis of these matrices.

Soft grade

These are more suitable for hard material, having a more open structure and high porosity which make them difficult to clog. They are capable of fast metal removal but have a high wear rate.

Hard grade

These are suitable for soft materials, having a more compact structure with low porosity and clogging only avoided by large grits. Slower work rates and less material removed between dressing.

Good tolerances and surface finishes can be achieved with grinding. Where higher finishes are required then honing or lapping can be used.

- Honing: Basically similar to grinding, but the tool is made from aluminium-oxide or silicon carbide. They are worked in a spiral motion and remove very little material.
- Lapping: Again similar to grinding except that the grinding mechanism is by grit embedded in a cast iron or copper lap or suspended in a slurry between the lap and the material being worked on.
- Abrasive belts: A continuous belt similar to emery paper. Grinding mechanism is by grit embedded in the belt material.

Polishing and buffing are also used to give a smooth lustre finish. They will not normally change the material profile as in effect no material is removed.

Other material removal processes

There are other material removal processes such as ultrasonic and electrochemical machining. These are mainly used for manufacturing tools as they tend to have low material removal rates. They are not described in detail in this textbook.

Problems 4.2.1

These all relate to the component shown in Figure 4.2.9. (Hint: You may wish to consult Section 1 in Chapter 7).

(1) Are there any problems in machining the profile of this component?

(2) Why may the component be produced on a lathe with the following errors:
- A taper on the 60 mm diameter?
- Ovality of the 100 mm diameter?

(3) Do you think it is best to precut the bar to the exact length? If not what extra length do you recommend be left on for later finishing?

(4) If you started with a bar measuring exactly 100 mm in diameter, how concentric do you think the 60 mm diameter will be to the 100 mm diameter?

(5) If you were required to produce the 60 mm diameter to a surface finish of 0.05 μm, what are your recommendations for a combination of metal removal operations?

(6) If you were requested to produce the component on a machining centre, i.e. utilizing milling cutters, how round do you feel you could produce the 60 mm diameter?

4.3 Conserving material by changing shape

In ideal material-conservation terms, all of the starting material should end up in the manufactured component without producing any waste. This section describes the processes which aim to achieve this ideal – casting and forging.

Casting

Casting processes have been around for thousands of years, but are still widely used today to produce intricate external or internal profiles. The process takes a fluid starting material and pours it into a mould cavity where it solidifies taking the shape of that cavity. There are two basic methods:

- Molten liquid starting material – solidifies by a freezing process.
- Slurry starting material – solidifies by chemical reaction or evaporation.

Molten liquid

These are the processes which we readily identify as casting. Solids produced have a structure which is crystalline, amorphous (shapeless) or a mixture of both.

A useful way to classify casting materials is using the temperature at which the material is fluid enough to flow into a mould. Materials are cast into moulds which have a higher melt temperature – the higher the difference the lesser the effect the process has on the mould life.

Material	*Temperature* (°C)
Tungsten	3400
Magnesia	2800
Molybdenum	2600
Alumina	2050
Silica	1710
Iron	1540
Nickel	1450
Mild steel	1415
Silicon	1410
Copper	1085
Aluminium alloy	600
Zinc	420
Lead	330
Nylon 66	270
Solder (50% Pb, 50% Sn)	215
Polystyrene	200
PVC	160
Low-density polythene	150

The very high melt temperature materials need to be processed in the solid state by powder processing (see page 169) because of the problem of finding durable mould material.

The low temperature polymeric materials require additional pressure to overcome their high viscosity and enable good filling. The pressure increases viscosity hence aiding mould filling. Metals are normally cast from the liquid and alloying can be carried out as part of the cycle. Pressure can be used when casting metals to aid penetration into cavities.

Mould production is a skilled sector of the casting process. To produce accurate and long-life moulds requires a high grade of material which in turn requires expensive manufacturing. We can reduce the cost of the mould materials at the expense of their life.

Disposable mould casting process

Mould production here is a two stage process.

First a pattern has to be produced identical in form to the finished component, except in one respect – size. The component shape is formed when the material is very hot, as it cools down and freezes it shrinks. The pattern has to be made larger to allow for this shrinkage. This pattern is used to make the actual moulds.

Depending on the number of moulds to be made, pattern material varies. For small runs, wood is a common material. For large runs metal patterns are used – the actual material used depending on the life required.

The mould is produced by forming an impression from the pattern. As a mould for the outside of the component has to be continuous, it is produced in two, or more, sections as in Figure 4.3.1. The rejoining of these sections can cause problems through misalignment and a visible joint line.

Internal surfaces are produced by using a core made from a similar material to the mould.

Processes which use disposable moulds are:

- Sand casting: The moulds and cores are made from compacting special sands. Surface finishes and accuracy are not high, although they can be improved by utilizing fine grained sand and coating the mould surface.
- Resin shell casting: This gives a better finish and accuracy. Produced by coating the exterior of the pattern with fine resin

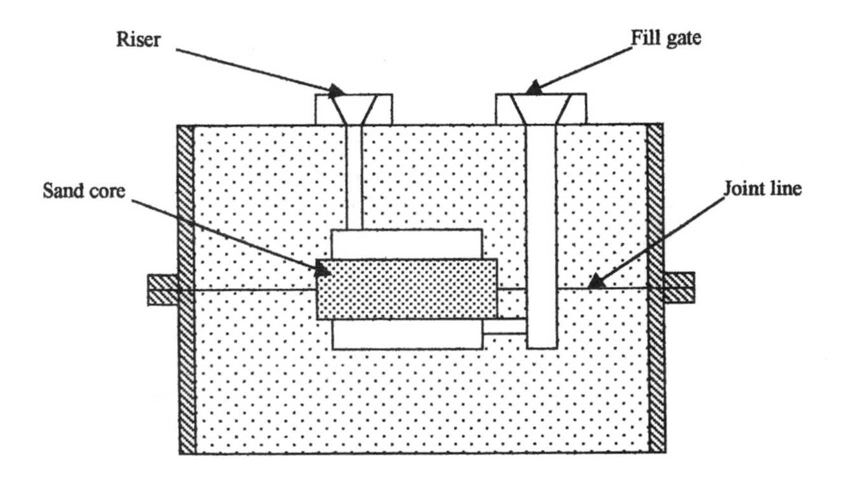

Figure 4.3.1 *Sand casting mould, ready for pouring*

coated sand. The thin coat is then cured by drying and heating to form a thin mould. When casting the mould is supported by either sand or metal shot.

- Investment casting: Also called the lost wax process, because a wax pattern is used. The pattern is dipped into a ceramic slurry repeatedly which builds up the depth of covering. The pattern is removed by melting it and pouring it out. This gives a mould with no joint line.

Disposable moulds are suitable for both small- and large-scale production.

Permanent mould casting process

Where long runs are required, it can be economic to produce permanent moulds, which are normally termed as dies. They can also improve quality aspects such as porosity, accuracy and surface finish as heat transfer rates are higher. Pressure can improve cavity fill and give better internal homogeneity, but requires more expensive machines. The processes are:

- Gravity die casting
- Low pressure die cast
- High pressure die cast

Die casting is quick but the dies are relatively expensive to make. The repeated contact with molten metal reduces their life. The need to remove the casting from the mould severely restricts the design. Die casting is therefore normally limited to fast production of large volumes of light components.

Centrifugal and slip casting

Used for large diameter discs or pipes and other hollow castings without using a core to form the internal profile.

Discs. In this process, the mould is centred on a revolving platform. The molten material is fed at the centre of rotation and the spinning process provides pressure to ensure the material flows towards the outside.

Pipes. The inside of the mould is profiled to give the outside shape of the pipe. Molten material is poured in and the mould is then swung round its axis forcing the molten material to the inner surface of the mould where it freezes.

Slip casting. A similar process to the pipe production using a hollow mould. This time the mould is in effect tumbled to ensure the material flows to all areas. Method also used extensively with slurries in industries such as pottery.

The solidification process

The internal structure of a casting is dependent on the number of crystals started and the speed of solidification. Internal unwanted cavities can be small pores or quite large. These are caused (propagated) by gas bubbles and flow related problems.

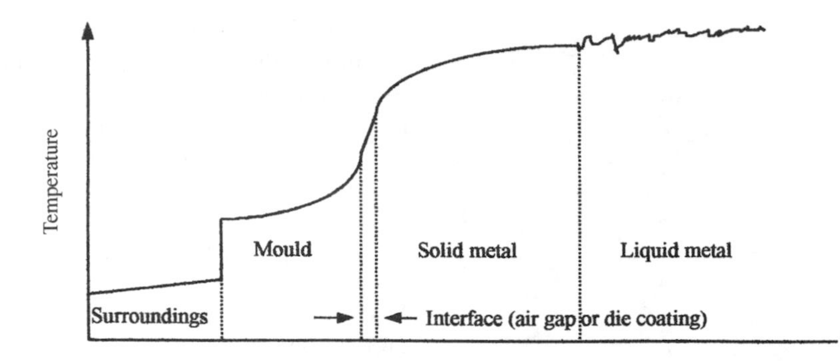

Figure 4.3.2 *Temperature profile of a liquid solidifying*

Solidification time is controlled by the speed at which heat can be diffused from the casting into the mould. Metal dies have high conductivity whilst refractory moulds such as sand are slow.

The resistance to heat flow can be seen from Figure 4.3.2. The thermal properties of the metal, the mould and any air gaps combine to determine the freezing rate.

In Table 4.3.1, we calculate the heat diffusivity of the mould using ($c \times p \times K$). The higher this value the quicker the mould can remove heat from the casting. This shows that the slowest heat extraction is from plaster and the highest from copper.

Table 4.3.1 Mould materials' heat diffusivity

Material	*Specific heat, c* ($J\ kg^{-1}\ K^{-1}$)	*Density, p* ($kg\ m^{-3}$)	*Thermal conductivity, K* ($Wm^{-1}\ K^{-1}$)	*Heat diffusivity* $c \times p \times K$
Plaster	850	1100	0.4	374 000
Sand	1150	1500	0.6	1 035 000
Iron	650	7900	40	205 400 000
Magnesium	1040	1700	150	265 200 000
Zinc	390	7200	120	336 960 000
Aluminium	850	2700	230	527 850 000
Copper	380	9000	390	1 333 800 000

A major factor in the heat dispersion is the relationship between the volume of the casting and its cooling surface area. This is expressed as the modulus, M, of the casting and is based on the ratio of its volume:surface area.

To calculate the cooling time we can use the data from Table 4.3.2 in the formulae:

$$\text{Cooling time, } t_f = \frac{p \times L \times M}{h\ (T_m - T_e)}$$

where:

h = heat transfer coefficient at interchange of mould and material
M = modulus of casting, i.e. volume/cooling surface
T = temperature (subscript $_m$ is the mould, $_e$ is the environment)

Table 4.3.2 Liquid metal constants

Metal	*Density,* p (kg m^{-3})	*Melting point,* T_m (°C)	*Heat of fusion, L* (kJ g^{-1})	*Specific heat, c* (J kg^{-1} K^{-1})
Iron	7020	1540	270	750
Aluminium	2400	660	390	1100
Copper	8000	1083	200	500
Magnesium	1590	650	360	1300
Zinc	6560	420	112	480

Looking at the above calculation to compare different metals in the same die arrangements and cooling from the same temperature, we would have h, M, T_m and T_e all being the same and the solidification time, t_f, is proportional to ($p \times L$). This works out for the above materials as a relative cooling time as shown in Table 4.3.3.

Table 4.3.3 Relative cooling time of a range of metals

Metal	*Relative cooling time*
Iron	1 895 400
Aluminium	936 000
Zinc	734 720
Magnesium	572 400
Copper	160 000

A magnesium casting will therefore cool about 20% quicker than a zinc one. Quick cooling times mean that the casting can be ejected from the mould sooner and hence production rates will be better. Zinc-based die castings remain more popular because of lower material costs and its lower melting temperature which reduces the die wear.

Chevorinov's rule for solidification also should be checked at the design stage for cooling:

- In pressure die and sand casting, the solidification time at any point is proportional to the *square* of the thickness of the section at that point.
- In gravity die cast, the solidification time is proportional to the thickness of the cross-section.

Fluid flow

To encourage a liquid to flow into a mould we need pressure. Often a head of metal is sufficient but other pressure sources are often used. There are two aspects to be considered:

- Surface tension: As the molten material flows through a narrow channel, the leading edge forms a spherical shape. The surface tension creates a pressure difference resisting flow. Therefore we must create sufficient pressure to enter thin portions. Conversely, we must also ensure that the size of sand particles is small enough to prevent penetration.

- Fluidity: The metal must be sufficiently fluid to fill all parts of the casting before solidification occurs. The fluidity depends on heat flow and the velocity of the material, which in turn is dependent on the pressure head.

Filling problems

A very critical operation during casting is transferring the metal from the furnace into the mould. Each casting process has different delivery problems.

Open-top gravity moulds. These are used for ingot production and concrete slabs. Main problem is exposure to atmosphere giving an opportunity for oxidation to occur whilst pouring or on the top surface.

Closed gravity moulds. These require a column of molten material to give a pressure head and as a feed to replace freezing material. This feeder is excess material – and is often recyclable although it does require energy to melt it down again. These moulds have an arrangement of feeding channels:

- Dross traps to ensure clean material is transferred to the mould.
- A tapered sprue to give a pressure head and maintain delivery during solidification.
- Wells to redirect material flow through 90°.
- Gates and runners to deliver material into the mould – often at several different points to aid flow before solidification.

Injection moulding. The main problem is excessive turbulence during filling which traps air. To reduce this the cycle should be closely controlled. Figure 4.3.3 shows two filling cycles – the multi-stage process reduces the turbulence.

Squeeze casting. In effect a liquid forging operation. Has the advantage of little fluidity needed, hence reduces a need for a special casting alloy. There is good contact with mould wall for cooling and fine microstructures can be produced.

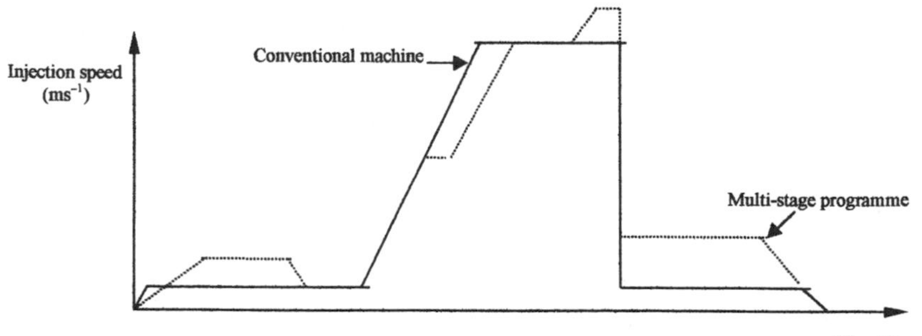

Figure 4.3.3 *Pressure die filling programme. (Reproduced courtesy of the Open University (author and publisher))*

Feeding of castings

Extra metal must be fed into a casting to compensate for any volume change during liquid cooling and solidification. (Note: Some metals such as Grey CI expand at solidification.)

After solidification this shrinkage is homogeneous and can be predicted and allowed for in the design of the mould. During solidification, however, the process of solidification itself can cause considerable volume change and can affect the flow of material to other still liquid sections.

There are four basic design rules involved in the feeding process:

(1) The material in the feeder must remain liquid until after the main casting solidifies. Apply a 20% safety margin by ensuring the relative volume/cooling surface ratios (M) are:

$$M_{\text{feeder}} = 1.2 \times M_{\text{casting}}.$$

(2) There must be sufficient liquid available to feed the casting during contraction.

$$V_{\text{feeder}} = V_{\text{casting}} \times (\beta/(E - \beta))$$

where
V = volume
β = contraction (Al = 7%, steel = 3%)
E = efficiency of feeder – typical is 14% for a cylindrical feeder

(3) There must be a path for the feed liquid to reach all parts. This means that the thickest section should be near to the feeder, with progressively thinner sections away. Use the inscribed circle technique (see Figure 4.3.4) to determine flow path.

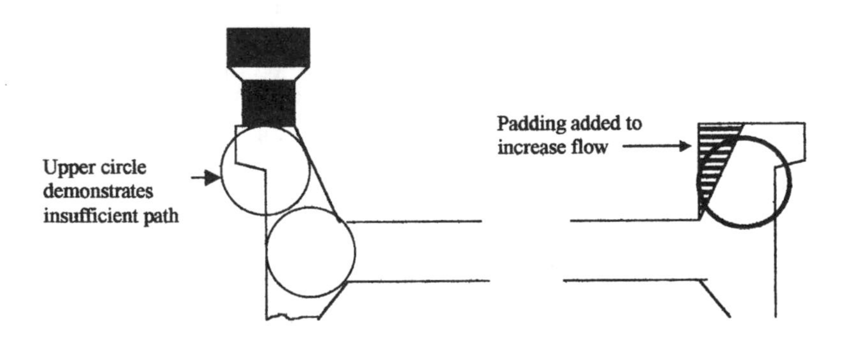

Figure 4.3.4 *Application of the inscribed circle method to determine sufficient flow paths. (Reproduced courtesy of the Open University (author and publisher))*

(4) The pressure head must be maintained to suppress porosity. Feeder height in sand casting should be 1.5 to 2 times the sprue diameter. In addition to the feed sprue, extra risers are often used spaced round the casting to form reservoir feeders to 'hot spots' in the casting.

Each of the molten casting processes have particular cost and production advantages as can be seen in Table 4.3.4.

Table 4.3.4 General production/cost of molten casting techniques

Process	*Die*	*Equipment*	*Labour*	*Pieces/hour*
Sand	Low	Low	Low–medium	<20
Shell	Low–medium	Medium–high	Low–medium	<50
Plaster	Low	Medium	Medium–high	<10
Investment	Medium–high	Low–medium	High	<1000
Permanent mould	Medium	Medium	Low–medium	<60
Die	High	High	Low–medium	<200
Centrifugal	Medium	High	Low–medium	<50

Casting problems

Planes of weakness due to crystal growth during cooling. These are caused by differences in thickness or sharp change of profile causing different crystal pattern growth. Inclusions also tend to be pushed into these planes.

Shrinkage during cooling must be allowed for in the mould design – and makes close tolerances impossible. May lead to cracking on different cross-sections or the formation of pipes (voids) in the centre of thick castings. There are three stages that must be allowed for as in Table 4.3.5.

Porosity if very fine and scattered does not cause much weakening, but if large will be a stress point. Porosity is caused by voids left by material solidifying, gases coming out of solution trapped in cooling material or segregation of different phases in the cooling material.

Polymer processing

The basic principles are similar to metal casting, but processes differ slightly due to a desired high molecular weight (leads to high viscosity) coupled to low thermal conductivity and degradation temperature.

These are usually overcome in two ways:

- Use low molecular weight material and polymerize them in the mould to increase their molecular weight. These are the *thermosets*. Any waste produced cannot be recycled.
- Use high molecular weight material and apply high pressure to create high shear rates. This reduces the viscosity but increases molecular alignment. These are the *thermoplastics*. A limited quantity of the waste produced can be ground down and recycled.

Table 4.3.5 Shrinkage of aluminium and steel casting

	Shrinkage for	
	Aluminium	*Steel*
Liquid stage per 100 K	0.8%	1.3%
Liquid to solid	3%	6.5%
Solid, down to room temperature	7.35%	5.5%

In addition it is possible to improve properties by fibre reinforcement, but cutoffs cannot then be recycled.

Process performance

Additional time must be added to the mould cycle to allow the polymer to cool sufficiently to be handled. In the case of material being polymerized, time is also required for curing, which may extend cooling time.

The high viscosity methods require high pressure to produce the necessary high shear rates and necessitate a higher specification of tools, and hence higher tooling cost. The basic processes are:

Low viscosity

- Reaction injection moulding
- Monomer casting

Mid viscosity

- Compression moulding
- Transfer moulding
- Melt casting

High viscosity

- Injection moulding
- Rotational casting

Material can be melted separately before casting or as it enters the casting process by a combination of heat and pressure.

Slurry casting

In these processes the material is fed into the mould without being heated beforehand. Heat may be applied afterwards to quicken evaporation or to aid diffusion.

Powder processing

Originally the reason for developing powder processing was to produce components in very high melt point materials. This process is very efficient in material usage and can produce to a high tolerance which has led to its spread to a variety of materials.

However, the granular material used produces a highly porous body which has to be sintered, i.e. heated and compressed, to reduce these voids. It is possible to make use of this porosity to infuse a lubricant for bearings. It is possible to sinter some low density, low stressed components by applying heat without pressure to the formed component whilst it is held in a mould. This still gives a high degree of porosity.

The main process is to press the powder at room temperature to give a weak 'green' compact. The containing mould is then not needed during subsequent sintering, which is at a higher temperature. A higher specification component can be produced by applying pressure and heat simultaneously. However, the cost to eliminate all voids is prohibitive.

Density variations occur especially when different cross-sections are produced in the component by unidirectional pressure (see Figure 4.3.5). Isostatic pressing gives better homogeneity.

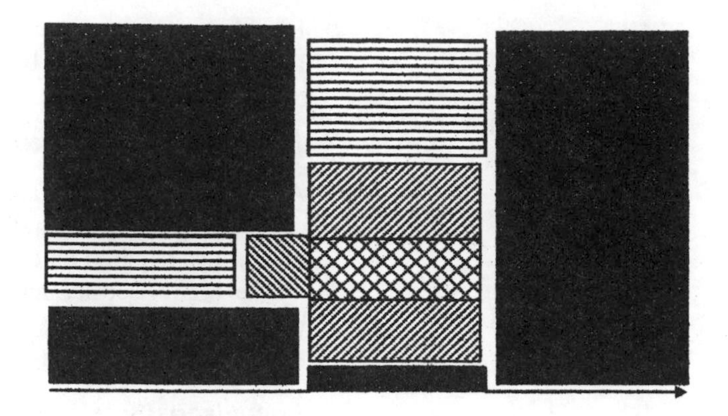

(a) Density differential in cross-hatched material due to combination of end and side action

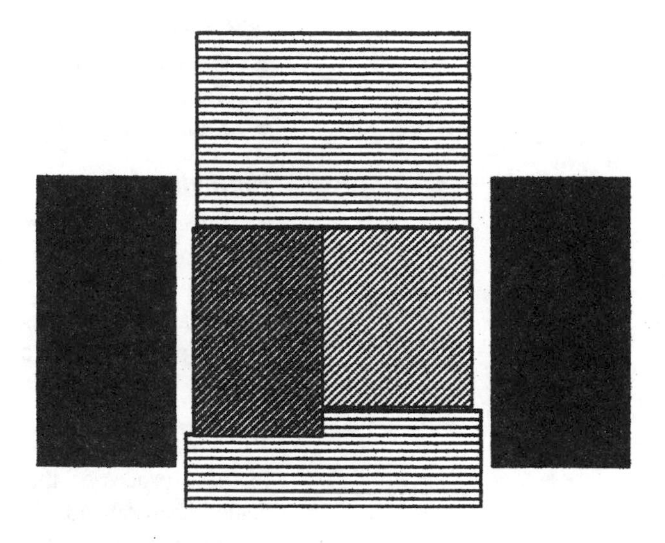

(b) Density differential due to differences in compression

Figure 4.3.5 *Density differentials in powder metallurgy products due to component design*

Sintering

This process allows diffusion between adjoining granules as in Figure 4.3.6 to form to increase internal cohesion. The degree of diffusion varies in relation to pressure and temperature. Density varies and long thin components are difficult to produce.

Production of tool inserts requires fine control of temperature during sintering. A high reduction in porosity has to be balanced with grain control.

(a) Granules touching
Small contact areas
Large connected voids

(b) Increased contact
Voids reduced,
channels reduced

(c) Full fusion
Voids small
and isolated

Figure 4.3.6 *Process of sintering*

Other slurry processes

These processes are used for a variety of materials, such as concrete, clay and other pottery materials.

These take place in similar moulds to the molten material centrifugal casting. The difference is that the input material is in the form of a slurry. The slurry is a suspension of the finished material in a liquid, often just water. The materials may chemically react, or combine as a mixture at the moulding stage.

The as-cast material is often in a very weak condition and requires careful handling until solidified. The solidification process is often simple evaporation to drive off the liquid, although isothermic reaction can also take place. This takes place either at normal room temperature, or by applying additional heat.

Forming

The forming of a product involves permanent deformation to change the shape of the starting material. The extent to which materials can be deformed depends largely on their temperature. Basically forming can be broken down into two main sectors, based on the geometry of the starting material:

- Bulk using a 3D piece of material.
- Sheet using a thin sheet blank.

Bulk processes involved

The different processes involved are described below:

- Upsetting: Cross-section of part of component is increased.
- Swaging: Cross-section of part of component is reduced.
- Forward extrusion: Material is forced through a shaped hole to give the outer profile of the cross-section. By using an internal mandrel, internal profiles can be produced. During forging, internal shearing can take place – leaving part of the input material 'dead' (see Figure 4.3.7).
- Backward extrusion: Different to forward extrusion in that the material is forced back over the pressing tool to form a deep cup shape. Can also have an outer profile imparted.

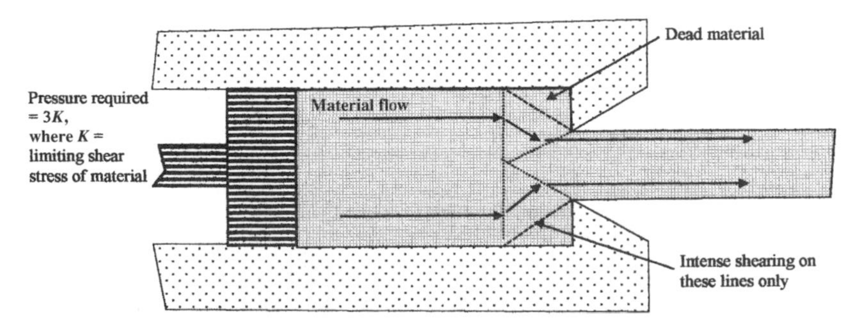

Figure 4.3.7 *Cross-section of extrusion process showing shear planes and dead material zones*

- Open die forging: Material is shaped between flat or simple contoured dies – depends on skill of operator for shape.
- Closed die forging: Material is shaped between finish contour dies. Excess material will produce a flash at the joint line of the dies.
- Coining: Similar to closed die forging but used as a final operation to impart sharp details and specific surface finish on component.
- Rolling: Used to produce specific cross-sections in a continuous operation. Process is an iterative one gradually changing the section. Extensively used for producing standards for stock material shapes such as tube, plate, bar, structural sections. Crystals are elongated in direction of rolling.
- Drawing: Changing the cross-section of a material by continually pulling through a series of changing size shaped holes. Commonly used for reducing wires and tubes. Can also be used to change cross-section shape as in converting a round tube to a square section.
- Shearing: A cutting process where the required length is separated by shearing of the bar.

Forging is often carried out as a multi-stage process to ease the shape transition. This may require intermediate heat treatment (annealing) processes to allow the internal structure to be destressed. In closed die forging, a flash is often produced where extra material is squeezed out at the joint between the dies, similar to the mould line in casting.

Hollow sections

Hollow sections such as tubes can be readily formed by similar processes to those used with solid starting material. However, because they are basically thin-walled cross-sections they will require close support or have to be heated to elevated temperatures to ease deformation as with sheet manipulation.

When being bent, to prevent collapse, hollow sections normally require internal support if the bend radius is less than three times their diameter. This internal support can be by sand filling, or coil springs on smaller diameters.

Sheet processes

We can consider that a sheet of material has relatively no thickness and exists mainly in a single dimension. The thickness is important as we shall see, but this image helps to visualize some processes as described below:

- Bending: Bending involves turning a portion of a sheet into another dimension. Bending can invoke a degree of springback due to residual stress. This can be corrected by overbending, stretching or by coining.
- Cutting/blanking, i.e. cutting a shaped hole in the material: Where the cut-out material is discarded, it is termed cutting. When the cut-out material is the component, it is termed blanking.

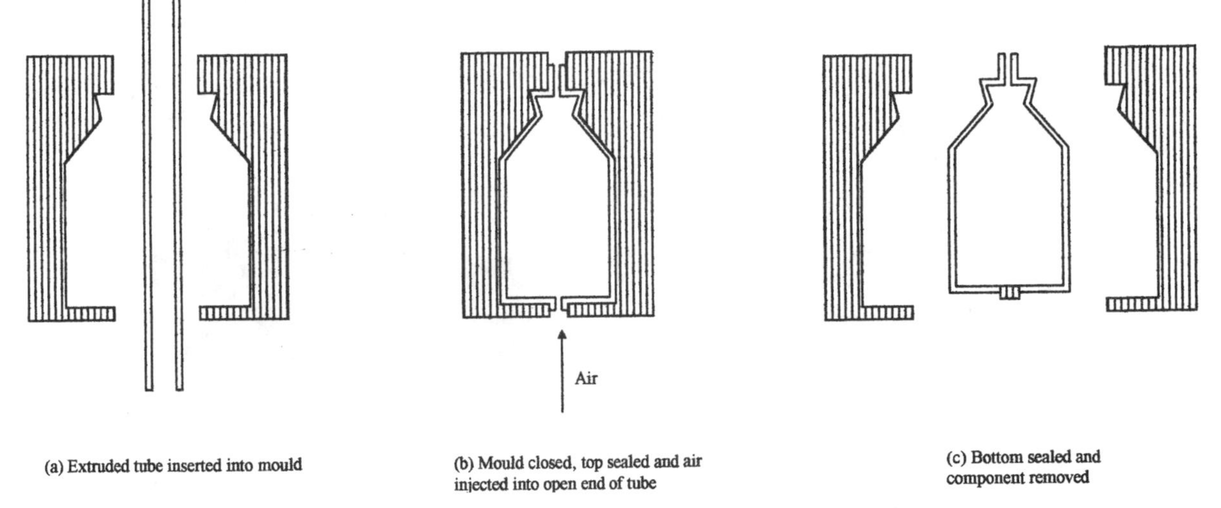

Figure 4.3.8 *Blow moulding operation*

- Stretching: Can just be subjecting the sheet to tensile force acting though its flat body, i.e. increasing the dimension. Can also deform into a second dimension by pressing sheet down onto a contoured surface, or by internally expanding a hollow thin-walled component, as in blow moulding a bottle (see Figure 4.3.8).
- Rolling: Used to form circular shapes from sheet plate. Large diameter pipes are produced by this method – sometimes in a spiral format.
- Deep drawing is a reverse bending operation followed by stretching. The strain in the flange is considerably different from the drawn material – being compressive hoop stress that can cause wrinkling of the flange material unless solid blank clamping is used (see Figure 4.3.9).
- Spinning or flow turning is sometimes used to turn a flat sheet into a cup or cone shape. The main difference between the techniques is that in spinning the starting diameter is reduced during shaping, but in turning it is maintained (see Figure 4.3.10).

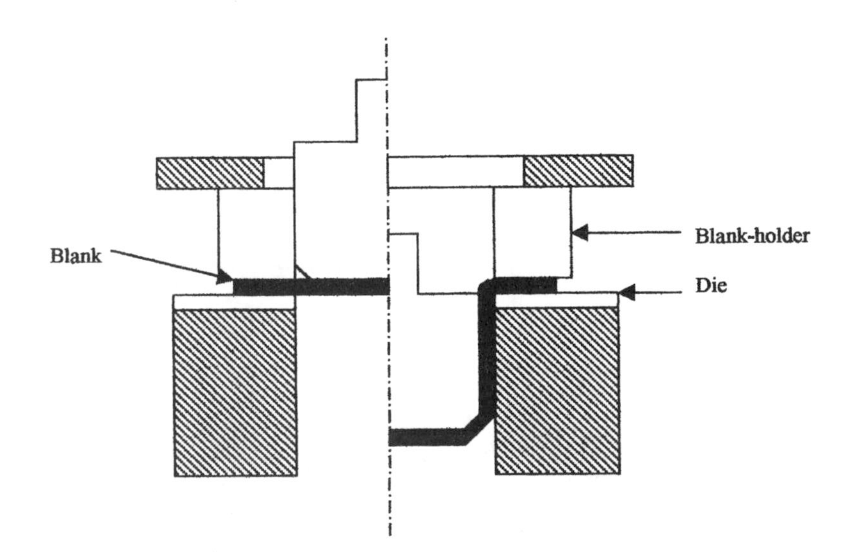

Figure 4.3.9 *Deep drawing technique*

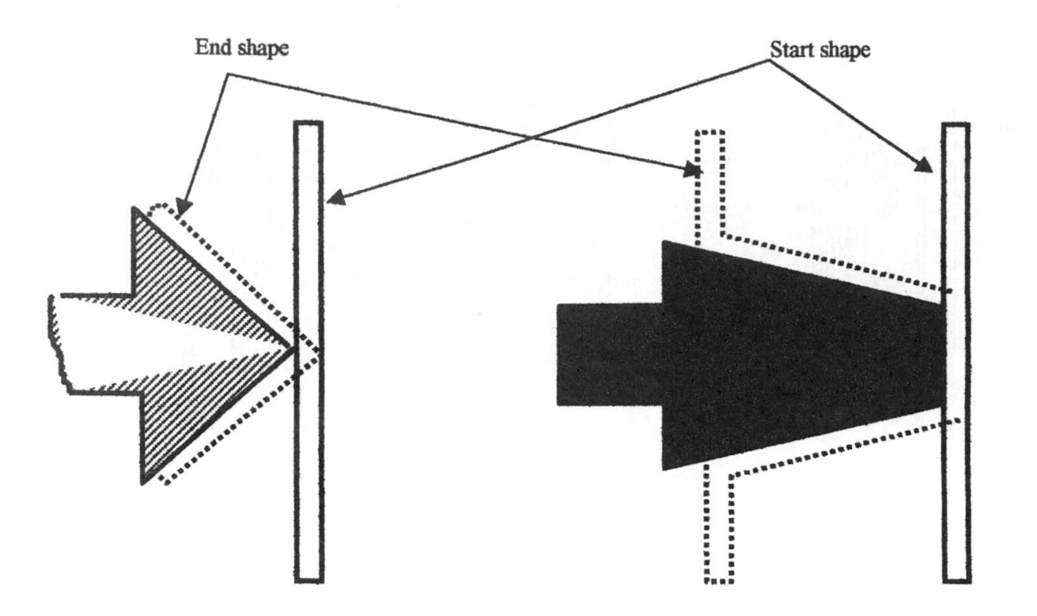

Figure 4.3.10 *Shaping by spinning and flow turning*

(a) Spinning
Material thickness remains constant
Start diameter reduces

(b) Flow turning
Material outer diameter stays constant
Thickness reduces, especially on taper portion

Deformation

To produce permanent deformation, the material must undergo plastic strain as in Figure 4.3.11.

Activity 4.3.1

Take a piece of soft cheese and press down on it in one direction. What happens to the shape of the cheese? Now take an eraser and repeat the pressing down. What happens to the eraser's shape? Why the difference?

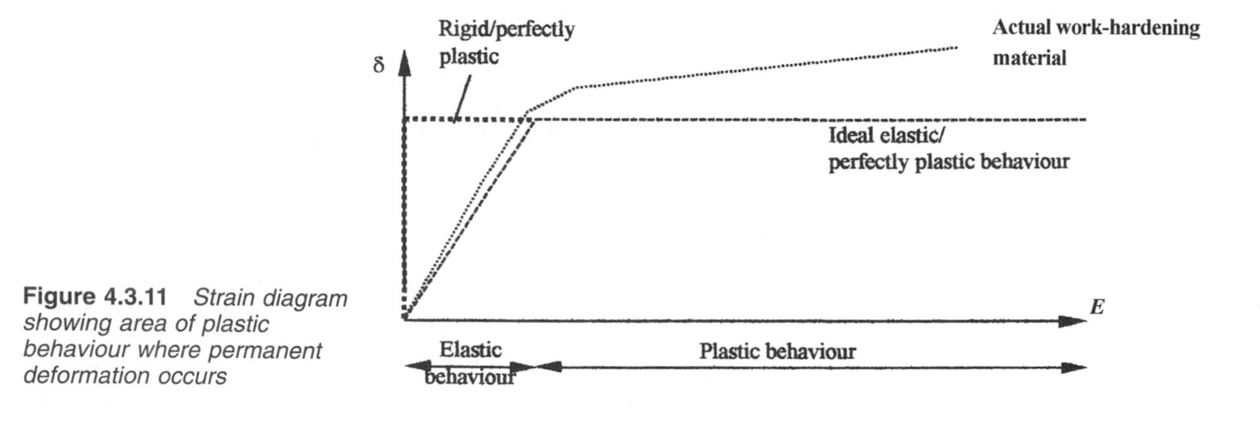

Figure 4.3.11 *Strain diagram showing area of plastic behaviour where permanent deformation occurs*

Stress provides a useful guide to predict the onset of plastic deformation, but there is no simple test to calculate a limit to the deformation before failure. The failure limit depends on the strain history, physical constraints, type of loading and temperature involved.

In a general stress system, there are three normal stresses and associated maximum sheer stresses. Yielding is a constant volume process and it is the shear stresses that control the direction and amount. When all stresses are equal, we have hydrostatic stress. If tensile stress develops then fracture may easily occur.

In sheet processing, the principal stresses normally are applied in two dimensions only with the material being free to move in the third direction (plane stress). Deformation in sheet processing is especially affected by grain orientation during prior processing, e.g. rolling.

The basic failure modes in sheet processing are:

- Folding, buckling, or wrinkling, under compressive forces. These can be at the inner radius of small radii bend, or on the flange of deep drawn material.
- Severe bending (small radius) causes a large degree of stretching on the outer radii and local tensile rupture.
- Different stretching due to basic crystalline structure produced in previous processes such as rolling.
- Necking from tensile force in stretching, especially in deep drawing.

If movement in all directions is not constrained in bulk processing, we can have different types of stressing within the material. Inner material can be in compression, whilst at the surfaces material will tend to bulge outwards. This causes tensile (hoop) stresses to build up and can cause failure. It is limited to some degree by friction between the material and the load face of the tool. In order to maintain complete *hydrostatic* stress, considerable all-round restraint is required and high process loads are needed.

Material variations and inclusions can be the starting point of failures, especially cracking. In addition, repeated deformation causes cold working hardening which increases the resistance to plastic deformation.

It is extremely difficult in bulk forming to predict the load required to complete plastic flow. The starting shape has a large bearing on the performance – hence the multi-staging of the change.

In extrusion, a shearing action occurs similar to that in single point cutting (see Figure 4.3.7). This can be simplified to examining only the areas of intense shear.

In practice, not all shear contributes to the final shape of the component. These can add to the pressure required. Friction also adds to this pressure, which depends on the materials in contact and any lubricant involved.

In complex geometry, simple shear is difficult to isolate and the more complex technique of finite element analysis has to be used. In 3D shapes this involves a considerable cost in computing resources. An alternative is wax modelling which can mimic the properties of a material, including work hardening, the relative strength of the tool and the lubricant's influence.

Effect of temperature

In relation to a metal's melting temperature, T_m, there are three ranges we can operate within – cold, warm and hot.

Cold forming ($<0.25\,T_m$)

Not all material can be readily cold formed. An ideal material has:

- Low yield strength and hardness.
- Low work-hardening rate.
- High ductility.
- Good response to heat treatment.

Materials which can be cold forged tend to be predominately mild steel, and alloys below 0.2% carbon, but it is also used for brass shell cases, aluminium containers, zinc battery cases and copper electronic components. Work hardening during forging can increase the tensile and yield strength of the material by up to 25% and 100% respectively.

Some materials may require subsequent heat treatment to give desired functional properties. Annealing will restore the original properties, or other treatments such as case hardening can be given.

Processes are normally limited to closed die, upsetting and extrusion. Components can be formed very close to their final dimension. It is easy to control input volume and hence there is seldom a need to allow for a forging flash.

Warm forging ($0.25–0.65\,T_m$)

Warm forging is an attempt to combine the good points of both cold and hot forging without the drawbacks of either. The stress required for deformation is reduced and ductility is increased. This temperature range enables higher carbon and alloy steels to be processed without scale formation.

Hot forging ($>0.65\,T_m$)

Hot forging is used to reduce the stresses involved and hence the power of the equipment required. This gain is counterbalanced by increased die wear (due to high operating temperature), and lower surface finish and tolerance control. Some advancement in finish and tolerance control has been made through hot die or isothermal techniques on aerospace materials.

Typical materials and their forging temperatures are shown in Table 4.3.6.

If time allows, recovery and recrystallization can occur during processing. Table 4.3.6 gives lower finishing temperatures when resistance to deformation is still high. Raising the temperature will reduce the stress involved but too high a temperature can result in structural problems. Strong oxidation of steels starts at 900°C, thereafter scaling causes surface defects and die damage.

Table 4.3.6 Metals and preferred forging temperatures

	Finish forging temperature (°C)	*Tensile strength at forge temperature* (MN m^{-2})
Mild steel	950	5
Aluminium (7079)	400	8
Stainless steel (403)	920	14
Magnesium alloys	380	17
Martensitic stainless steel	945	25
Austenitic stainless steel	945	28
Nickel alloys (Incoloy 901)	960	38
Titanium alloys	950	39
Tungsten alloys	1390	51

Formability

There are some tests which give an indication of a material's performance in forming:

Bulk formability tests

Uniaxial compression (see Figure 4.3.12). Various material height/ diameter ratios and effect of lubrications can be tested. Surfaces can be hatched to give a detailed measurement of the deformations endured.

For complex/heavy reduction, a good indication of formability can come from torsion tests, i.e. the number of revolutions before failure.

Lubrication reduces the friction between the material and the die during forging. This friction can severely restrict deformation by giving a drag on the surface. The lubricating techniques used at the different forging temperatures are:

Cold Forging. A phosphate soaping is applied by passing the raw material through a sequence of the following processes:

- Dip in acid pickle in 15% H_2SO_4 at 60–70°C for 5–15 min to clean and etch.
- Wash.
- Dip in zinc phosphate solution at 80–90°C for 3–7 min to give a crystalline surface coating.
- Wash.
- Dip in sodium stearate solution at 80°C so that we have the reaction zinc phosphate > zinc stearate.

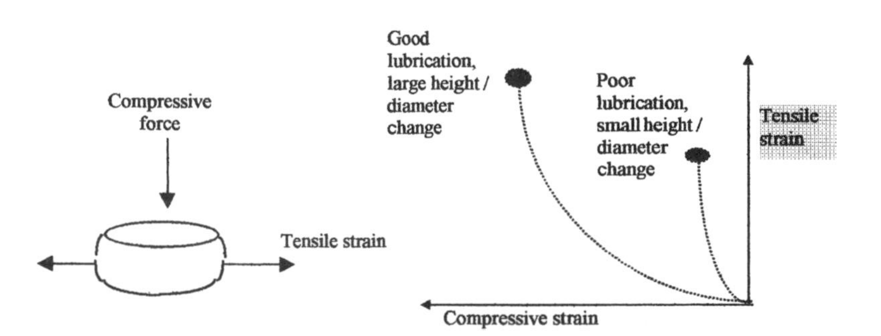

Figure 4.3.12 *Uniaxial compression test*

The resultant coating is a zinc phosphate bond to the die with a surface coat of zinc stearate and sodium stearate to act as the lubricant.

Warm and hot forging. Dies sprayed with spherical graphite solution. Glass lubricant is used on hot extrusion of steel (1100°C).

Sheet formability tests

A tensile test gives simple 'uniaxial' tension performance but cannot simulate tool friction. Hardness is another indication of yield stress.

Another technique used is the circle grid strain analysis. This involves etching an array of 2 mm circles on a sheet as in Figure 4.3.13. The sheet is then subjected to a variety of bulge tests. The circles deform into ellipses where the major axis corresponds to maximum principal surface strain and the minor axis the minimum surface strain.

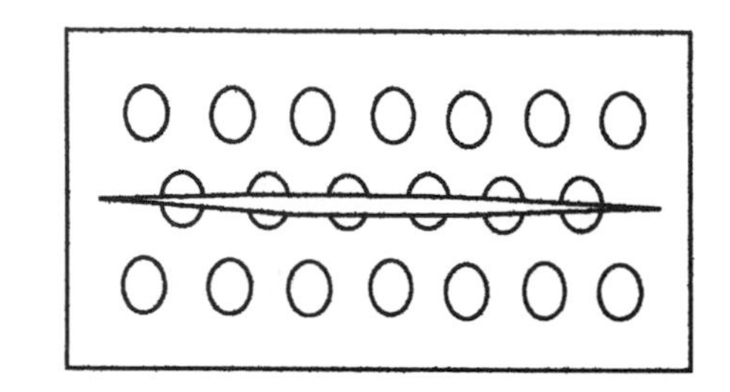

(a) Circle grid analysis

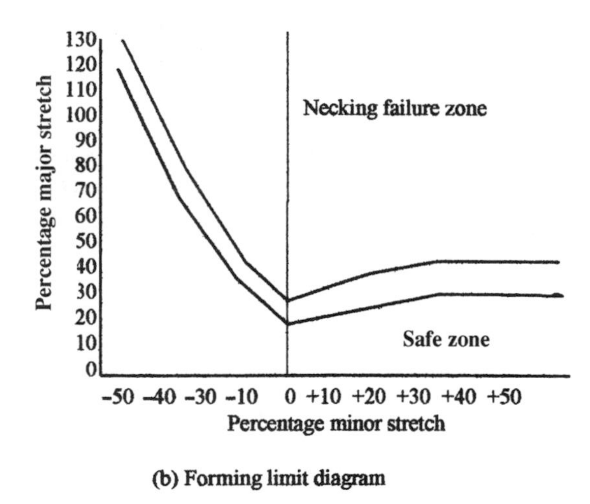

(b) Forming limit diagram

Figure 4.3.13 *Grid circle test and forming limit diagram*

These can be graphed to give a forming limit diagram (FLD) as in Figure 4.3.13. The higher the position in the safe zone, the better the formability. The FLD necking failure zone rises with thickness and falls with yield strength. FLD values can aid tool, process and component design.

Sheet. Relatively low loads make lubrication pressure resistance less of a problem. To reduce tool wear sometimes coatings such as titanium nitride are used.

Types of forming machines

Bulk forming plant

For this operation, the deliverable load during the forge operation is critical. Some machines deliver only a single blow (energy and stroke restricted) in one cycle – and rely on repeated blows. Others apply a constant squeeze action (load restricted).

Hydraulic presses give long die times and lower strain rates, but greater chilling in the component. This has led to the development of heated dies, or applying very fast rates of production which calls for automatic transfer devices. Sintered materials are used for high production punches to prolong tool life.

Sheet forming plant

The cubic size of the die space is critical with loads less so as forces involved are low. Factors used in plant selection are:

- The degree of variable control over the stroke of the press.
- The stroke length available in the press.
- Daylight space between the top and bottom die.

Often only the bottom half of a tool set is used to define the geometry of the component in sheet forming, giving an opportunity to reduce the cost of the upper half of the tool. The latter is sometimes replaced by a rubber pad or a flexible diaphragm, but this has a shorter life.

Sheet processing is often carried out in a linked series of operations on the one press termed progression tooling.

Problems 4.3.1

These all relate to the component shown in Figure 4.2.9. (Hint: You may wish to consult Section 7.1 in Chapter 7.)

(1) If you had 10 off to produce of this component, what are the advantages of producing it as a sand casting followed by machining?

(2) If you had to produce this component in cast iron by centrifugal casting, how much material would you require to leave a hollow core of 40 mm diameter?

(3) If you had 100 off this component to produce, list the advantages and disadvantages of three casting methods.

(4) If you had 10 000 off this component to produce, list the advantages of two forging methods.

(5) If you had to produce this component from sheet, plate or tubing material, describe two methods you may use.

4.4 Joining materials

Most products are not a single homogeneous piece but an assembly of a number of separate components. This section deals mainly with the permanent, or semi-permanent types of joints.

There are many reasons for joining together different components into one complete assembly:

- Functional movement of parts is required. This may be occasional (e.g. in disassembly for maintenance) or continuous (e.g. in a shaft rotation).
- Different material/component characteristics may be needed at different points. Examples include electrical circuits and a hard wearing tool tip held in a tough holder.
- Convenient size of components relative to finished assembly, e.g. individual pipes in a cross-country pipe system.
- Saving material or processing. An example is forming a large box section using standard plate material.

The method of joining can be mechanical or chemical:

- Mechanical joints include nails, screws, rivets, clips, etc. These rely on deformation to produce the holding force. Where joints need to be disassembled later, mechanical joints are mainly used. These can cause problems in manufacture because the increase in the number of components increases the logistic and material handling functions. This has led to a variety of self-contained fasteners. Mechanical joints are not described within this section.
- Permanent chemical joints are where two components are joined by a layer of a solid material that adheres to both. This includes glue, cement, solder, braze and weld. Where joints are meant to be permanent, chemical joints often are preferred, as fewer components are required.

  Where surfaces do not match at the microscopic level, inter-atomic forces cannot aid joining. To make intimate contact a liquid is used to readily contact (wet) each surface. As a liquid has no shear strength, it must be solidified by:
  - Freezing, i.e. a lowering of the material temperature to one at which it no longer remains liquid.
  - Evaporation of some of the liquid content leaving behind a solid.
  - Reaction: A chemical action which combines materials so that the resultant material is solid.

This section deals only with the chemical processes.

Joint type

A joint which is under a compression force is normally safe. Shear forces can be compensated for, but tension is normally undesirable. Problems in joints come from stresses created through distortion, incompatible materials and errors in the process. Cavities, inclusions, poor adhesion and sharp edges are principal stress raisers. These must be avoided or compensated for in the design.

The basic types of joints are:

Butt Joint (see Figure 4.4.1(a)). Requires the jointing material to be similar in strength to the host material. Bend tests and NDT (non-destructive testing) predominate their assessment.

Lap joint (see Figure 4.4.1(b)). Uneven distortion results in peel stress dominating. Long overlaps increase results by reducing stress, but these have a maximum effect.

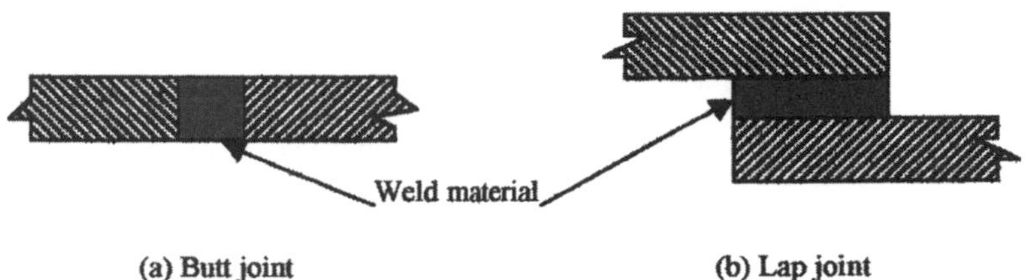

Figure 4.4.1 *Main joint types*

More elaborate joints can be designed for even stress distribution or imparting a compressive force. These cost more, therefore they should be used only where cost effective. Redesign can often allow complex stresses to be eliminated allowing conventional joints to be used.

Metal-to-metal joints

In the past, the only way to join metals was by blacksmiths hammering two red hot pieces together. Otherwise mechanical joints, such as riveting, had to be carried out.

Low temperature (185–250°C) solder, a tin and lead compound, was developed first. This has some adhesion to the base metal but generally lacks its strength. Oxide films on the base material have to be removed by fluxes to give a reasonable bonding surface. Soldering remains the key method in electronic circuitry joints and as a gas or liquid seal at low pressures where high forces are not involved as in domestic water systems.

The ability to produce higher concentrated localized heat meant that the higher melting metals could be joined, initially through brazing (800–1000°C) which gave a higher strength joint than soldering, but still did not melt all the base materials being joined.

Initially welding was very similar to brazing using a filler piece of material melted to fill the gap between two adjacent pieces of material, using first oxyacetylene then electric arc as the power input. The main improvement on brazing lies in the input metal being similar to the base metal and full fusion takes place.

Modern welding is a complex process which requires a high degree of skill. In many shapes this is difficult to automate except in highly repetitive work such as long seams and spot welding. In heavy plate welding, long lengths are automated.

The concentration of heat in welding causes distortion and high stresses due to local thermal expansion and high temperature gradients in the host metal. It may be necessary to heat adjacent sectors of the host material prior to welding, and carry out post-welding heat treatment, especially in the heat affected zone (HAZ).

The metallurgical composition of the molten weld has to be controlled. This included removing oxides from surfaces and protecting the molten pool from oxygen in the environment and hydrogen in water vapour. The slag chosen to do this melts at a lower temperature and is less dense than the molten metal itself, therefore forming a protective shield. An alternative method is to shield the molten metal in inert gas (TIG, MIG and MAG). This was developed initially to weld aluminium with its high affinity for oxygen, but is now widespread for other materials.

The welder has to ensure:

- Correct fit-up – root gap, root face, included angle and alignment.
- Good access giving visibility and freedom of movement.
- Position: Flat and down-hand welding is easy and gives good quality. In vertical and overhead welding it is difficult to control the pool of molten metal.
- Continuous power and feed metal input.
- Correct pre- and post-weld heat treatment.

Case study

Extracts from a general fabrication specification for carbon steel pipework up to 0.40% carbon

Heat treatment

All mild steels (up to 0.26% carbon), having a design thickness of up to 19 mm, do not normally require heat treatment after welding.

All low carbon steels, mild steels over 19 mm thick and where the distance between welds is less than four times the minimum thickness will require heat treatment after welding as follows:

- Up to 115 mm outside diameter: Stress relieved or normalized where access prevents full stress relieving being carried out.
- Above 115 mm outside diameter stress relieve.

Preparation of weld joint

The preparation of weld joints is as shown in Figure 4.4.2. The joint faces will preferably be machined, but may be prepared by thermal cutting, chipping or grinding. If thermal cutting is used, all scale and slag will be removed from the weld faces prior to welding. No preheat is required to these materials before thermal cutting.

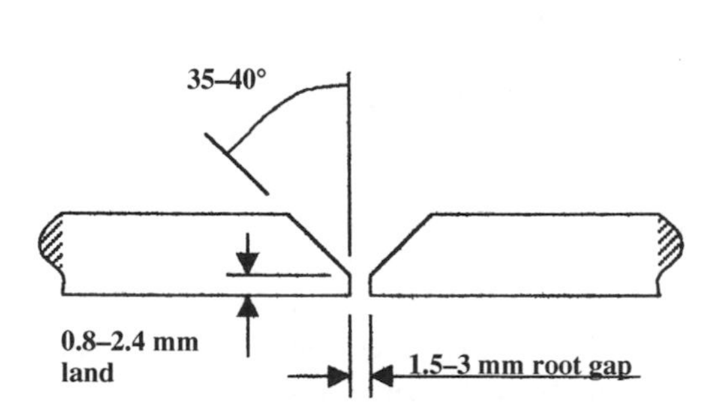

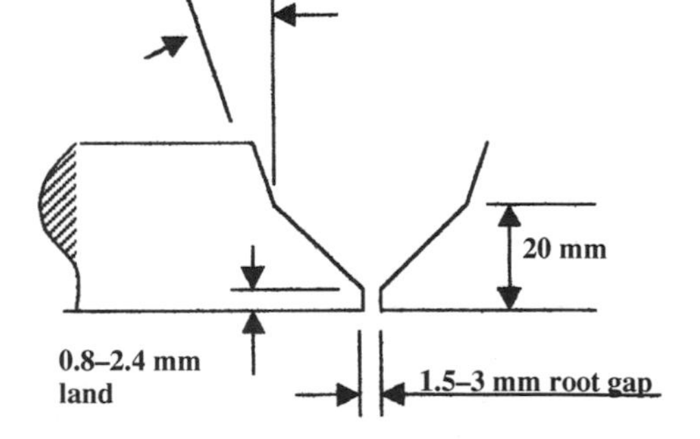

Figure 4.4.2 *Weld preparation and weld gap at set-up*

Set-ups

The joints will be set up as per Figures 4.4.2 and aligned as Table 4.4.1 as accurately as practical within the existing commercial tolerances on pipe diameter, thickness and out-of-roundness.

All welding will be in accordance with a procedure approved under one of the following:

- ASA Code for Pressure Piping B31.1
- ASME Boiler and Pressure Vessel Code Section IX
- BS 2633 Specification for Arc Welding of Ferritic Steel Pipework

Welding procedure

Electrodes. Welding electrodes, filler wire or combination of filler and flux will be such that sound weld deposits of compatible analysis to the base materials are made. Coated electrodes will conform to BS 639 or comparable American Standards, such as ASTM Filler Metal specification SA-316 F.1–4.

The following preheating will be applied to all welds:

- No welding will take place below 0°C.
- Up to 0.26% carbon up to 19 mm thickness – none required.
- Up to 0.26% carbon above 19 mm thickness – 100°C.
- 0.26%–0.40% carbon all thickness – 150°C.

Preheating will be applied to the area local to and including the weld preparation and the temperature maintained during the entire operation.

On completion of welding, the joint is to be wrapped in dry insulating blankets to ensure slow cooling, unless immediate post-weld heat treatment is carried out.

Table 4.4.1 Fabrication alignment

Bore		*Maximum permissible difference in internal diameter*				
		Process other than TIG		*TIG welding*		
				Close butt welds		*Root gap and oxyacetylene run*
Over	*Up to and include*	*Backing ring*	*No backing ring*	*Fusible insert/ no filler wire*	*With filler wire*	
mm	*mm*	*mm*	*mm*	*mm*	*mm*	*mm*
–	100	0.40	0.75	0.40	0.75	0.75
100	300	0.40	1.50	0.40	0.75	1.50
300	–	0.75	2.25	–	1.20	2.25
			Maximum out of alignment at the bore			
–	100	0.40	0.75	0.40	0.75	0.75
100	300	0.40	1.50	0.40	0.75	1.50
300	–	0.75	1.50	–	0.75	1.50

Post-weld heat treatment:

- Up to 0.26% carbon up to 19 mm thick – none required, over 19 mm thick – stress relieve.
- 0.26–0.40% carbon – up to 12 mm thick – none required, over 12 mm thick – stress relieve.

Radiography. When specified by customer, radiographic inspection of butt welds will be carried out to BS 2910. The first butt weld of each operative will be radiographed with 10% of the remainder.

Any one of the following may be a cause for rejection:

- Any crack.
- Any zone of incomplete fusion.
- Lack of root penetration exceeding 12 mm in length, or a total length of 25 mm in any 300 mm length of weld.
- Slag inclusions or gas pockets exceeding 6 mm in length, or a total length of 25 mm in any 300 mm length of weld.
- Total area of porosity projected radially through the weld will not exceed 6 mm^2 (equivalent to three areas 1.6 mm diameter) in any 600 mm^2 of projected weld area.

Spot weld

Welding two thin sheets together represents a problem in controlling the heat input to prevent burning away the parent material. This method of joining thin sheets of material together reduces this problem. The process involves forcing the two materials into close contact and passing a localized electric current through them, as in Figure 4.4.3. The joint produced is not continuous, therefore it is not pressure tight.

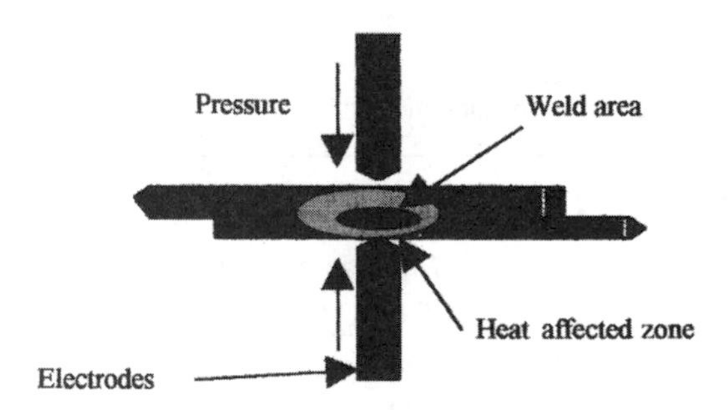

Figure 4.4.3 *Spot welding. (Reproduced courtesy of the Open University (author and publisher))*

A pressure-tight joint can be produced by the resistance seam weld process. This is basically a series of spot welds overlaying each other. It is produced by rotating wheel electrodes with an intermittent current passing through.

Friction weld

A clean process where high rotational speeds produce plasticity in host metals being rubbed together (see Figure 4.4.4). They are then quickly pressed (forged) together, expelling air. A large flash is produced which requires trimming.

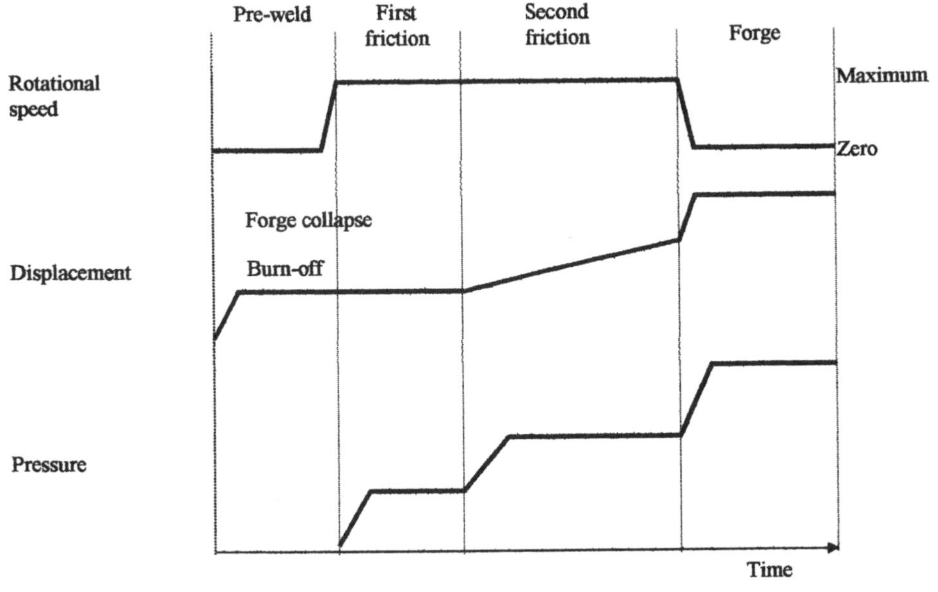

Figure 4.4.4 *Friction welding cycles. (Reproduced courtesy of the Open University (author and publisher))*

Useful in volume production for dissimilar materials:

- Steel: Aluminium (anode hanger in smelter).
- Stainless steel: Carbon steel transition tubes.
- High speed steel drill tip: Carbon steel shank.
- Rear axle assemblies.

High energy beam welding

An oxyacetylene flame injects 10 kW into 100 mm^2 and a TIG arc concentrates the same energy into 10 mm^2. An electron beam gives a further magnitude increase in energy concentration and hence reduces the volume to be heated, with the reduction in side effects. However, the system must work in a vacuum, which is a disadvantage with large components. It may sometimes be advantageous with small components, however, because of the tight control possible.

Laser beam welding does not require a vacuum for operation and gives similar advantages to electron beam welding. Inert gas can be used to shield the molten pool. Laser welding is economical in micro-electronics assembly, but lack of reliable large lasers has held back the spread of this process until recently.

Weld faults

A cross-section of a butt weld and the type of weld problems that arise are shown in Figure 4.4.5. Some radiography views of the typical weld faults are shown in Figure 4.4.6. Many of these can only be established by examining the internal structure of the joint – radiography and ultrasonics are used:

- Porosity: Caused by damp rods in first 50 mm of a run, until they are heated up beforehand. In small amounts not a problem as spherical and not stress inducers.
- Lack of penetration: Severe potential stress point problem if root run does not fill weld gap.

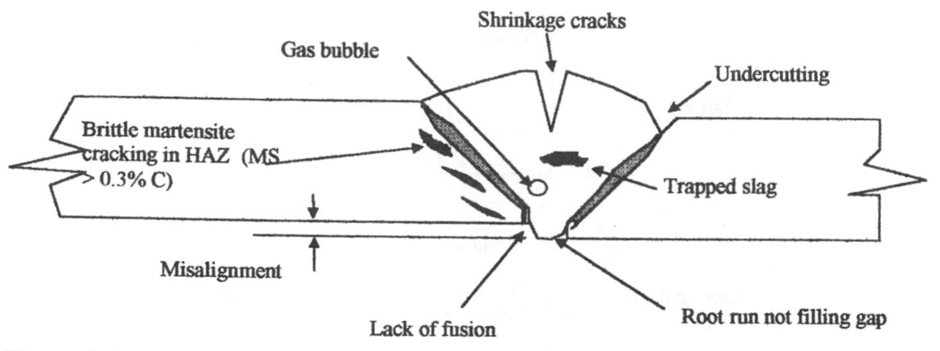

Figure 4.4.5 *Cross-section of weld showing potential faults*

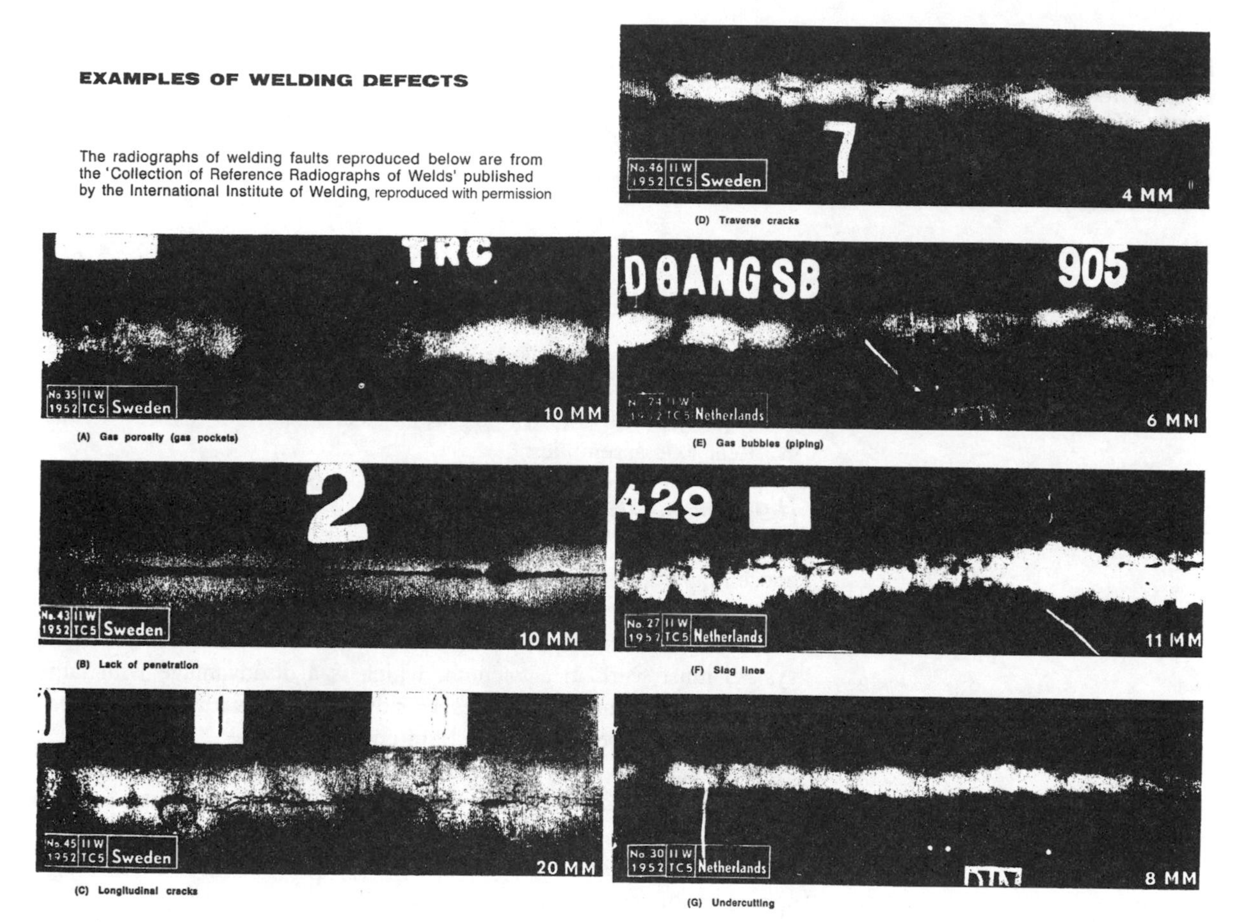

Figure 4.4.6 *Examples of welding defects: (a) gas porosity (gas pockets); (b) lack of penetration; (c) longitudinal cracks; (d) traverse cracks; (e) gas bubbles (piping); (f) slag lines; (g) undercutting. (Reproduced with permission of the International Institute of Welding)*

- Longitudinal cracks: Poor design or processing, e.g. lack of preheat or poor support against distortion.
- Traverse cracks: Broken runs without reheating between runs leading to thermal stresses.
- Gas bubbles: Caused by incorrect selection of rods, or poor preheating.
- Slag: Trapped by poor operations.
- Undercutting: Caused by too high temperature (i.e. current). Reduces the cross-section of weld.

- Incomplete side wall fusion (not shown): Caused by damp rods, low temperature or lack of surface preparation. Difficult to detect.

As we saw in the case study (page 182), welding has to be carried out under strict control to enable high quality products to be produced.

Glue

Glue is a chemical jointing method which does not provide primary bonding, as in welding, yet allows dissimilar materials to be permanently joined. The qualities required of a glue are as follows.

Wetability

The glue must be in continuous contact with the surfaces it is joining. It must, in effect, 'wet' the surface to allow secondary chemical bonds (i.e. adhesion) to be made between the glue and the base material substrata (see Figure 4.4.7). Wetting is essential to give complete surface contact. The wetability is a function of the surface energies involved.

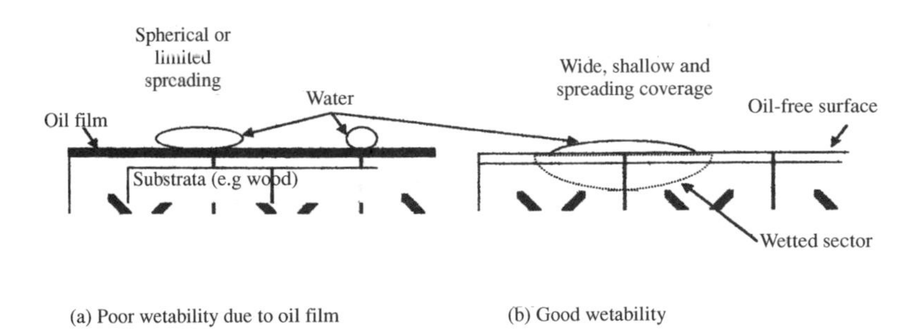

Figure 4.4.7
Wetability of surfaces

Therefore to enhance wetability we can:

- Use glue with low internal cohesion.
- Use a similar chemistry for the glue and the materials being joined.
- Ensure the joint surface offers secondary bonding possibilities. In some cases this may mean applying a surface treatment to it.

Problems arising are:

- Some surfaces are oxides which are weak in themselves or in their adherence to their base metal. Some polymers, however, have their surface deliberately oxidized.
- Some polymer surfaces have concentrations of constituents such as plasticizers, antioxidants, diluents, etc. Solvent extraction can remove these.
- Air, moisture or vapour can be trapped due to incomplete wetting. This can be minimized by pressure during application and curing. A continuing decrease in volume during assembly is ideal.

A good glue therefore should:

- Fill completely the space in the joint.
- Be tough enough to withstand stress concentration.
- Match the stiffness of the substrata to transmit stress.
- Be durable against:
  - Vibrations.
  - Thermal variation.
  - Chemical resistance.

The qualities which may also aid in the selection of a glue include:

- Setting time
- Solvents used
- Removability
- Cost
- One/two part
- Cold/hot setting
- Toxicity
- Waterproof
- Rigid/flexible
- Strength
- Range of adherents
- Fungicidal
- Shelf life
- Ease of application
- Impact adhesion

Surface preparation

In gluing, surfaces are often roughened prior to application. In metals, this is primarily to give a clean oxide film, not to give an interlocking effect. However, embedding protruding fibre from the substrata into the adhesive can be effective.

Glue constituents

Base/Binder. The main constituent which has to hold the substrates together. Chemically classified as epoxy or cryanoacrylic. In melt and freeze glue this may be the only ingredient, e.g. fish-glue applied at 100°C and set at ambient temperature.

Solvent/Dispersant. Evaporation setting glues require a liquid phase which will dissolve and disperse the binder. Water is often used, but damp and mould growth may undo these. Organic solvents have been developed to replace water.

Plasticizers. Added to reduce the brittleness of the set base. Normally similar to solvent with a lower volatility – however, this may still evaporate over time leading to age embrittlement.

Hardeners. Reactive glues require a chemical action to turn the base to a solid. This may be achieved by mixing two chemicals (liquids) together to form a new solid phase. Others require the presence of a catalyst (hardener) to undergo the phase change. Some reactive glues use heat as the catalyst.

Fillers. Cheap material incorporated to give increased bulk. Some fillers can counteract shrinkage and reduce stress effects.

Flexibilizers. Crack blunting devices are often added to polymers to contain crack spreading and release strain energies. Similarly small particles of rubber can be introduced into an epoxy and increase impact and fatigue resistance.

Design problems

The main problem with glued joints is that the glues are normally brittle and weak under tensile forces. Joints therefore have to be designed so that the joint is in compression or in sideways shear.

In addition the joints often deteriorate with age and can be subject to attack by water or chemicals.

Problems 4.4.1

These all relate to the component shown in Figure 4.2.9.

(1) What would be the disadvantage of producing the component by welding two solid cylinders together, i.e. 100 mm diameter by 90 mm long and 60 mm diameter by 100 mm long?
(2) What difference would it make to your answer to 1 if the loading was very light and the cylinders were glued together?
(3) How would you design the joint in your answer to Problem 2 to minimize tensile forces?
(4) What do you think the advantages would be if you joined the two cylinders using the friction weld process?
(5) Do you think a variation of the spot weld technique could be used if the cylinders were made out of sheet material?
(6) If the cylinders were made out of a plastic material, what advantage would there be in producing a fabrication rather than moulding the component complete?

5 Materials selection

Summary

The design process for any product will involve many people with skills from CAD to manufacturing. The materials engineer has a central role in this process although he or she can sometimes be neglected or contacted as an afterthought. In this chapter, we will begin by looking at the role of the materials engineer and then proceed to identify the functions of the product. Once the functions have been defined, the properties required of a material can be identified and the search for suitable materials can begin. We will look at the external pressures on the product designer, coming from the manufacturer, the user and the environment. Some methods for narrowing down the choice of materials will be considered, as will strategies for combining material properties. From Section 5.2 onwards, we will look at detailed properties of common engineering metals. The effect of composition and heat treatment on the properties of materials needs to be understood so that best use can be made of materials. It will also help to avoid costly mistakes being made.

Objectives

By the end of this chapter, you should be able to:

- identify the functions of a product;
- extract property criteria for materials;
- develop clear strategies for selecting materials;
- understand the nature of materials and the influence of composition and heat treatment on their properties.

5.1 Role of the materials engineer

Products can be classed in the following way:

- New products developed to meet a completely new market, such as the personal computer, the mobile phone and even the surfboard – new technology will dominate the product but there may be a need to convince the customer of its value.

- Existing products redeveloped for a changing market or function, such as the car or telephone – new technology will be used to upgrade the function but with a great deal of competition; the product cost will dictate its viability.
- Existing products with little redevelopment to meet unchanging market or function, such as food products (e.g. loaf of bread) and household tools (e.g. the screwdriver) – in this case, markets may saturate as there are more products than the customer can consume. The customer has to be persuaded by your product through good marketing, low cost and new ideas – for example, a new shape of loaf.

Often the materials engineer is called in to deal with problems with the product. For example, one of the first plastic kettles was manufactured from acetal (also called polyoxymethylene, or POM). This material is an engineering polymer, easily moulded with the required engineering properties such as strength, stiffness and chemical and temperature stability beyond 100°C. However, the temperature of the heating element is considerably higher than the design temperature and the polymer combusts easily. It is quite easy for the materials engineer to see the need in terms of the *primary functions* of the product only.

The choice of a material is dependent on many factors and different people in the development process. The creator of the concept may select a material for its looks and texture. A bright plastic kettle developed to compete with a stainless steel one must offer a variety of colours as well. The manufacturer will want to use a material for which tooling costs and quality needs are easily met. The customer would expect the same reliability as from a steel kettle and at lower cost.

In making a material choice, the manufacturer has to satisfy several requirements, which we could define as follows:

Primary requirements (referring to the function of the product)

- The kettle must be strong enough to carry the water.
- The kettle must withstand some mishandling in the form of accidental knocks.
- The body must be electrically insulating.
- The material of the kettle must not affect the quality of the water.
- The material of the kettle must not react chemically with the water and change its properties.
- The material of the kettle must not distort either due to heating or by absorbing water.

Secondary requirements (additional criteria for the manufacturer to consider):

- The kettle must have a handle so that it can be carried easily and safely.
- The number of stages in the manufacturing process must be minimized.
- The speed of manufacture must be maximized.
- The quality of the product must be consistently high.
- The product must meet test standards such as BS in the country where it is sold.
- Any tests for safety must be carried out.

Additional requirements (to get and keep customers):

- What competition is there and how does the product differ from the competition?
- What is the target price for the product?
- How easily will the product sell?
- How can the market for the product be further stimulated by design?
- What is a suitable life for the product?
- How is the product likely to fail?
- What are the consequences of failure?

All these statements need to be converted into precise criteria that will lead to the selection of materials and the manufacturing process. The overriding concerns of a manufacturer are:

- Is there a market for the product?
- Is the market big enough to cover the cost of investment in new technology?
- Can the product be made for a price that will provide the required level of profit?
- Can existing manufacturing processes be used to make the product?

From these criteria, a set of requirements is passed back to the design stage:

- Ensure that the product will meet the functional requirements for the stated minimum period.
- Incorporate features in the product that will attract customers.
- Ensure that the manufacturing process will deliver a product of the required standard and at minimum cost.
- Select the cheapest material that will meet the stated specifications.

This will lead to a specification for the material for each part in the component.

Product criteria

When a designer looks for a suitable material to meet a particular need, he or she will often be looking for a precise value to incorporate into an analysis. For example, if the component is a hook required to carry a load of 10 kN, the criteria for selecting the material will be that it has sufficient strength. Properties, such as strength and others we have already met in Chapter 3, have to be defined so that they are understood all over the world since products are often made from components from one part of the world and sold and used in another part of the world. Hence methods for testing various properties are defined. This is the task of organizations such as British Standards in Britain, DIN in Germany, American Society for Testing Materials (ASTM) in the USA and ISO in Europe generally. Such standards will specify the dimensions of test pieces as well as other conditions under which the test should be performed. Even when tests are conducted to standards, there could be variations because of the nature of materials and the manufacturing process. The property range of the material can also increase during the manufacturing process. Impurities in the composition could also significantly alter the properties. Hence the designer must verify the source of the data and the conditions under which the data was obtained.

The designer must also understand how the materials respond and why they do so. The designer should be aware of methods that can be used to improve the performance of a material. It would be foolish to rely solely on a database or a textbook of values without an appreciation of the implications. The designer should also be very clear about the conditions under which the component and therefore its materials operate.

In the example of the hook designed to carry a load of 10 kN, if the environment was a loading bay at a sea port, the desirable properties could be:

- It must have the highest strength possible so that it did not fail under load.
- It must have the highest stiffness possible so that it did not extend too much.
- It should be relatively light.
- It must cope with the corrosive marine environment.

On the basis of these criteria, a trivial solution would be diamond. Not only is this solution extremely expensive, but also impossible to manufacture. Much less effort will be wasted if the designer is able to be more precise as follows:

- It must have a strength of at least 100 MN/m^2.
- It must have a stiffness (Young's modulus) of greater than 100 GN/m^2.
- It must have a mass of less than 50 kg.
- It should withstand alternating wet and dry conditions and sea water.

With a better understanding of the market for the hook a more realistic, cost-effective solution such as steel or aluminium alloy could be chosen.

Activity 5.1.1

Write down some clear criteria for the following products:

- A screwdriver.
- A compact disc.
- A picture frame.

Types of properties

A designer should consider the whole life of the product and therefore of the materials making the product. This life includes:

- The processing of the material.
- The manufacture of the component.
- The operation/use of the component.
- The environment in which the component is used.
- What happens when the component is finally scrapped.

In this section we will consider only the life of the component. Relevant properties need to be identified and addressed.

Properties relevant for manufacture

- When in the molten form, material is allowed to flow into a cavity to take up a final shape. The viscosity of the melt will change with temperature and so a suitable temperature must be specified for the process. This is important for casting a metal, for extrusion and injection moulding of plastics, and for forming glass. It is also important for the sintering of ceramics. If the temperature is too high, too much energy is used and the process is unnecessarily expensive. In the worst case, the material could also degrade. If the temperature is too low, the process will simply not take place! The temperature must be sufficient to allow additions to mix in the case of polymers and for complete filling of the mould.
- In the solid form, the material must be malleable. The total deformation possible before failure occurs must be known. This can vary significantly with the rate at which the load is applied. The degree to which a metal work-hardens or a polymer cold-draws must be known. How much surface hardening or softening is caused by a process and is the surface liable to cracking? These are questions that need to be answered before the process is applied.

Properties relevant to operation

- Mechanical behaviour: The yield stress is the highest value that a designer can afford to have in a material to avoid permanent damage to the shape of a component. In practice, the ultimate tensile stress (or strength) is easier to obtain. The modulus of elasticity (or stiffness) is a measure of the amount of elastic deformation in a material under a given stress. The ductility, fracture toughness, impact strength and hardness are all measures of the suitability of the material for different applications. All of these properties can change with temperature or rate at which load is applied. Particularly important are the effects of fatigue, caused by the cyclical changes in load, where fatigue strength and endurance limit can be defined. At temperatures approaching the melting point of a material, creep, which is a property that makes materials deform like toffee, becomes important.
- Electrical and magnetic properties: Electrical resistance or resistivity is an important property for a conducting material. This determines the amount of energy lost in transmitting a current

through the material, and the heat gained as a result can degrade the material in the long run. Insulators will not allow electrical current flow as the name implies, but the designer must be aware of the limit before the material breaks down. Electrical currents have associated magnetic fields, defined by the magnetic field strength, B. Under the influence of a changing magnetic field, electrical currents will be induced in metals and semiconductors.

- Chemical properties: The interaction between materials and the environment on the surface can lead to damage by oxidation in air or aqueous corrosion in damp or wet environments. Two dissimilar metals in contact with each other may cause one to corrode in an aqueous environment. An inert, protective coating may be necessary in each case.
- Thermal properties: Properties such as thermal conductivity, heat capacity, coefficient of linear expansion, glass transition temperature and maximum operating temperature need to be considered before selecting materials.
- Other physical properties such as density can prove vitally important when selecting materials for lightweight applications. Optical properties such as refractive index and the photoelastic effect are important for particular applications.

Properties relevant to the environment

- Chemical reactions with the environment can lead to degradation by either oxidation or aqueous corrosion. Ultraviolet light can lead to chemical changes in the surface and infrared radiation can lead to thermal degradation. All of these factors modify the surface layers and so make the material more prone to early failure under normal operation. Where flammability is a potential hazard, the choice of material must include information on whether it will burn for a given amount of oxygen in the environment. This is defined as the limiting oxygen index for the material.

During manufacture, the state of the material may be altered by the temperature variations around the component. The outer surface of a thick section may cool much faster than the core. In a steel casting, this would lead to greater hardness and strength on the outside than on the inside. In the extreme case, this variation could lead to cracks within the casting that may not be spotted without some further testing.

Methods for materials selection

Any material choice must therefore go through a process as follows:

(1) Apply the primary functional requirements to all materials grouped appropriately – make a limited selection of materials.
(2) Judge materials against factors such as availability of material and its relative change, cost of material and its variability, meeting of secondary criteria such as the effect of the environment and customer preference.
(3) Acquire materials and conduct tests on properties and processability.

(4) Repeat earlier stages if all materials prove unsatisfactory or review selection criteria.
(5) Manufacture prototype products and conduct long-term tests.
(6) Manufacture product and sell to the customer.
(7) Monitor performance of the product during operation by recording failure data from customers.

In the first stage, the vast range of materials available needs to be reduced to a few sensible choices. This can be done in the following way:

- From the analysis of product function, create a hierarchy of property needs.
- Look at the materials specifications of a similar product.
- Look up a database or a table of materials and their properties and compare several materials.
- Compare the performance of each of the selected materials against primary, secondary and other requirements.
- Reduce the selection and talk to materials manufacturers and suppliers about the selection.
- Obtain data on the properties of specific grades, price, availability and reliability of delivery.
- Make the final decision and work closely with the supplier.

Often the priority list can become complicated, as several properties are equally important. Hence a combination of properties needs to be considered together. Figure 5.1.1 illustrates the steps in the process of selection once the properties have been specified. The manufacturing requirements of the product need to be added to the list. The effect of the product environment and the impact of the failure of the product need to be considered. Finally, manufacturers will in future have to consider how to deal with products that reach the end of their useful life.

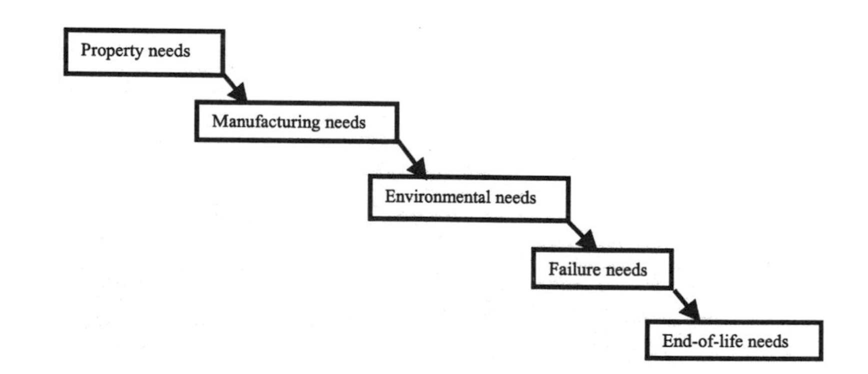

Figure 5.1.1 *Hierarchy of demands on materials*

For example, consider the design of a small ladder for use in the home. The specifications of the customer may be as follows:

- It must raise a person by a height of at least one metre.
- It must be light enough to be carried around.
- It should be stored away neatly.
- It should require little or no maintenance.
- It should be safe to use.

For the designer this implies that, the ladder should:

- conform to British Standards (BS7377);
- be manufactured from suitable materials with the lowest density;
- be foldable for storage;
- be protected from environmental deterioration such as corrosion;
- form a stable and safe structure when in use.

A typical solution will be made from a tubular aluminium alloy frame. The density of aluminium alloy is approximately one-third that of steel although it is more expensive. The amount of material and therefore the weight of the ladder is reduced by using a tubular construction. When the ladder is set up with its frame inclined to the vertical, bending causes the most significant stresses. In this case the deflection of any beam in the structure is related to the flexural stiffness, EI, of the material, where E is the Young's modulus and I is the second moment of area.

Review – second moment of area

This can be regarded as a measure of the resistance of a section to bending. When a straight length of rod or tube or any other profile is flexed about its centroidal axis as illustrated in Figure 5.1.2, the second moment of area, I, is given as shown below:

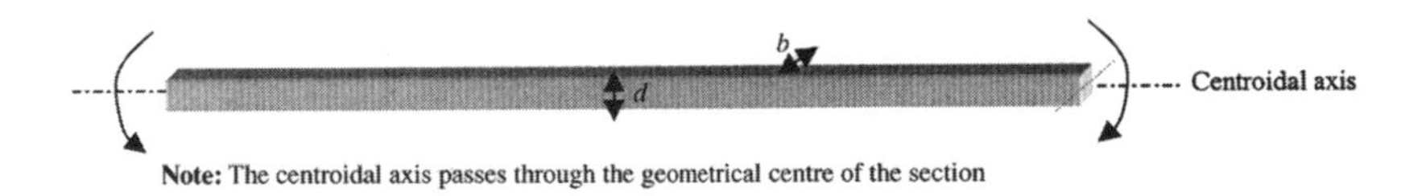

Figure 5.1.2 *Bending of a straight slender beam*

- For a cylindrical rod section, $I = \pi D^4/64$ where D is the diameter of the rod.
- For a tubular section, the second moment of area is just the difference between that for the rod and the missing inner piece of the rod. Hence, $I = \pi D^4/64 - \pi d^4/64$ where D is the outer diameter and d is the inner diameter of the tube.
- For a rectangular section, $I = bd^3/12$ where b is the width of the cross-section across the direction of bending, and d is the depth, that is to say the dimension in the direction into which the section will be bent.

The second moment or areas of some more complex sections can be worked out by similarly adding or removing standard shapes from the section.

Example 5.1.1

Given a 2 metre long aluminium alloy tube of outer diameter 3 cm and inner diameter 2.6 cm, determine its second moment of area.

- What would the diameter of a solid rod have to be to have the same flexural stiffness?
- How do the weights of the two structures compare?

Since the material is the same, the flexural modulus will be the same for the rod and tube, if the second moment of area, I, is the same. For the tube,

$$I = \pi \times 3^4/64 - \pi \times 2.6^4/64 = 1.7329\,\text{cm}^4.$$

A rod with the same second moment of area will have a diameter,

$$D = (I \times 64/\pi)^{1/4} = (1.7329 \times 64/\pi)^{1/4} = 2.44\,\text{cm}^4$$

Weight of the aluminium alloy rod,

$$W_r = \rho\pi l D^2/4 = 2700 \times \pi \times 2 \times 0.0244^2/4 = 2.52\,\text{kg}$$

Weight of the aluminium alloy tube,

$$W_t = \rho\pi l D^2/4 - \rho\pi l d^2/4 = 0.95\,\text{kg}$$

Hence, for the same stiffness, the rod diameter will be 81% (about five-sixths) of that of the tube, but its weight is 165% of (more than two and a half times) that of the tube.

Once the basic shape and dimensions of the ladder have been determined, secondary functions need to be considered. The stability of the aluminium frame can be improved by increasing the separation of the legs near the base. This will mean that the aluminium extrusions will need further processing. Rubber mouldings can be introduced and slotted into the aluminium frame to increase the coefficient of friction at the base. An added feature may be a restraint built in to limit the separation of the rear legs from the front legs in a four-legged ladder. For further comfort and safety of the user, the frame needs to be extended to rest the arms. You will find on a visit to any DIY store or a search though a home products catalogue that there will be small variations in stepladders although they will all meet the same basic functional requirements. The following specification comes from a catalogue:

Frame: tubular steel with white enamel finish
Top step height 1.06 m
Ribbed steps to increase grip
Ribbed plug-in feet for better grip with the floor
Overall open height = 1.68 m
Maximum loading = 150 kg
Special locking device to prevent ladder collapsing

Activity 5.1.2

In the earlier example, determine the dimensions of a square section of aluminium alloy if it had the same value of second moment of area as the tubular section. What would the dimensions be if a hollow square section was used to give the same second moment of area, given that the wall thickness should be 0.2 cm?

Compare the weights of each section if the length is again 2 m.

Performance criteria

When assessing the needs of a new product, the designer is able to identify some restrictions or constraints that will help to make the selection of materials easier. For example, in the design of the ladder, the specification of a maximum load applied on any step is a functional requirement that automatically places a design limit on the framework. Other functional requirements include the requirement for stability and safe grip when standing on the ladder. Another constraint may be that the ladder must not exceed 1.5 m in height and that each step must be 20 cm wide (or wide enough for a person to stand comfortably on a step). These are geometrical requirements that again limit the design. Another useful criterion is a measure of the efficiency of the structure. This could be a measure of the weight of ladder needed to carry a given load, i.e.,

$$\text{Efficiency} = P_L/W_L \quad (5.1)$$

where P_L is the load carried by the ladder and W_L is the weight of the ladder. Clearly, the higher the value of this factor, the greater would be the efficiency and therefore the attraction of the material. In an aircraft structure, the weight of each component adds a significant penalty to the performance of the aircraft and so this measure of efficiency is very valuable. In the case of an automotive vehicle, the cost of the product may be more important and so the efficiency may be given by,

$$\text{Efficiency} = P_P/C \quad (5.2)$$

where P_P is the load carried by the ladder and C_P is the weight of the ladder.

In terms of the mass of material, we can rewrite this equation as,

$$\text{Efficiency} = P_P/(C_M \times M_P) \quad (5.3)$$

where C_M is the unit cost of the material and M_P is the mass of product.

We begin to see how parameters can be combined to give a better means of narrowing down the search for a material or combination of materials.

Activity 5.1.3

Define an efficiency factor for the following products:

- An ordinary door hinge.
- A bicycle pedal.
- An electrical resistor.

At this stage, it would be useful to refer to a database of properties. A very useful tool has been the development of maps or charts. These provide a visual representation of all classes of materials relative to each other, allowing an initial selection of materials. For example, in Figure 5.1.3, we have a map of the range of Young's modulus values against the density of the material.

The chart contains all materials from cork and polymer foams up to the stiffest and densest. A quick survey of the chart shows that materials can be conveniently grouped by their class. Some clusters are compact and distinct as in the case of engineering ceramics. Others such as the metals and common ceramics overlap. The most dense materials are metals and the least dense are the woods and polymers. The properties of wood depend significantly on the direction along which they are

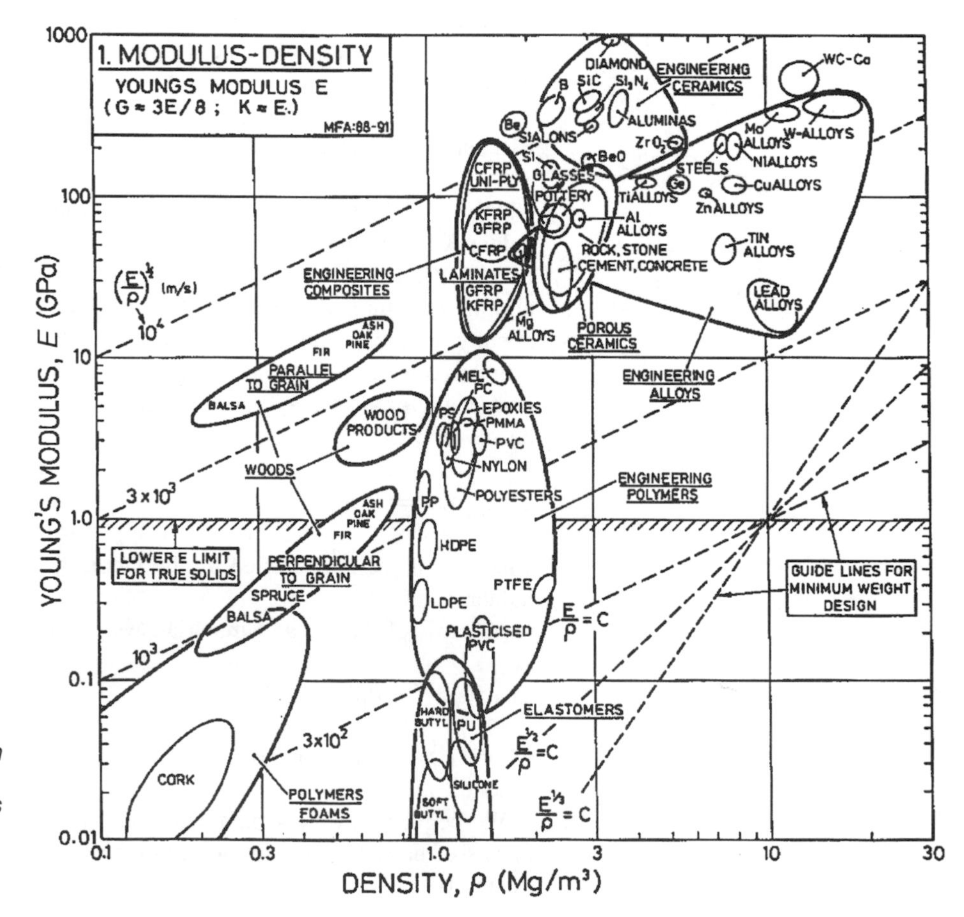

Figure 5.1.3 *Relative position of materials comparing Young's modulus with density* (*Reference:* M. F. Ashby. *Materials Selection in Mechanical Design.* Pergamon 1992)

measured. The properties of the composites are also dependent on the fibre direction and the proportion of the fibres in the composite.

In the design of a ladder, where the stiffness of the material is important, a simple analysis would involve considering a line parallel to the density axis, taken from E = 1000 GPa and moving to lower stiffness, gradually uncovering materials. Diamond is the stiffest material but would not suit our purpose. The first material to suit our purpose would be steel. Aluminium alloys are seen to have about one-third the modulus and therefore would give three times the deflection for the same section dimension. However, aluminium alloys are also about one-third as dense. In principle therefore there is little difference between the two materials as, using a thicker section can compensate for the lower modulus of the aluminium alloy. As more material would be needed, we must also consider the impact on the cost of the product.

The particular ratios are determined by evaluation of the mode of loading and are summarized in Table 5.1.1.

In Table 5.1.1, the materials' properties are defined as follows:

E = Young's modulus, G = shear modulus, ρ = density and σ_s = tensile strength

If a line is drawn parallel to the one defined by E/ρ = constant, it will pass through both the steel and aluminium alloy regions. In this case constant C, is approximately equal to 5000. A superior material would have a higher value of C and hence looking along a line parallel to C, we can see that some woods can perform equally well. Most ceramics and polymer composites give better values of C. The line indicates that on the basis of modulus and density, there is no difference between them.

For household applications, both materials are stiff enough for the load to be carried. Therefore the size of the ladder frame is more likely to be determined by the dimensions needed to keep the ladder stable. The surface appearance and the cost of the ladder will also ultimately determine how well each type of ladder would sell. In another example, ladders for use in sewers in the Middle East were made from glass fibre reinforced polyester (GFRP) – this is the standard form of glass fibre reinforced plastics. Looking at Figure 5.1.4 we can see that GFRP can be as good as aluminium alloys in modulus but have about half the density. GFRP is also very stable in corrosive environments. Hence in the sewer, where the ladders are used infrequently during the

Table 5.1.1 Materials parameters for design by component shape and mode of loading. (Reference: M. F. Ashby, *Materials Selection in Mechanical Design*. Butterworth-Heinemann, 1995)

Component shape and loading	*Parameter to maximize for maximum modulus*	*Parameter to maximize for maximum strength*
Bar in tension	E/ρ	σ_s/ρ
Bar in torsion	$G^{1/2}/\rho$	$\sigma_s^{2/3}/\rho$
Beam in bending	$E^{1/2}/\rho$	$\sigma_s^{2/3}/\rho$
Buckling limit for a column	$E^{1/2}/\rho$	σ_s/ρ
Cylinder under internal pressure	E/ρ	σ_s/ρ

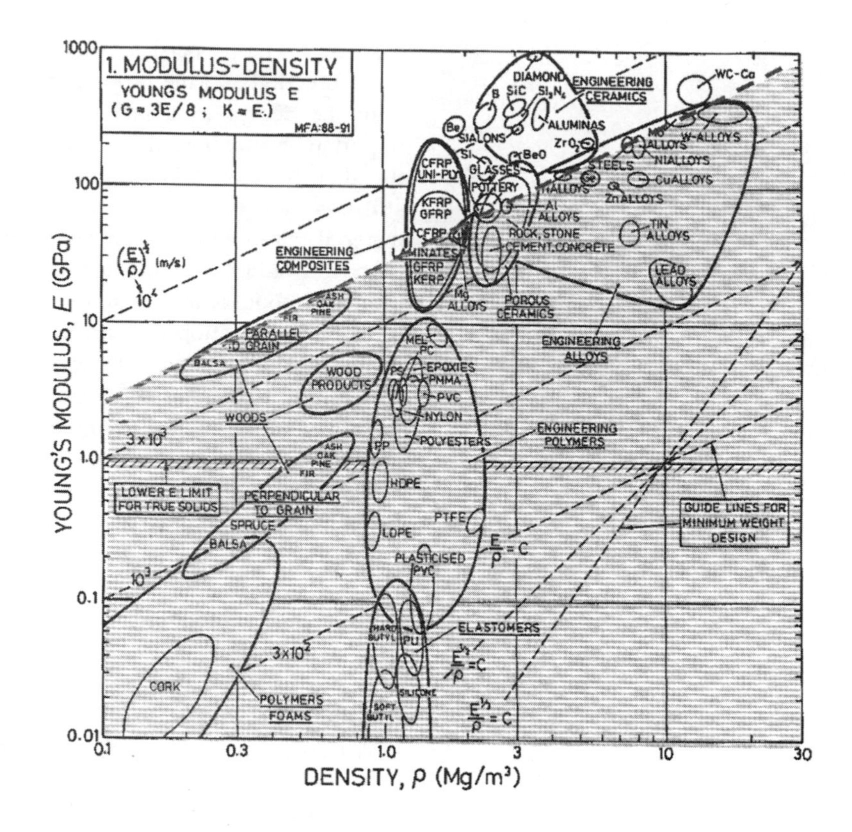

Figure 5.1.4 *Selection of materials superior to steel and aluminium alloys in tenson*

maintenance of the sewer, the corrosion resistance would be the most important criterion after modulus and strength. Because of the lower density, the cost of transporting the ladders from Britain where they were manufactured is reduced.

Note on properties of composites

The properties of composites are very dependent on the nature of the fibres, their orientation and the proportion of fibres in the composite. Fibres can take the following form:

- They can be continuous or unbroken and laid along a specific direction. They could also be wrapped around a shape forming a spiral pattern.
- They can be woven into a fabric. The properties depend on the type of weave pattern and the direction across the fabric in relation to fibre direction.
- They can be discontinuous or short. In this case they could be between 0.1 mm and a few mm long but as they are a few μm in diameter, they can still be classed as fibres.
- They could be in the form of a mat of short randomly oriented but entangled fibres.

The properties of the fibre are usually best along the direction of the fibre length and so the best composites will have continuous fibres oriented along the appropriate direction. As the properties of the fibre and the matrix holding the fibres together are usually widely different, the proportion of fibres will also significantly change the properties.

For example, for glass fibre, $E_f = 70\,\text{GN/m}^2$ whereas for polyester, $E_p = 3\,\text{GN/m}^2$. Hence, in a composite where the fibres are continuous and parallel to the main direction of the component, the modulus, E_c, of the composite is given by,

$$E_c = v_f \times E_f + v_p \times E_p$$

where v_f = proportion by volume of fibres in the composite and v_p = proportion by volume of matrix in the composite.

For example, if the composite is 20% fibres by volume, then the matrix must make up 80% by volume of the composite. Hence, $v_f = 0.2$ and $v_p = 0.8$. Therefore,

$$E_c = 0.2 \times 70 + 0.8 \times 3 = 16.4\,\text{GN/m}^2.$$

We could work out the density of the composite by using the same equation by replacing each modulus term by the equivalent density term.

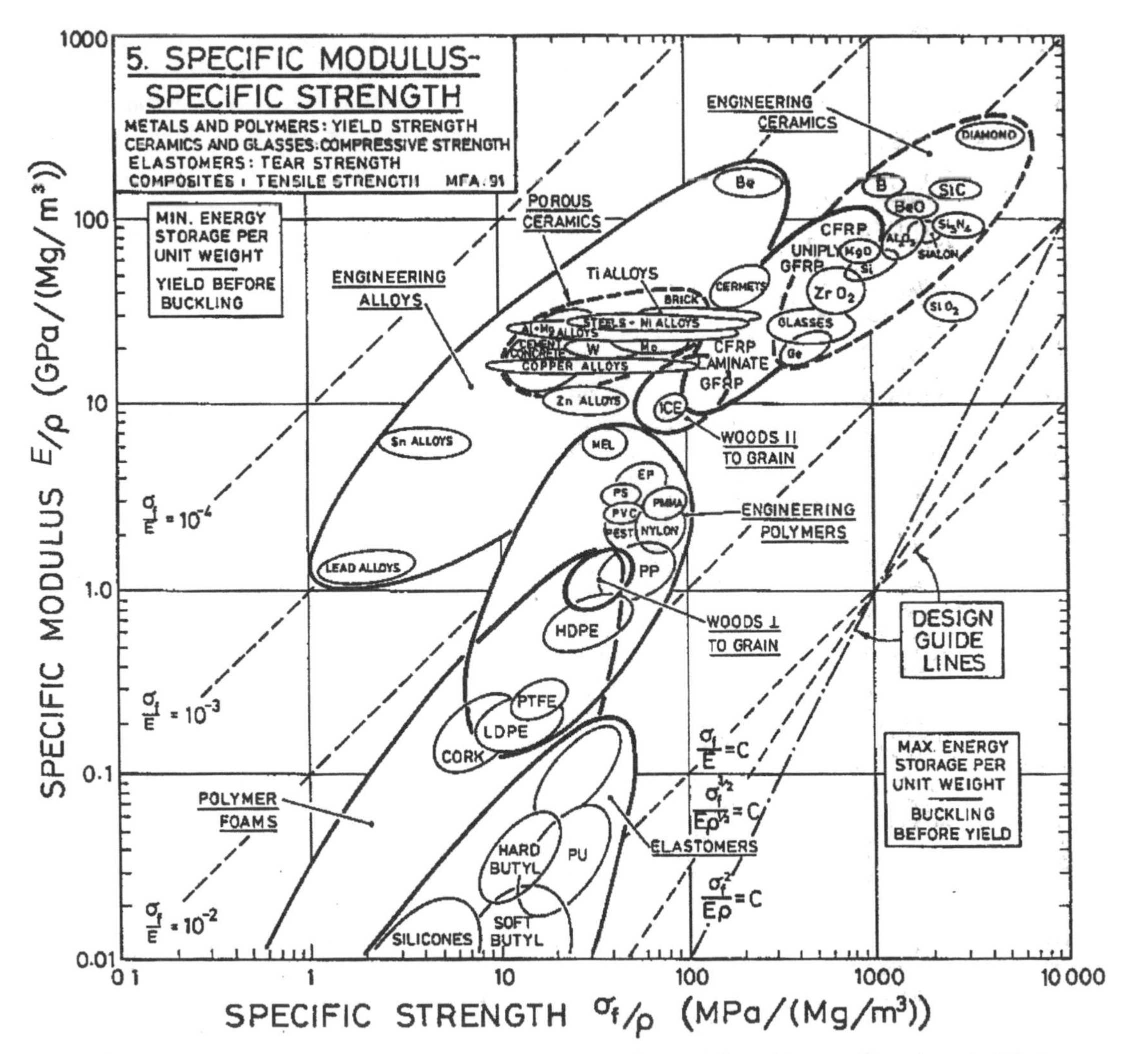

Figure 5.1.5 *Relative position of materials comparing specific modulus with specific strength.* (*Reference:* M. F. Ashby. *Materials Selection in Mechanical Design*. Butterworth-Heinemann, 1995)

In the case again of a ladder on an aircraft where the mass of any component has to be kept to a minimum, a three-way comparison between modulus (to minimize deflection), strength (to prevent mechanical failure) and density (to minimize material usage) is needed. This can be achieved by scanning the chart comparing specific modulus against specific strength (Figure 5.1.5) where,

Specific modulus = modulus/density

and,

Specific strength = strength/density

In this case the polymer composites based around carbon fibre (CFRP) and glass fibre (GFRP) overlap with the best class, the ceramics. Where the cost is less of an issue, carbon fibre-reinforced epoxy will make an excellent choice. Ceramics are impractical from a manufacturing point of view and also suffer from poor impact strength. The recently installed toughened glass floor at the Science Museum shows that glass of the appropriate geometry does have the required strength and stiffness.

Please note that the values given for materials are taken from tensile tests but the conditions under which they were obtained are unclear. This is normally the case with values specified in textbooks and even in databases. The data would need to be verified once a small selection of materials has been made.

At this point, we will look at some common engineering materials particularly so that we can identify how their properties are determined.

5.2 Plain carbon steels

Steels are the most important material available to a structural engineer, be it for the body of a container such as a car or for reinforcing concrete in modern buildings. They are highly versatile and are available in hundreds of compositions. They are essentially iron–carbon alloys although there are always other additions as well. A small proportion of carbon atoms can occupy the space between iron atoms in an iron crystal without severely distorting the structure. This is called an *interstitial solution.*

They are characterized by good strength and stiffness in tension and compression and have high ductility. They can be joined by welding and can be machined easily. Their properties can be controlled by varying their composition and by heat treatment. They are also relatively cheap.

During the process of manufacture, impurities are introduced in the steel. They can be expensive to remove completely. They include:

- Sulphur: Harmful because it may form brittle iron sulphide making the steel brittle.
- Manganese: Beneficial because it forms manganese sulphide in preference to iron sulphide. MnS is ductile and so the steel is not made brittle. Mn also increases the strength. Most steels contain 0.7% Mn.
- Phosphorus: Harmful and should be kept to low proportions.
- Silicon: Beneficial in small quantities.

Classification of plain carbon steels

Steels can be classified into a number of different groups. There are three main groups:

- *Low carbon steels* containing up to about 0.25% carbon by weight: They have moderate strength and excellent fabrication properties and are used in great quantities for bridges, buildings, ships and other vehicles. The great majority of all steel used (over 90%) falls in this group.
- *Medium carbon steels* containing between 0.25% carbon and 0.60% carbon by weight: They have greater strength and hardness but less ductility. They are therefore also more difficult to fabricate than the low carbon steels. They are used to manufacture products such as gears and rails, which need better wear resistance.
- *High carbon steels* containing between 0.60% carbon and 1.4% carbon by weight (although steels with greater than 0.8% carbon are rare): They have the highest strength and hardness but the least ductility and toughness of the three groups. They are used to manufacture springs, dies and cutting tools where high hardness and wear resistance are important. Better materials have generally taken their place.

Note: In all our future discussions we will assume that the proportions of elements in any alloy are given in terms of the weight of material rather than volume.

Microstructure of steels

Iron is polymorphic and forms different crystal structures over different temperature ranges as was seen under 'Polymorphism (allotropy)', Chapter 3. The structural changes also exist when carbon is dissolved in the iron structure. The dissolved carbon forms interstitial solid solutions. The proportion of carbon that exists in the BCC ferrite form is much less than can dissolve in the FCC austenite form. We can see this in the thermal-equilibrium diagram for the iron–carbon system (Figure 5.2.1).

When the solubility limit of carbon in either ferrite or austenite is exceeded, a compound, Fe_3C, forms between the excess carbon and iron. This is called iron carbide, or *cementite*. It has a composition of 6.67% carbon by weight. The complete iron–carbon thermal equilibrium diagram is more complex than the idealized diagrams we have looked at up to now. However, it helps to show the different forms taken by the steel.

You should notice straightaway that the melting point of the material falls as the amount of carbon is increased up to about 4.3% carbon. Furthermore, unlike a pure metal, the material does not usually change state at one temperature. Instead, the liquid transforms to solid over a range of temperatures. There is just one composition at which the liquid behaves like a pure metal and changes state at one temperature. This is the *eutectic composition* (with 4.3% carbon). We will look at this composition in more detail as well as two others. A *eutectoid* is similar to a eutectic except that there is a transformation from one form of solid

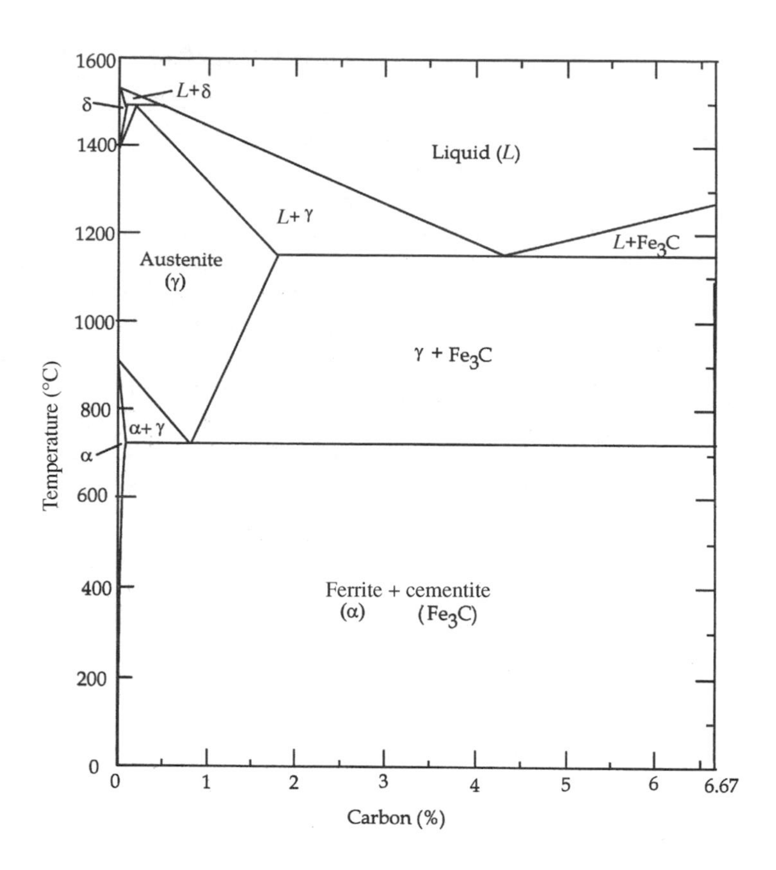

Figure 5.2.1 *Thermal-equilibrium diagram for the iron–carbon system*

to another at one temperature. An alloy cooling through a eutectoid will behave exactly as an alloy through a eutectic except that the changes will be slower in the solid state. The third composition to consider is one that forms a *peritectic*. Consider an alloy with 0.3% carbon cooling from the liquid state. δ ferrite forms in the liquid but as it goes through the peritectic, all of the material will form the single phase of austenite (γ). It will then cool through the eutectoid to give primary ferrite and a eutectoid phase of $\alpha + Fe_3C$. The eutectoid phase, which appears as a series of dark lines in a bright background when viewed using an optical microscope after polishing and etching, is called *pearlite*.

Note on studying the microstructure of metals

We can see the structure of metals using the optical microscope after suitably preparing the surface. This process involves first cutting a piece of the metal across the section to be examined. This is illustrated in Figure 5.2.2. The section is then mounted in a thermosetting polymer resin to give a product that can be handled easily as it is gradually worked and polished using successively finer grades of silicon carbide paper to give a smooth surface. The final polishing is carried out using diamond paste in the case of steels and cast irons or another solution. The finished surface should be flat and have the reflectivity of a good mirror so that it can be examined at high magnification using the optical microscope. The first stage of examination, however, is often just viewing with the naked eye. This is followed by examination under the microscope. Flaws and grain boundaries may be seen. However, the most interesting structure is seen after the surface is *etched*. This

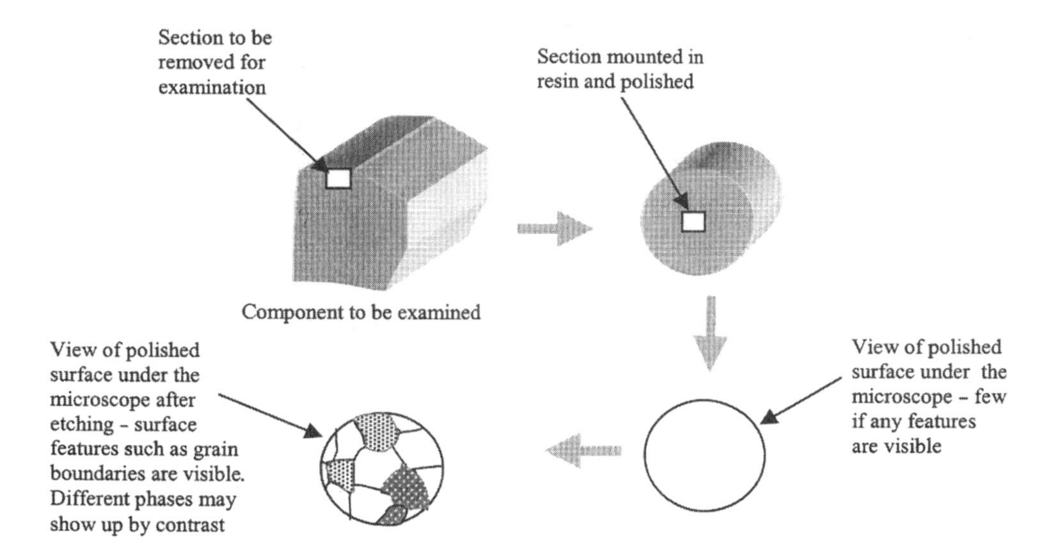

Figure 5.2.2 *Preparation and optical microscopy of a metal*

involves the exposure of the surface to a dilute solution of acid. The particular medium will depend on the metal. For example, for steels, about 2% by volume of nitric acid in industrial alcohol (called Nital) may be used. The acid preferentially attacks certain phases and so increases the contrast between phases and at grain boundaries. The process of etching is a delicate operation requiring good control of the time of etching as insufficient exposure of the surface to the etchant might not reveal any features, but more importantly, too much exposure could lead to destruction of the characteristics being examined. Etching is also valuable in removing the surface deformation that inevitably takes place as the section is cut and polished and so should reveal the actual structure in the section.

Activity 5.2.1

Look at some aerial photographs of your town or a region that you recognize and try to relate the features that you know, such as buildings, roads, rivers and woods to the features that you can see in the picture. How easily can you match features? Now look at a satellite image of an area that you are less familiar with. How easily can you identify features? What pitfalls are there in predicting what you see?

Heat treatment of steels

An important advantage with the use of steels is that a simple heat treatment process can control the properties of many of them. There are three main heat treatment processes. For all three the steel must be heated to the austenite region and allowed to remain there for long enough for it to change to a completely austenite structure (soaking).

The steel can be cooled in one of three ways:

- Very slow cooling, for example in a furnace. This is the process of full annealing.
- Intermediate rate of cooling, for example in still air. This is called normalizing.
- Rapid cooling, for example by plunging into cold water. This is called quenching.

To study these heat treatment processes only the eutectoid part of the thermal equilibrium diagram (Figure 5.2.3) need be considered.

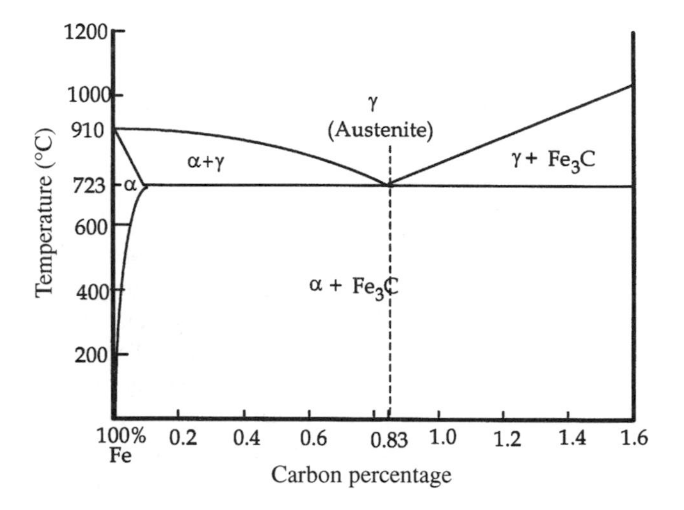

Figure 5.2.3 *Eutectoid region of the iron–carbon thermal equilibrium diagram*

A eutectoid steel contains 0.83% carbon. The *lower critical temperature* is constant for all compositions of steels. It has a value of 723°C. The *upper critical temperature* for all steels is the line above which only austenite is stable. Consider the heat treatment of a medium carbon steel containing, for example, 0.6% carbon.

Full annealing

The steel is heated to the austenite (γ) region and soaked. *It is important to realize that the steel has not melted. The steel remains solid during these heat treatments.* The 0.6% carbon steel is then cooled very slowly, by switching the furnace off and allowing it to cool inside the furnace or by transferring it to a special 'soaking pit'.

The steel is completely austenite as it cools slowly to the upper critical temperature. At the upper critical temperature, primary ferrite starts to form, and increases as the steel cools between the upper critical temperature and the lower critical temperature. Ferrite can only dissolve a low amount of carbon and so the carbon has to diffuse from the ferrite crystals to the austenite. The carbon content of the austenite increases.

At the lower critical temperature (723°C) the carbon content of the austenite is 0.83%. The temperature remains constant while the austenite changes to ferrite and cementite. This new combination is called *pearlite*. The final structure consists of ferrite regions and pearlite regions. Within the pearlite, the cementite forms short, almost parallel bands separated by ferrite and can often be seen clearly when viewed under an optical microscope, after polishing and etching. Ferrite is soft, weak and ductile whereas pearlite is hard, strong and brittle. A mixture of the two, as a result of full annealing, gives a strong, ductile and tough material.

The properties of the steel will be affected by the volume ratios of ferrite to pearlite. A steel with a higher carbon content will have more pearlite and will be stronger and harder than one with a lower carbon content. A fully annealed steel with 0.83% carbon will form pearlite only. Steels with more than 0.83% carbon will have a structure of cementite and pearlite. They will be hard and inevitably brittle.

Normalizing

The cooling rate is much faster now. There are three main effects:

- There is less primary ferrite because the steel cools much quicker between the upper critical temperature and the lower critical temperature.
- Ferrite is supersaturated with carbon because the carbon does not have sufficient time to diffuse from the ferrite to the austenite.
- The laminations of the pearlite are much finer.

In consequence, a normalized steel will be harder, stronger but less ductile than a fully annealed steel of the same carbon content.

Quenching

The austenite starts to transform ferrite but the cooling rate is so fast that there is no time for the carbon to diffuse out of the ferrite. The ferrite crystal structure is therefore heavily extended. Some of the distortion is relieved by a shear process and the structure takes a body-centred tetragonal form, called *martensite*. Under the microscope, martensite appears as a needle-like or acicular phase. It is very hard and brittle. The hardness depends on the carbon content as illustrated in Figure 5.2.4.

Low carbon steels (less than 0.3% carbon) are not often quenched because they are usually required in the ductile condition.

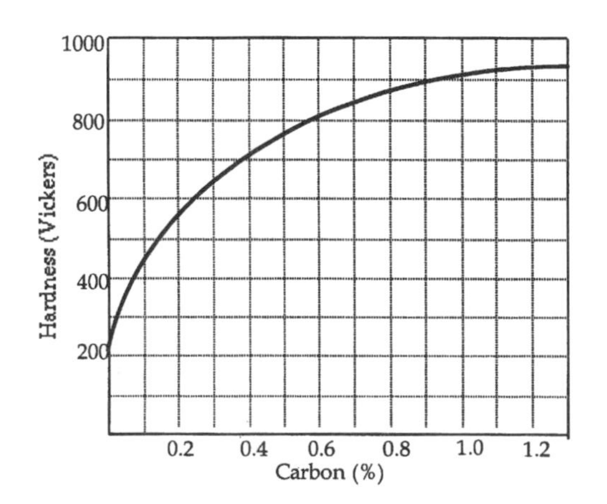

Figure 5.2.4 *Variation of hardness achieved by quenching with % carbon*

Tempering

Martensite is very hard, and potentially has high strength. However, the presence of excess carbon leads to *residual stress* within the component as well as a loss of ductility. The material would therefore fracture at relatively low tensile stresses with no prior warning. To avoid such failure, it is *tempered* to increase its toughness. This process involves reheating the steel to a temperature below the lower critical temperature. The thermodynamically stable phases at that temperature are ferrite and cementite, not martensite, and so there is a tendency for the martensite to change to ferrite and cementite. That does not take place at room temperature because there is insufficient energy in the metal. The cementite forms as tiny spherical particles. Tempering can be carried out at temperatures of 200–650°C. The higher the tempering temperature the coarser are the cementite particles and the softer and tougher the steel. The cementite particles only become visible under the microscope at temperatures of about 400°C. The structure, which consists of cementite particles in a ferrite matrix (background phase), is called *sorbite*.

The properties of quenched and tempered steel can be controlled to give a wide range of properties by varying the tempering temperature. They are usually treated so that they are stronger and harder than steels treated by normalizing. A qualitative comparison of the properties achieved by the various heat treatment processes is indicated in Table 5.2.1.

Table 5.2.1 Comparison of properties achieved by various heat treatments

	Fully annealed	*Normalized*	*Quenched and tempered* †
Strength	Low	Moderate	High
Ductility	High	Moderate	Low
Hardness	Low	Moderate	High
Toughness	Low	High	Moderate

† = Properties depend on tempering temperature

Stresses developed during quenching

The rapid rates of cooling during quenching lead to the development of high stresses in steels. This is aggravated because expansion occurs when the face-centre cubed (fcc) austenite is transformed into body-centre (bct) martensite as illustrated in Figure 5.2.5.

Some regions within a component treated in this way may retain stresses (residual stresses) that could lead to unexpected failure of the component in normal use. Some quenched parts may become distorted or even cracked, especially if there are complex shapes with sharp corners or abrupt changes of cross-section. Some strategies that can be adopted to minimize this build-up to residual stresses are given below:

- Design the component to avoid sharp corners if possible. They can occur, for example, at the base of a screw thread and in key-ways.

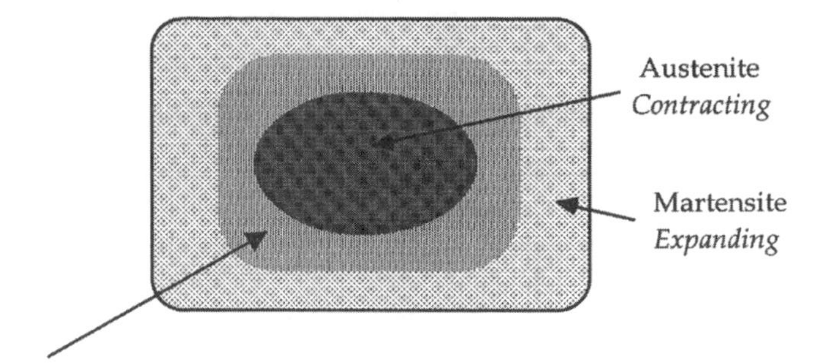

Figure 5.2.5 *Changes in structure across a section during high rates of cooling*

A generous radius should be applied. Also avoid abrupt changes in cross-section where possible.

- In industry steels are usually quenched in oil rather than water. This will cool the steel less quickly and so reduces some of the deleterious effects of water quenching. However, the steel will not be so hard as when quenched in water. Different types of oil are available which will give different cooling rates. Water-soluble polymer quenching agents are available which have some advantages over oil, but they are expensive.

During heat treatment the steel must be soaked in the austenite region of the thermal equilibrium diagram so that it changes to austenite completely. Heating it to a temperature well above the upper critical temperature could reduce the time required to soak the steel. However, that should be avoided because the austenite crystals would grow and become large. On cooling, ferrite would precipitate in the coarse austenite crystals along certain planes with the remaining austenite forming pearlite at the lower critical temperature. This structure tends to have poor ductility and is called a widmanstatten structure.

Activity 5.2.2

Have you noticed how a bar of soap can sometimes develop cracks and even split after it has been used for a while? This is an example of the presence of residual stresses. When the bar of soap was first compacted into shape, it would have contained more moisture than it has in the dry state. Describe the stages by which stresses may develop. What causes the bar to crack?

Consider now a solid steel cylinder of diameter 20 cm and length 2 m, which is quenched from the austenitic region to ambient temperature. Describe the stages in the process by which stresses develop and explain why they may lead to internal cracks.

5.3 Cast irons

Plain carbon steels are the most important engineering alloys because of their good all-round properties. They can be cast to shape and some steel castings are made, but there are some problems. Steel has a high melting temperature, which requires an advanced furnace and the input of considerable energy to melt it. There is considerable shrinkage when steel castings are produced because of the large temperature difference between the solidification temperature and room temperature. Steel castings tend to have an overheated structure. Whereas steels contain up to 0.8% carbon and exceptionally a little more, cast iron is the same form of alloy with more than 2% carbon as can be seen in Figure 5.3.1.

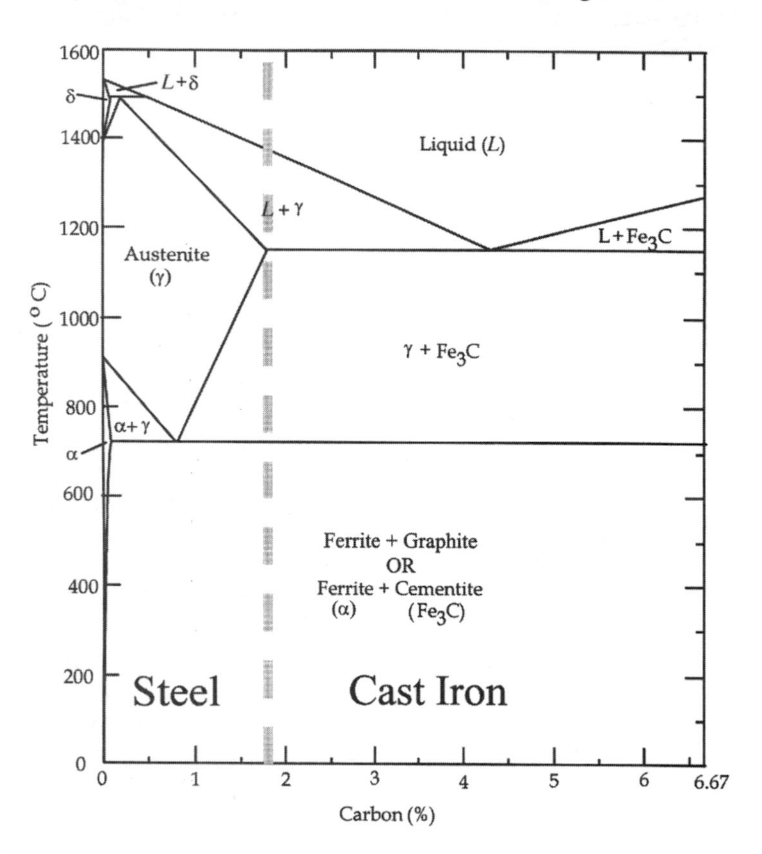

Figure 5.3.1 *Iron–carbon thermal equilibrium diagram showing cast iron and steel compositions*

Cast iron is more easily produced because of its lower melting temperature. The liquid can be poured into a mould and solidified readily. Yet, they have many of the advantages of ferrous alloys. There are basically four groups of cast iron called:

- Normal, flake graphite
- Spheroidal graphite
- Malleabilized
- Special

Flake graphite (grey) cast iron

Depending on the impurity content, particularly silicon, the Fe–C equilibrium diagram can be changed so that graphite rather than cementite is formed with the austenite, as the melt solidifies. This is indicated in Figure 5.3.1. A high manganese content encourages the formation of cementite rather than graphite. In steel there is usually sufficient manganese and a low silicon content to ensure that no graphite is formed. Other impurities or alloying elements will encourage either graphite or cementite formation but silicon and manganese are the two elements used for control. The graphite takes the form of flakes.

As the lower critical temperature (723°C) is reached, a eutectoid reaction occurs. Following rapid cooling from the melt, ferrite (α) and cementite (Fe_3C) will form. This combination is usually known as *white iron*. If the cooling process is slowed down, then ferrite and graphite will form below the lower critical temperature. In this form, the final structure consists of graphite flakes in a background of ferrite and pearlite. Graphite has very little strength in tension and is brittle. The flakes of graphite also do not bond with the surrounding metal and so their sharp-pointed tips leave the equivalent of cracks in the metal, which act as stress concentrators. Under tension a crack can easily spread from one graphite flake to the next causing failure at low loads. Flake graphite cast iron is therefore brittle. The fractured surface is grey in colour because of the graphite, and so it is called *grey iron*.

To summarize:

- Grey cast irons have low viscosity when molten and so flow easily to fill complex mould shapes.
- They have low shrinkage on solidification because the normal contraction is balanced by the formation of graphite.
- They can be melted in simple furnaces compared with steel.
- They are easy to machine because the graphite flakes cause discontinuous chip formation.
- They provide good damping of sound and vibration.
- Cast iron components are relatively cheap because of some of the above advantages and because the material is inherently cheap as the level of purity needed is less than that for steel.

The cylinder block for an internal combustion engine (ICE), for example, is made from cast iron because of most of these advantages. It would be difficult to manufacture by any method other than casting. Phosphorus is a common impurity in grey cast irons. It can be present up to 1.5%. It reacts to form a eutectic phase called phosphide eutectic. Phosphorus increases the fluidity of the molten iron while phosphide eutectic increases the wear resistance of the solid material. Cylinder liners for the ICE are made from phosphoric grey cast iron. The hard phosphide eutectic phase reduces wear while the soft graphite is worn away to leave channels for the lubricating oil. Grey cast irons are specified in BS 1452.

A *balanced cast iron* can be produced by suitable control of the silicon content. When a casting of this material is produced, fast cooling results in *white cast iron* and slow cooling results in *grey cast iron*. ICE camshafts are made from a balanced iron. The camshaft is cooled slowly in the mould to give a grey iron. The lobes of the camshaft are cooled quickly by putting metal 'chills' in the mould to extract heat at a rapid rate. This makes the lobes very hard and wear resistant.

Spheroidal graphite cast irons ('SG.' or 'nodular cast irons')

The disadvantage with grey cast irons is that they are brittle. With *spheroidal graphite cast irons*, the shape of the graphite is changed from flakes to small spheres. This is done by purifying the molten cast iron and just before casting adding a small quantity of magnesium. The graphite spheres do not have the same stress concentration effect as the sharp-pointed flakes and so SG iron is quite tough, but retains most of the advantages of grey cast iron. It is quite an expensive material because of the high level of purification and quality control required. However, SG iron components are still cheaper than steel, because casting is a cheap method of manufacture. In the example of the mass production of ICE crankshafts, casting with SG iron is cheaper than forging with steel.

Heat treatment of SG irons

Graphite spheres form at 1137°C and at normal cooling rates and pearlite will form as the background phase at 723°C. If the iron casting is fully annealed by soaking at between 850 and 900°C and very slowly cooled, ferrite and graphite will form rather than ferrite and cementite. The SG iron can also be heat treated like steels and properties similar to steels can be produced:

- Normalized (as cast): They have a pearlitic structure and will be strong and tough, but not very ductile.
- Annealed: They have a ferritic structure and will be less strong but more ductile.
- Partly annealed: They have a mixed ferritic/pearlitic structure where the strength and ductility can be controlled.
- Quenched and tempered: They have a tempered martensite structure and will have high strength and low ductility.

Spheroidal graphite cast irons are specified in British Standard BS 2789. They are given grades of two numbers that specify their tensile strength and ductility measured as % elongation to failure. Examples of such grades are illustrated in Table 5.3.1.

Table 5.3.1 Grades of cast iron given in terms of material properties

Tensile strength/ductility (MN m^{-2})/(% elongation)	*Matrix phase*
370/12	Ferrite
420/12	Ferrite
500/7	Ferrite/pearlite
600/3	Ferrite/pearlite
700/2	Pearlite
800/2	Pearlite or tempered martensite

Other cast irons

- *Malleabilized cast irons* are produced by an old method. They are cast irons that have the advantages of grey irons, but are tough and ductile. The irons used contain 0.6–1.0% carbon. These solidify as white irons when they are cast. The castings are subsequently annealed at 900–950°C for long periods of time and then cooled slowly. The cementite in the white iron decomposes to give ferrite and graphite, as it should according to the thermal equilibrium diagram. According to the process of manufacture, the resulting materials are described as *blackheart* or *whiteheart*, each having slightly different properties and so leading to different applications.
- *Meehanite*: Just before casting a finely powdered ceramic is mixed into the molten iron (inoculation). The particles of ceramic act as nuclei for the formation of graphite. The final structure contains very fine graphite flakes, which give improved mechanical properties, particularly strength.
- *Silal* has a high silicon content (5%). This gives a structure with very fine, disconnected graphite flakes. Such a material has better dimensional stability at elevated temperatures.
- *Ni-resist* contains 15% nickel, 2% chromium, 6% copper, giving very high corrosion resistance.
- *Ni-hard* has low silicon content and also contains 4% nickel and 2% chromium. A very hard structure can be produced, especially at a rapidly cooled surface, during casting. Its uses include rolls for rolling mills.

Problem 5.3.1

The load-bearing base of a measuring instrument is to be made from cast iron. The instrument needs to be isolated from mechanical vibrations and have moderate impact resistance. The base needs to be at least 1.5 m long, 20 cm wide and 1.5 m tall. The surface should have good wear resistance and a smooth finish. Determine which cast iron is best for the application, giving reasons.

5.4 Stainless steels

Looking at the full thermal equilibrium diagram for the iron–carbon system in Figure 5.2.1, we can see that pure iron as it solidifies first takes the δ form, then the γ form and finally the α form. Of these, the δ and α forms are body-centre cubed (bcc) whereas the γ form is face-centre cubed (fcc). This polymorphic change in the iron and the steel structure is responsible for many of the property changes in steel. Normally, steel can only take the body-centre cubed form at room temperature, but we will now look at methods for changing even the structure of steel.

Iron alloyed with a bcc element

If iron is alloyed with a bcc element the bcc form is stabilized. The effect is to alter the temperatures at which polymorphic change takes place. Elements such as chromium, tungsten, vanadium, molybdenum

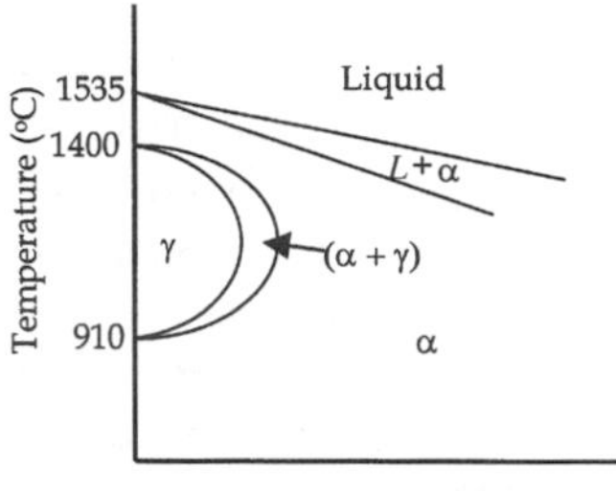

Figure 5.4.1 *Effect of a bcc element on the thermal equilibrium diagram of steel*

and silicon have this effect. Their thermal equilibrium diagram is illustrated in Figure 5.4.1.

When chromium, which is a passive metal, is alloyed with steel, the alloy will also be passive if it contains more than 13% Cr. A protective film of chromium oxide (Cr_2O_3) will form on the surface as illustrated in Figure 5.4.2. (The process of passivation is described under 'Aqueous corrosion', Chapter 6.)

The film forms and will be repaired if damaged as long as there is freely available oxygen in the environment. The film separates the metal from the environment so that no further corrosion occurs. If there is no freely available oxygen in the environment the film may be damaged and break down. This is an example of a stainless steel. Different types of stainless steel are formed by varying the chromium content above 13%.

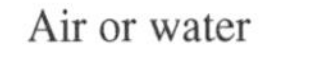

Figure 5.4.2 *Formation of a protective coating of Cr_2O_3 on the surface of Fe–Cr alloy*

Ferritic stainless steel

The carbon content of this steel is kept low (0.1%) but there is 17% Cr in the alloy. This gives rise to good corrosion resistance. Ferrite is ductile and so these steels can be rolled into thin sheet easily and then pressed to shape. Their strength remains low and they cannot be heat treated in the same way as plain carbon steels because there is no austenite range as illustrated in Figure 5.4.3. Typical uses of ferritic stainless steel include car trims (although being replaced by plastics) and kitchen utensils and work surfaces.

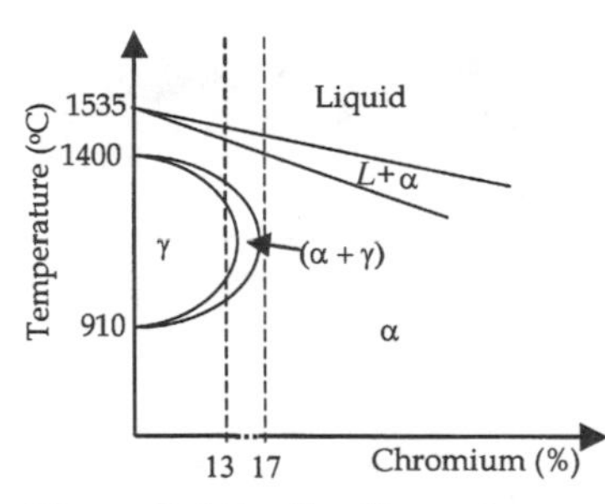

Figure 5.4.3 *Ferritic and martensitic stainless steels*

Martensitic stainless steel

These are steels alloyed with about 13% Cr and up to 0.25% C. With the slightly greater carbon content, austenite formation is extended to slightly higher chromium content and so on cooling, the iron passes through the γ phase (Figure 5.4.3). This form is cheaper than ferritic alloys. Their corrosion resistance is not as good, but is much better than plain carbon steels. They form martensite on cooling in air making them hard and strong. It follows that they cannot be cold worked and are difficult to weld. Hence they are usually cast to shape. Typical uses of martensitic stainless steel include pumps for corrosive slurries and surgical and dental instruments.

Iron alloyed with an fcc element

If iron is alloyed with an fcc element the fcc form is stabilized. Elements such as nickel, manganese, carbon and nitrogen have this effect. Their thermal equilibrium diagram is illustrated in Figure 5.4.4.

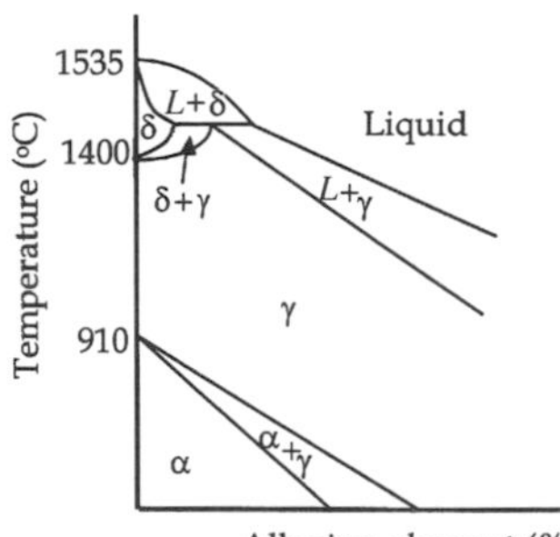

Figure 5.4.4 *Effect of an fcc element on the thermal equilibrium diagram of steel*

Austenitic stainless steel

These steels are Fe alloy with 18% Cr and 8–10% nickel (Ni). They are more expensive than the other stainless steels, but have better corrosion resistance. Ni stabilizes the austenite phase to make the steel ductile (Figure 5.4.5). These steels are strong and tough even at sub-zero

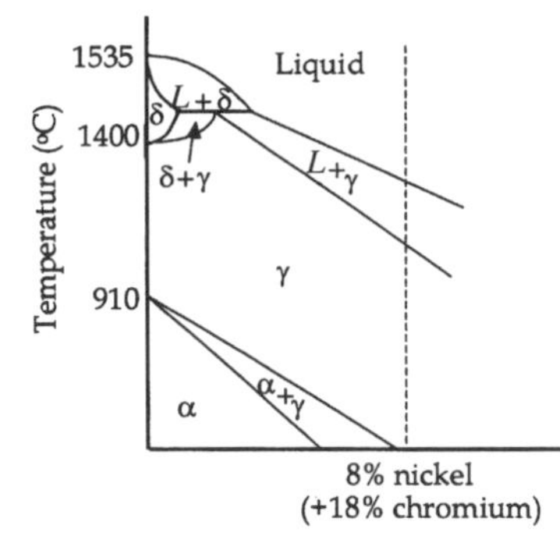

Figure 5.4.5 *Thermal equilibrium diagram for Fe alloy with 18% Cr and 8% Ni*

temperatures, and can be readily welded. They cannot be heat treated like plain carbon steels. They work-harden very rapidly and are non-magnetic. Typical uses of austenitic stainless steel include large-scale chemical vessels, high quality domestic items and artificial implants in the human body.

Problem 5.4.1

Stainless steels appear to have the characteristics of most steels as well as being rust free. What could prevent their use in the manufacture of car bodies? Give more than one reason with explanation.

5.5 Aluminium

Properties of aluminium

Aluminium and its alloys are the most important and widely used group of non-ferrous metals. Almost pure or unalloyed aluminium is widely used. It has the following properties:

- Low density: The density is 2700 kg/m^3 compared with 7800 kg/m^3 for steel. It is possible to make alloys which have high strength-to-weight ratios. This is important for the manufacture of vehicles, especially aircraft. The stiffness-to-weight ratio is not significantly better than other materials.
- High electrical conductivity: Aluminium can carry as much current as a copper conductor of twice the mass. Aluminium is used where lightweight conductors are required. The main limitation on the use of Al as a conductor is that it is difficult to solder.
- High thermal conductivity: This is useful for heat exchangers. It is being used increasingly for automobile radiators.
- Low melting temperature (660°C): This is useful for making castings and means that Al alloys can be used for pressure die-casting. The disadvantage is that the maximum service temperature for Al alloys is relatively low.
- High ductility: This makes it easy to fabricate components.
- Good corrosion resistance: Aluminium is a *passive* metal (see 'Aqueous corrosion', Chapter 6). It reacts with oxygen in the air to form a thin, invisible film of oxide over its whole surface, which protects it from further corrosion. The thickness of the naturally occurring oxide film can be increased by *anodizing* (see Chapter 6). This gives increased corrosion protection and increased resistance to wear, e.g. for the jet intakes of aircraft.
- Low tensile strength: The tensile strength of pure aluminium is 45 MN/m^2. Much of the 'pure' aluminium used contains about 0.5% iron, which increases strength, but reduces ductility and corrosion resistance, e.g. saucepans.
- High coefficient of thermal expansion: This can cause problems for some applications, e.g. in internal combustion engines.

Aluminium alloys

The addition of small amounts of other elements helps to increase strength although other properties could be adversely affected. There are two main groups of alloys:

- Casting alloys as defined in BS 1490.
- Wrought alloys as defined in BS 1470–1475.

These two groups can be further divided into those that can be heat treated and those that cannot. In the case of *non-heat treatable* alloys, strength is increased by solute hardening, for example with magnesium or by the formation of second phases, such as with silicon. The strength of wrought alloys is often increased by work hardening. These alloys can be annealed after cold work. The *heat-treatable* alloys are strengthened by a special heat treatment process called *precipitation hardening*. Alloys of this type are often referred to by their trade name, *Duralumins*. There are many other alloys of this type. The process of precipitation hardening in aluminium alloys is described below.

κ (kappa) is a solid solution of copper in aluminium, which can be seen in the thermal equilibrium diagram (Figure 5.5.1). When the solubility limit of copper in aluminium is exceeded, a compound θ ($CuAl_2$) forms.

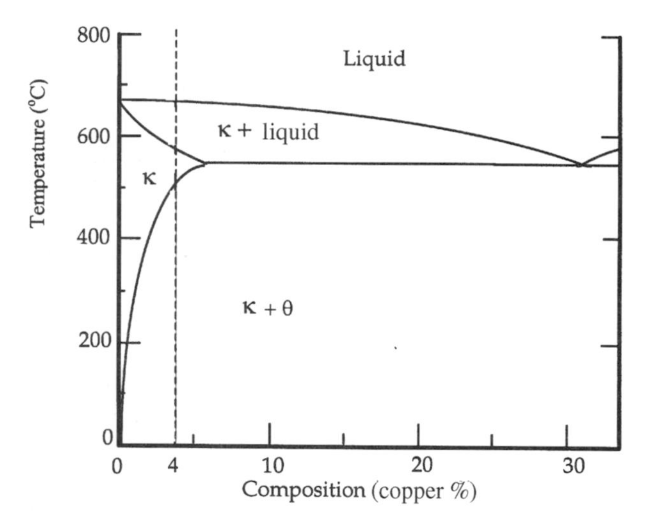

Figure 5.5.1 *Thermal equilibrium diagram for the aluminium–copper system*

Now consider an aluminium alloy containing 4% copper, identified by the dotted line in Figure 5.5.1. There are two stages to the heat treatment. The first stage is called *solution treatment*. In this process, the alloy is heated to the κ region and soaked so that the alloy becomes completely κ. It is then quenched. In this process, the normal formation of θ is suppressed by the rapid cooling rate. As a result, at room temperature, the structure will consist of a supersaturated solid solution of copper in aluminium. This is a *substitutional solid solution* of 4% copper atoms dissolved in aluminium. The copper atoms will form a completely random arrangement on the aluminium crystal lattice. This is illustrated in Figure 5.5.2. The strength of the alloy will have increased to some extent by *solute hardening* because the copper atoms are a different size to the aluminium atoms.

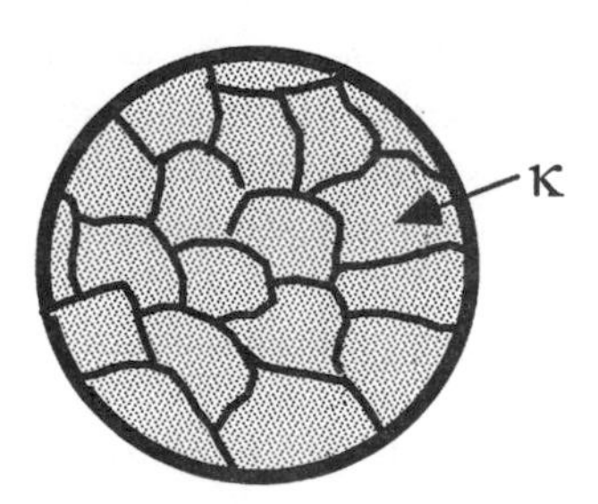

Figure 5.5.2 *View of the κ phase in Al–Cu alloys supersaturated with Cu*

The second stage is called *precipitation (or ageing)*. The alloy is left for some time either at room temperature or reheated to a fairly low temperature (up to 250°C) to achieve this effect. The thermodynamically stable phase, θ ($CuAl_2$), will now tend to form as small particles and the internal strain energy of the material decreases. Strength and hardness increase as the θ phase forms. The precise degree of increase is a function of temperature and time.

To form $CuAl_2$ there must be regions where there are copper atoms in such concentrations that there is one copper atom for every two aluminium atoms. The copper atoms diffuse through the aluminium crystal lattice to tiny sites, which are rich in copper atoms. There will be very many of these sites in each crystal. The higher the temperature, the faster will be the rate of diffusion of the copper atoms. There will be severe distortion of the aluminium lattice at the copper-rich sites because the copper atoms are a different size to the aluminium atoms. The distortion of the aluminium lattice causes a *strain energy field* to be set up. The copper-rich sites are still part of the aluminium lattice so this is called a *coherent precipitate* (Figure 5.5.3). Resistance to dislocation movement causes the increase in strength. There are many tiny particles of 'precipitate' and the dislocations must cut through these if plastic deformation is to take placc.

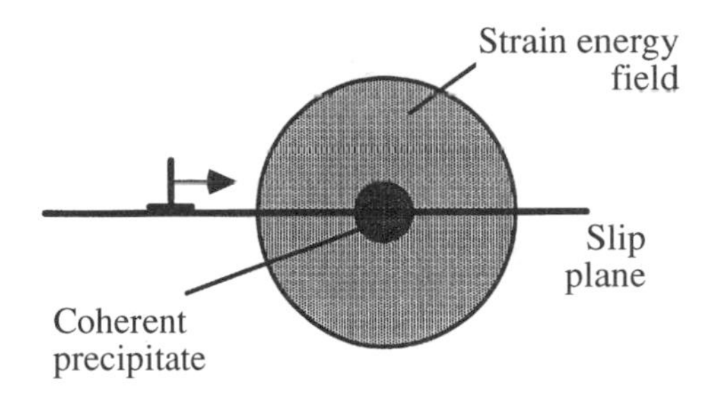

Figure 5.5.3 *Strain energy field around a coherent precipitate*

This resistance causes the proof stress to be raised. The larger the precipitate and its associated strain energy field, the more resistance there is to dislocation movement and the higher the strength. Strength also increases with time at the ageing temperature. However, as long times are involved to achieve the maximum strength, the process can be expensive. If the temperature is too high, the strength can decrease after some time, as the particles of θ become larger. This effect is called *overageing*.

The maximum strength is reached at about 200°C with Al with 4% Cu. If the alloy is then cooled to, and used at, room temperature, it will not overage. This is because, as the particles become larger, the strain energy associated with the copper-rich region increases until it becomes so high that the precipitate of $CuAl_2$ breaks away from the aluminium lattice. The aluminium lattice is no longer distorted and so the strain energy field is reduced. It is then easier for dislocations to cut through the precipitate and so the strength is reduced. The alloy will not overage if it is cooled to room temperature when it has been fully strengthened. Care must be taken to avoid overageing when precipitation-hardening alloys are heated during their service life.

Precipitation hardening was first used on aluminium–copper alloys. Other aluminium alloys were developed which could be treated in a similar way. Precipitation hardening is useful because it can give a high increase in strength with the addition of small quantities of alloying elements in, for example, nickel alloys and steels. It always involves the two stages of solution treatment followed by precipitation. Whether or not the alloy could actually be precipitation hardened would have to be determined experimentally. It would depend on, for example, the rates of diffusion of the two metals in each other.

Alloying elements in aluminium

Silicon is important for alloys that are specified for making castings, using all casting processes. As can be seen in Figure 5.5.4, when the eutectic alloy (11.7% silicon) is used to make castings, the usual lamellar structure is produced. The structure is coarse, making the alloy brittle. To prevent that, a small quantity of sodium can be added just before casting, which results in the lamellar eutectic structure being fine and less brittle. This also moves the eutectic point to a higher silicon content (14%) and lowers the melting temperature slightly (dotted line on thermal equilibrium diagram).

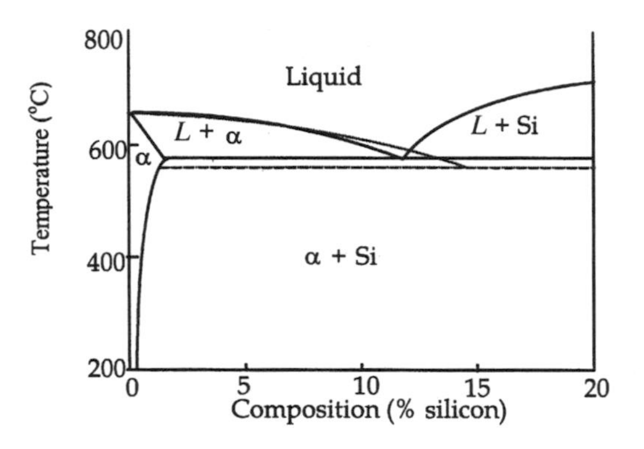

Figure 5.5.4 *Thermal equilibrium diagram for the aluminium–silicon system*

The eutectic alloy, which has a low melting temperature and solidifies at one temperature, is particularly suitable for pressure die-casting. The alloys are suitable for casting because they have:

- high fluidity when molten, and
- low shrinkage on solidification because the normal shrinkage is balanced by expansion when silicon forms from the melt.

Alloys with more than the eutectic composition of silicon will have primary silicon in them. This is hard, and also rather brittle. However, the coefficient of thermal expansion is reduced, which makes these alloys suitable for use in internal combustion engines.

- *Manganese* increases the strength of the aluminium by solute hardening. The strength can be further increased by work hardening. Typical uses of this alloy include hollow-ware such as saucepans.
- *Magnesium* also increases the strength in the same way as manganese, but rather more effectively. The ductility remains high and corrosion resistance is increased. This alloy is therefore especially suitable in marine environments, for example for the superstructure of ships.
- *Copper* is the most important alloying element responsible for precipitation hardening and can give a high increase in strength.

Corrosion resistance is, however, reduced and copper alloys are not suitable for anodizing. This can be overcome by using a thin coating of pure aluminium on the surface (ALCLAD) or an alloy that does not contain copper. Copper-based alloys are very important for structural parts in aircraft and general engineering because of their high strength coupled with the low density of aluminium alloys.

- When *magnesium* and *silicon* are used together, precipitation hardening is caused by the formation of Mg_2Si. The strength is increased but not as significantly as with the Al–Cu alloys. The corrosion resistance is not so adversely affected as with the Al–Cu alloys and such alloys are used for structural components in road transport vehicles.
- In the case of *nickel*, precipitation hardening is caused by the formation of $NiAl_3$. This is more resistant to overageing than the other alloys and so it can be used at moderately higher temperatures. It is used in internal combustion engine cylinder heads and pistons.
- *Zinc* gives a high increase in strength when used in combination with copper and magnesium. The alloy becomes difficult to fabricate and suffers from corrosion and overageing problems. Such alloys are used with care in some aircraft alloys. Zinc is regarded as an impurity for other applications.
- Aluminium–*lithium* alloys are precipitation-hardening alloys giving a good strength increase. They also have increased modulus of elasticity, which is unusual. They have a lower density than other aluminium alloys. Thus, they have high strength-to-weight and stiffness-to-weight ratios. They have great potential in aircraft construction.

Problem 5.5.1

Alloys used for aircraft construction are highly developed and can have complex compositions. The alloy used for the airframe of Concorde, for example, has to be strong and able to resist overageing at the moderately elevated temperatures that exist because of frictional forces. In this case, the aluminium alloy used, known as RR58, has the following additions made to the host metal (Table 5.1):

Try to explain the purpose of each element noting the proportions being used.

Table 5.5.1 Composition of aluminium alloy RR58

Element	*Proportion* (%)
Copper	1.8–2.7
Magnesium	1.2–1.8
Silicon	0.15–0.25
Iron	0.9–1.4
Nickel	0.8–1.4
Manganese	0.2
Titanium	0.2
Zinc	0.1

5.6 Copper

Properties of copper

- High electrical conductivity: Pure copper has a high electrical conductivity and so is widely used for electrical conductors. The conductivity increases with purity.
- High thermal conductivity: Copper is used in radiators, heat exchangers and boiler tubes where heat energy has to be transmitted efficiently. Cu alloys resist fouling in sea water and so are especially useful for heat exchangers using sea water.
- Good corrosion resistance:
  - In water: Therefore copper is used in environments containing sea water or tap water and in the brewing and chemical industries.
  - In air: Cu corrodes slowly, forming a protective green 'patina' on the surface.
- High ductility: This is useful because copper can be drawn into wire for electrical applications. It can be made into tubing, which can be easily bent for water supplies.
- Copper is *easily joined* by brazing and soldering.
- Copper has *low strength and hardness* and so needs to be alloyed for applications where the applied stress is significant.

Copper–zinc alloys (brasses)

Brasses have good corrosion resistance and are suitable for decorative finishes. The two common types are α *brass*, which contains up to 39% zinc and α–β *brass* which contains 39–45% zinc. The thermal equilibrium for copper–zinc systems is illustrated in Figure 5.6.1.

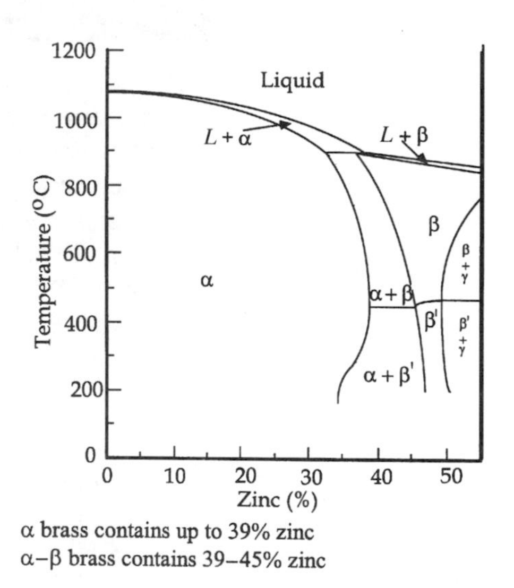

Figure 5.6.1 *Thermal equilibrium for copper–zinc systems (brasses)*

Brasses fall into two categories, which are determined by the nature of the phases formed in the metal at room temperature and therefore their properties. First, α *brasses* are those that form a single, α phase. Their strength increases as the zinc content is increased because of solute hardening. Ductility increases to a maximum at 30% zinc as illustrated in Figure 5.6.2.

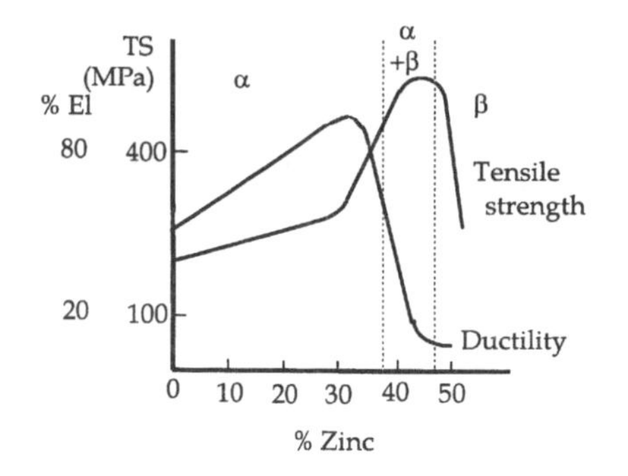

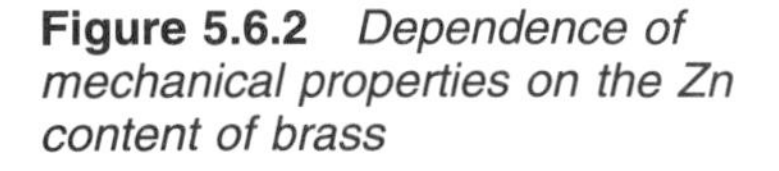
Figure 5.6.2 *Dependence of mechanical properties on the Zn content of brass*

A well-known form called *cartridge brass* consists of 30% Zn in copper. Copper with 35% Zn is more widely used because it is cheaper. Typical uses include condenser tubes and decorative components made by deep drawing.

The second category is called α–β *brass*. They are stronger than α brass but have low ductility as can be seen in Figure 5.6.2. The corrosion resistance is also reduced. Copper with 40% zinc is called *Muntz metal*. It gives a widmanstatten structure on casting because α precipitates from β (see phase diagram). This tends to be brittle. Hot working breaks up the widmanstatten structure and makes the alloy tougher. It will also be anisotropic. β which is a disordered solid solution, may change to β′, an ordered solid solution, on very slow cooling. Typical uses of Muntz metal include condenser and heat exchanger plates.

Free cutting brass can be made by the addition of 2% lead to brasses. *High tensile brasses* contain small additions of manganese, tin, aluminium or iron to increase strength and corrosion resistance. A ship's propeller is an example of a high tensile brass.

Copper–tin alloys (tin bronzes)

The addition of tin to copper produces the valuable class of materials known as bronzes. Their thermal equilibrium diagram is illustrated in Figure 5.6.3.

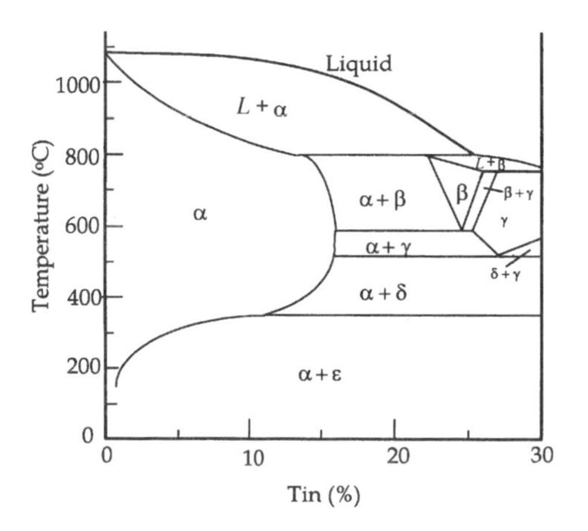

Figure 5.6.3 *Thermal equilibrium diagram of copper–tin systems (bronzes)*

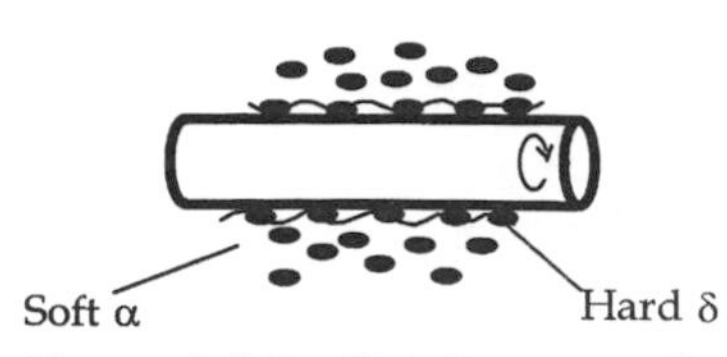

Figure 5.6.4 *Relative wear of the α and δ phases in α–δ alloy bearings*

The α *bronzes* are ductile and have better strength and corrosion resistance than α brass. They are, however, more expensive. Five per cent Sn in copper is used for springs and steam turbine blades. Three per cent Sn and 1.5% Zn added to copper is used for so-called 'copper' coins. α–δ *alloys* can be cast to make bearings. In use, the soft α phase wears away leaving the hard δ phase to support the rotating journal. The δ provides good wear resistance. The gaps where the α has worn away provide channels for good oil flow and retention (Figure 5.6.4). The ductile α provides good shock resistance.

Many bronzes contain some phosphorus. They are then called *phosphor bronzes*. The phosphorus forms copper phosphide (Cu_3P). This is a hard phase which improves the wear resistance of bearings because it acts like the δ phase α–δ bronzes containing some zinc are called *gun metals*. They are used for pumps and valves operating at high pressures.

Copper–aluminium alloys (aluminium bronzes)

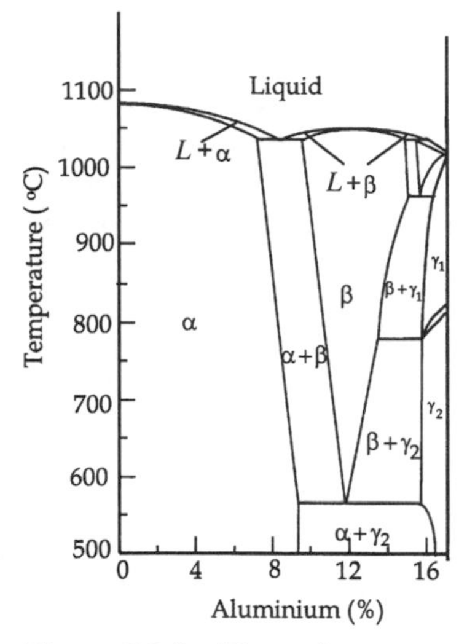

Figure 5.6.5 *Thermal equilibrium diagram of copper–aluminium systems (aluminium bronzes)*

Aluminium bronzes have good corrosion resistance and are suitable for marine environments. They are also resistant to high temperature oxidation. They have strengths similar to tin bronzes but are cheaper. Their thermal equilibrium diagram is illustrated in Figure 5.6.5.

α alloys are ductile and can be cold worked. They are used to make jewellery and condenser tubes. α–γ_2 alloys are cast into ships' propellers and valves. There is a eutectoid at 565°C. This can give a martensitic structure on fast cooling, which can be tempered at 500°C to give a useful fine-grained structure.

Activity 5.6.1

The Bronze Age was an important stage in the development of mankind. Explain the main reasons for the development of bronze and identify the main uses of bronzes.

Further reading

Ashby, M. F. *Materials Selection in Mechanical Design*. Pergamon Press, 1992

Callister, W. D. *Materials Science and Engineering*. John Wiley & Sons, 1999

Alexander, W. & Street, A. *Metals in the Service of Man*. Penguin, 1998

6 Materials and the environment

Summary

The term environment is used in a variety of often confusing ways. From the design point of view, it sets the conditions under which the object or component operates. It is therefore the set of conditions under which the materials chosen have to meet the functional requirements. This includes the temperature, humidity, chemistry and indeed the user. The environment itself responds to the process by which materials and products are manufactured, used and disposed of. In this chapter, we will first examine in detail the chemical conditions under which metals degrade. This will be followed by a discussion on measures that can be taken to minimize or protect against deterioration. In the second part of the chapter, we will examine the effect on the environment of materials and products. Beginning with the consumption of energy and therefore their sources in producing materials, we move on to look at the ways of managing materials once they are discarded after their primary use. With this appreciation, we can hope to design products where materials are considered as reusable resources.

Objectives

By the end of this chapter, you should be able to:

- understand the electrochemical mechanisms and conditions responsible for the degradation of metals;
- identify techniques available to the designer to avoid or minimize the effects of corrosion;
- generate awareness of the effects on the environment of the life cycle of materials from extraction from raw materials to disposal at the end of their useful lives;
- identify strategies for effective management of the waste materials;
- enable the designer to design with recyclability in mind.

6.1 Degradation of materials in the environment

Most solid materials react with the environment and new compounds or products develop on their surface. In the case of metals, this is referred to as corrosion. In the case of polymers and ceramics, we talk of degradation. In all cases, there is a *chemical reaction* between components in the environment, especially oxygen and the solid material. Through the reaction, the material returns to a lower state of energy.

In nature, metals occur as ores where they exist in combination with other elements. The natural state is in the form of oxides, sulphides, carbonates, etc. The extraction process raises the potential energy of the material as illustrated in Figure 6.1.1. It is also very expensive and unfortunately, on first exposure to the atmosphere, the metal usually attempts to return to the lower energy state!

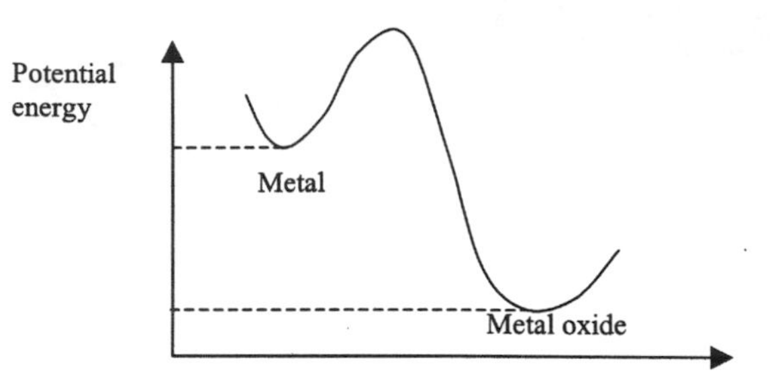

Figure 6.1.1 *The equilibrium states of a metal and its oxide*

In the United Kingdom, some reports show that 1 tonne of steel is converted into rust (iron oxide) every 90 seconds. The energy cost of producing the 1 tonne of steel is 42 000 MJ.

Aqueous corrosion

Electrochemical action is the presence of water which acts as the medium for the reaction, providing the energy to overcome the 'hump' in potential energy. An example is the rusting of iron and steel. In this case, iron combines with water and dissolved oxygen to form rust.

$$\underset{\text{iron}}{4Fe} + \underset{\text{water}}{2H_2O} + \underset{\text{oxygen}}{3O_2} \rightarrow \underset{\text{hydrated ferric oxide}}{2Fe_2O_3 \cdot H_2O} \tag{6.1}$$

The rusting reaction has two stages. In the first stage, *oxidation* takes place at the anode. In this process, electrons are given up by the metal atoms as they go into solution in the water. Iron becomes positive ions.

$$4Fe \rightarrow 4Fe^{2+} + 8e^- \tag{6.2}$$

In the second stage, *reduction* takes place at the cathode. The electrons released in the first stage are combined with oxygen atoms and water molecules to form negative hydroxyl ions.

$$4H_2O + 2O_2 + 8e^- \rightarrow 8OH^- \tag{6.3}$$

(Ordinary water contains about 8 ppm (parts per million) of oxygen molecules.)

Note on symbols

The letters in the chemical formulae refer to the symbols of each element as defined in the periodic table. A number before each symbol signifies the number of the atoms or molecules in the term. For example,

$4H_2O$ is the same as 4 molecules of H_2O. Each molecule consists of 2 hydrogen atoms (H_2) combined with one oxygen atom (O). In the $8OH^-$, the basic unit is the OH which consists of one oxygen atom and one hydrogen atom. The negative sign after the OH signifies that this is a negatively charged ion with one more electron than needed to have charge balance. Similarly the $4Fe^{2+}$ signifies 4 positive Fe ions each of which is missing 2 electrons.

Eventually, the ions in (6.2) and (6.3) will combine to give the brown rust, $Fe_2O_3 \cdot H_2O$, that we have all seen. When written in this form, we can see why this is called 'hydrated ferric oxide'. The medium or *electrolyte* in this case is water containing dissolved charged species such as H^+ and OH^-. Pure water does not allow the reaction to take place being a poor conductor of electricity. It is important to note therefore that a complete electrical circuit is needed, consisting of conducting metals for the electrons to flow and a conducting fluid to allow positive and negative ions to complete the electrical circuit. The anode is the site on the metal where positive ions are released into the fluid and the cathode is the site where the excess electrons are able to combine with other positive ions in the fluid. The anode and cathode may exist as separate and permanent sites on the metal although they could equally be arbitrary, mobile sites. Atmospheric corrosion mostly involves water and oxygen and is therefore called *aqueous corrosion.*

Electromotive series

Only metals corrode since only they can provide electrical conduction. The tendency of a metal to lose material and corrode can be measured as an electrical potential. In a standard experiment, the electropotential of metals is measured by creating a cell as illustrated in Figure 6.1.2. In one half, the metal being tested, e.g. copper, is suspended in a solution containing copper ions. The copper rod is connected to another conducting metal, platinum, suspended in the other half of the cell. This second half is the reference material and hydrogen gas circulating around the platinum rod acts as this material. The two halves are separated by a semi-porous membrane that allows ions to flow to complete the electrical circuit while maintaining the different material concentrations in each half of the cell. A voltmeter connected between the two metals allows the potential between the two halves to be measured. The resulting hierarchy, called the electromotive series, is given in Table 6.1.1.

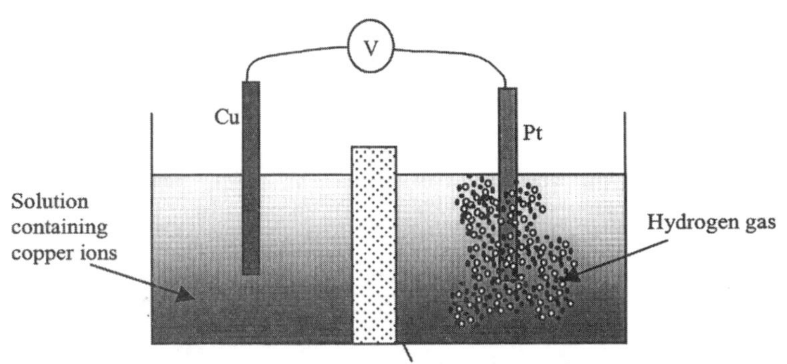

Figure 6.1.2 *Experiment to generate the electromotive series*

Table 6.1.1 Electromotive series giving the reduction potentials for metals

Metal	*Electrode potential (volt)*
Gold	+1.42
Platinum	+1.20
Silver	+0.80
Mercury	+0.80
Copper	+0.35
Hydrogen	0.00
Lead	–0.13
Tin	–0.14
Nickel	–0.25
Cadmium	–0.40
Iron	–0.44
Chromium	–0.71
Zinc	–0.76
Aluminium	–1.67
Magnesium	–2.38
Sodium	–2.71

The sign of the potential indicates that when compared to hydrogen, the metals with positive potentials will produce a reduction reaction and the metals with negative potentials will produce oxidation reactions. The series is valuable when choosing material combinations in design since in a given pair, the metal with the lower potential will corrode. For example, if copper and iron were used in contact with each other, there will be a potential difference between the two metals of,

$$+0.35 - (-0.44)\ \text{V} = 0.79\ \text{V}$$

The copper will become the cathode and the iron the anode and electrons will flow from the iron to the copper. The iron will therefore corrode. The potentials measured are particular to the environment of the experiment. If tap water was used, then different potentials would be measured. Useful comparisons can be made in any fluid and a relative table obtained. One very useful series for sea water is illustrated in Table 6.1.2.

Table 6.1.2 Galvanic series in sea water

Graphite
Stainless steel (passive)
Monel metal (70% nickel, 30% copper)
Nickel (passive)
Bronze
Copper
Aluminium bronze
Nickel
Brass
Tin
Lead
Stainless steel (active)
Cast iron
Plain carbon steel
Aluminium and its alloys
Cadmium
Zinc
Magnesium and its alloys

Activity 6.1.1

Look at components around the home and garden and look for evidence of corrosion. Make notes on the materials used, how they are joined and the environmental factors that could lead to corrosion.

If you ever take a walk by a port, you would not have failed to notice how easily the hull of a ship in contact with water will show signs of rust. Even closer to home you may have identified examples such as a wrought iron gate. In these cases, although all the metal may be one material, there will be separate sites on the same surface where anodes and cathodes have developed. This is possible under the following conditions:

- Different phases such as cementite and ferrite in a cast iron or steel will have different electropotentials. The cementite is usually cathodic with respect to the ferrite.
- Areas of high stress and worked edges are usually anodic.
- Grain boundaries and inclusions can be either anodic or cathodic.
- Inert deposits or dirt particles can starve the metal surface of oxygen and therefore become anodic.

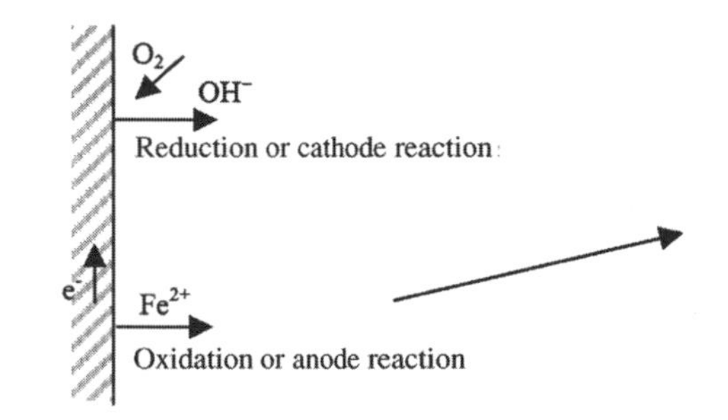

Figure 6.1.3 *Corrosion mechanism*

Considering the diagram in Figure 6.1.3, the reaction at the anode will be according to equation (6.2). The reaction at the cathode could be according to equation (6.3) in waters which are chemically neutral (i.e. where pH ~ 7) such as sea water. In acidic solutions (i.e. low pH) hydrogen gas is generated as shown below:

$$H^+ + H^+ + 2e^- \rightarrow H_2 \qquad (6.4)$$

Note on pH

The pH is a measure of the acidity of a solution. It is a measure of the concentration of the positive ions, H_3O^+ in the solution given as a logarithm of its concentration. Those solutions with a high concentration of the negative ions, OH^- (and therefore low concentration of H_3O^+), are said to be basic. As far as we are concerned, pH = 1 or 2 suggests a highly acidic solution (very high concentration of H_3O^+ ions) and a pH = 13 or 14 indicates a highly basic solution (very high concentration of OH^- ions). In both cases, the environment can promote corrosion. A neutral solution is one where pH = 7. Milk and tap water have a pH approximately equal to 7. Beer and tomato juice are slightly acidic at a pH of 5 and detergents are more basic with a pH of 10.

When the two metals from our earlier example, iron and copper, are connected together in an electrolyte as illustrated in Figure 6.1.4, a current will flow and the metal with the lower electropotential will lose material.

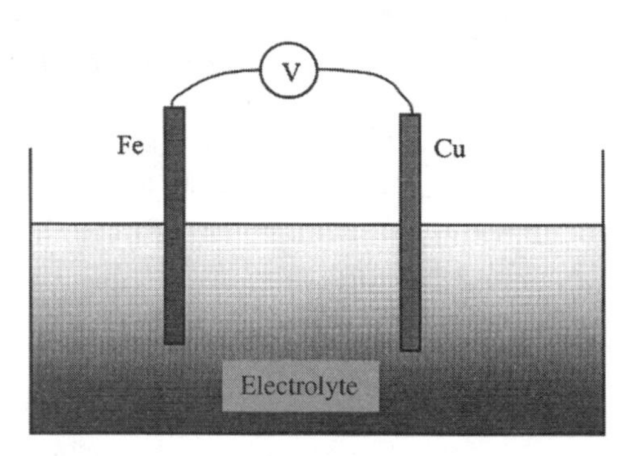

Figure 6.1.4 *Daniell cell*

In this case, iron will corrode as it loses electrons. Its electrical potential will rise in the process. 'Maps' showing the relationship between potential and pH have been produced to show the likelihood of corrosion. Such maps are called *Pourbaix diagrams*. An example for iron is illustrated in Figure 6.1.5. We can identify three regions on this map:

I – Immunity region. In this region, the electrical potential difference is too small to cause iron ions to be lost. Iron is therefore in a stable state. There is no tendency to corrode.

II – Corrosion region. In this region, the electrical potential difference is high enough to maintain the corrosion current through the metal and electrolyte.

III – Passive region. In this region, although corrosion occurs, the oxide film forming on the metal surface strongly adheres to the surface. This acts as a barrier to further corrosive attack and so first slows the reaction down and eventually blocks it altogether.

Pourbaix diagrams can be useful but they have severe limitations in their use in practice. As an example of their use, consider iron in water at the point * in Figure 6.1.5. Corrosion takes place under this condition. Corrosion can be prevented by one of two methods:

- By lowering the electrical potential to the immunity region. This is done by *cathodic protection*, which will be described under 'Design' below.
- By introducing an alkali into the water to increase the pH and push the iron into the passive region. This method is used for protection against corrosion in sealed systems such as boilers.

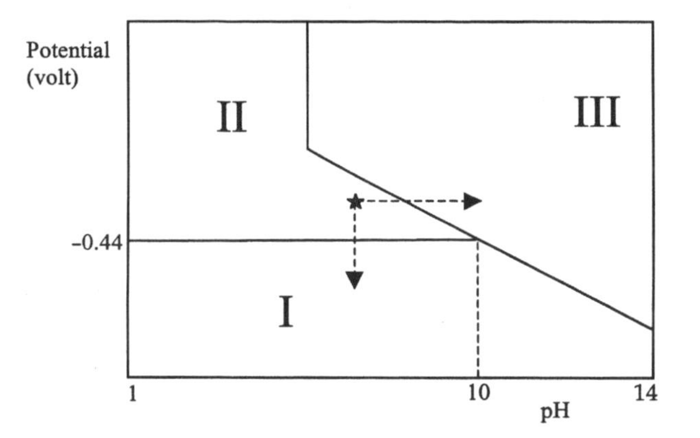

Figure 6.1.5 *Pourbaix diagram*

The Pourbaix diagram will show whether corrosion will take place. However, the rate at which corrosion takes place is more important. For example, zinc is lower in the electropotential series than steel and therefore more unstable chemically. However, it corrodes at a much slower rate than steel in air and therefore a small amount of zinc can be used to protect a larger amount of steel.

The more useful measure of the corrosion is the *intensity*, i, of attack, which is defined as the rate per unit area. This can be measured in units of:

- mm/year – giving an indication of the loss of thickness on a sheet, for example, or
- $g/m^2/day$ – giving an indication of the mass lost per unit of surface per day.

When mild steel corrodes, it does so in a uniform manner. The surface, however, may become roughened because of local variations in the rate of corrosion. If the intensity of corrosion is known, then it can be designed for. For example, the component can be made thicker to take account of the intended life.

Example 6.1.1

A steel tank containing aerated water is found to corrode at a rate of 47.4 mg per square decimetre of surface per day. How long will it take for the wall thickness to decrease by 0.40 mm?

Solution

$$\text{Corrosion rate} = 47.4\ \text{mg/dm}^2\ \text{day} = \frac{47.4 \times 10^{-6}\ \text{kg}}{(10^{-1}\text{m})^2\ \text{day}}$$

$$= 47.4 \times 10^{-4}\text{kg/m}^2\ \text{day}$$

$$\text{Density of steel} = 7800\ \text{kg/m}^3$$

The mass lost in a thickness of 0.4 mm

$$= 0.4 \times 10^{-3}\text{m} \times 7800\,\text{kg/m}^3$$

$$= 3.12\,\text{kg/m}^2$$

Therefore time taken for wall thickness to reduce by 0.4 mm is

$$= \frac{3.12\,\text{kg/m}^2}{47.4 \times 10^{-4}\text{kg/m}^2\text{ day}}$$

$$= 658.2\text{ days}$$

Such data can be found in design guides. Manufacturers of corrosion-resistant materials also provide such data although this may give a limited choice of materials. However, it must always be remembered that small changes in the environment such as the temperature and pH can have a very significant effect on the corrosion rates. The rate of corrosion is increased by:

- Agitation of the water: This allows dissolved O_2 to reach the cathode surface more rapidly (ships corrode faster when moving).
- Higher temperatures: There is faster diffusion of O_2 at higher temperatures (ships corrode faster in warm waters).
- Increasing the amount of dissolved oxygen: Dissolved oxygen levels fall rapidly with depth of water.

The presence of pollutants in the atmosphere can further influence the rate of corrosion. The following are sources of such corrosion:

- Inert materials such as dust and dirt particles can trap moisture on the metal surface.
- Oxides of sulphur generated by biological decay (e.g. of seaweed) and the burning of fuels (such as coal and oil) especially around power stations and even domestic central heating systems.
- Salts such as sodium chloride in marine environments and deposited on roads in winter.
- Ammonium sulphate in industrial environments.

Crevice corrosion

In the restricted space between two metal components joined together, particularly aggressive corrosion can take place. For example, two plates joined together by rivets can provide an environment for corrosion (Figure 6.1.6). As oxygen in the water between the plates is

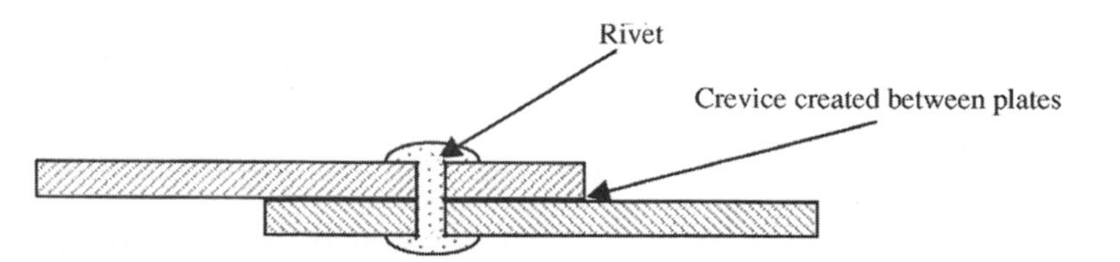

Figure 6.1.6 *Crevice corrosion between plates*

used up, an anode is set up in this area in preference to a cathode. There is still plenty of dissolved oxygen reaching the outer surfaces of the structure, which becomes the cathode. Very intense corrosion takes place in the region between the plates. This is a common type of corrosion called *crevice corrosion* and can be avoided by good design.

Some metals are resistant to corrosion because they form an invisible, inert film on their surface. This protects the underlying metal from further attack. Metals such as Al, Cr, Ti and Pt all form protective coatings. In the case of chromium, the coating has the composition Cr_2O_3. As was seen in Section 5.4, the protective qualities of stainless steels come from a film of Cr_2O_3 on the surface of the alloy. The film is formed by an anodic reaction, but then forms a barrier to further corrosion. As long as the environment is oxidizing (i.e. there is plenty of oxygen) the passive film is self-repairing. This means that if the passive film is damaged by wear, it is immediately formed again. If oxygen is not readily available as is the case in deep sea conditions, the passive film cannot form. Damage may happen by either mechanical or chemical action and the self-repairing action may not work. Similarly, in crevices, the protective films can be destroyed by H^+ and Cl^- ions, for example. The anodic reaction can therefore continue in the crevice.

In some circumstances, passive metals can suffer from localized or pitting attack. It occurs especially in neutral solutions ($pH \sim 7$) containing chloride ions, such as sea water. Stainless steel, which is normally passive, can suffer *pitting corrosion* in deep sea environments where there is also a lack of oxygen. The chloride ion breaks down the passive film by entering into it. The passive film is now a cathode and the exposed metal becomes an anode and therefore active. In this situation, there is a large cathode and a small anode and so intense attack by pitting takes place. The rate of penetration by pitting cannot be predicted accurately. It can result in rapid failure even though very little metal has corroded. Adequate allowance cannot be made in design and so this is a particularly dangerous form of corrosion.

In crevices, there are usually the right conditions for pitting to occur. Therefore passive metals are especially prone to crevice attack. The passive nature of the film can be maintained if oxygen is allowed to reach the surface. Agitation of the water helps to reduce the tendency to corrode.

Design

We have already seen some measures that can be taken to minimize the potential for corrosion. They will now be examined in more detail.

- By careful design, horizontal surfaces, crevices and other water and dirt traps can be avoided. Make sure that the area of the cathode is much smaller than the area of the anode. The process of corrosion is controlled by the availability of O_2 or other positive ions at the cathode, which can combine with any excess free electrons on the metal surface. If this process is slowed by reducing the cathode surface area, then metal ions at the anode are less likely to be lost.
- Avoid horizontal surfaces or troughs where water and sediment can collect. Cathode reactions locally under the sediment will use up the

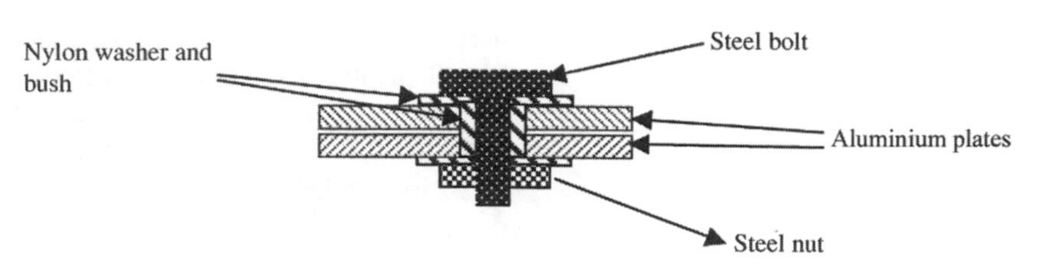

Figure 6.1.7 *Insulation of metals from each other*

oxygen in the region. As oxygen flow to this region is now restricted, the only reaction in this area becomes the anode reaction. This area then becomes a preferential anode. Cathode reactions will continue in the 'clean' areas of the tank bottom. Since the cathode area is much larger than the anode area, intense corrosion will take place under the sediment. Even when the composition across the sheet is the same, separate anode and cathode areas can form because of the differences in oxygen concentration. This is called an *oxygen concentration cell.*

- Insulate metals from each other. For example, when two aluminium plates are held together by a steel nut and bolt (Figure 6.1.7) corrosion of the aluminium can be prevented by separating the metals using a nylon bush and washer.

Activity 6.1.2

Look at common engineering products that have been in use for several years and identify locations where corrosion has taken place and give reasons for the corrosion.

Examples of products:

- The bodywork of a lawnmower.
- The bodywork of a car.
- Steel bicycle frame.
- Railings in the street.

Selection of materials

Where corrosion is a problem, if possible, avoid using different metals in contact with each other. If this is not possible, then use metals that are close together in the galvanic series. Special alloys such as stainless steel may also be used. This adds to the cost of materials and of manufacturing. However, it may prove more economical to use cheaper materials such as mild steel with coatings despite the need to replace the component more often.

Inhibitors

These are substances that are added to the environment to reduce the rate of corrosion. For example, in the car radiator system, antifreeze is added to prevent freezing of water. This, however, is highly corrosive and therefore inhibitors are also added to the liquid. In another example, toothpaste containers used to be made from aluminium. To prevent chemical reaction between the aluminium and the toothpaste, inhibitors were added to the toothpaste. The inhibitor formed a solid film on the inside surface of the aluminium to act as a barrier between the container and the toothpaste.

Anode inhibitors

The corrosion of iron is relatively easy. This reaction cannot be stopped as the surface remains exposed in aqueous environments. However, there are several compounds such as phosphates, sodium chromate and sodium benzoate that can form a thin, invisible film on the surface of iron or steel by reacting immediately with the Fe^{2+} ions that are produced. Sodium chromate is the most effective for steel although it is also very toxic. The film is extremely effective. However, if there is insufficient inhibitor, then a complete film may not form. The reduced area of the anode could accelerate the corrosion process. Sodium benzoate is effective for steels even when the amount of inhibitor is inadequate. This is used in car antifreeze mixtures.

Cathode inhibitors

Soluble salts of metals such as Zn^{2+}, Mg^{2+} and Ca^{2+} work by using up the available oxygen in the water. Hydroxides of the metal precipitate at the cathode surface. In hard waters, calcium carbonate is precipitated at the cathode:

$$Ca^{2+} + OH^- \rightarrow Ca(OH)_2$$

$$Ca(HCO_3)_2 + Ca(OH)_2 \rightarrow 2CaCO_3 + 2H_2O$$

The films formed are relatively thick and stifle the corrosion reactions. They are not as efficient as anodic inhibitors, but can reduce the corrosion rate to acceptably low levels. They are safe in that they cannot lead to increased localized corrosion. Calcium and magnesium salts are present in many natural waters and help to slow down the corrosion process in components such as mains water pipes. Zn^{2+} ions are, however, toxic.

Vapour phase inhibitors

These are substances which are added to the atmosphere in sealed containers to prevent steel corroding. The inhibitor is of low vapour pressure so that it forms a vapour that condenses on the metal surface. Corrosion is prevented as the film formed on the surface is protective. Complex organic solids such as dicyclohexylamine nitrate are used. Wrapping paper can be impregnated with this type of inhibitor and will provide adequate protection to metals when covered with it. Razor blades and other steel precision instruments, for example, were protected by this method. Sodium benzoate can also be used for this purpose.

Cathodic protection

In an electrochemical cell, the anode corrodes and the cathode is protected or, at worst, has reduced corrosion. The position of the two metals in the electrochemical series will determine which of the two is the anode. Cathodic protection works by making the metal to be protected the cathode in an electrochemical cell. The system requires another conductor to act as the anode, a conductor between the anode and cathode and an electrolyte. The electrical potential is lowered into the immunity region, I on the Pourbaix diagram (Figure 6.1.5), and so the tendency for corrosion is lowered. There are two methods of applying cathodic protection.

Impressed current system

The example in Figure 6.1.8 shows a metal structure located underground. This could be a steel water pipe, for example. The pipe is connected in an electrical circuit including a d.c. supply and an anode.

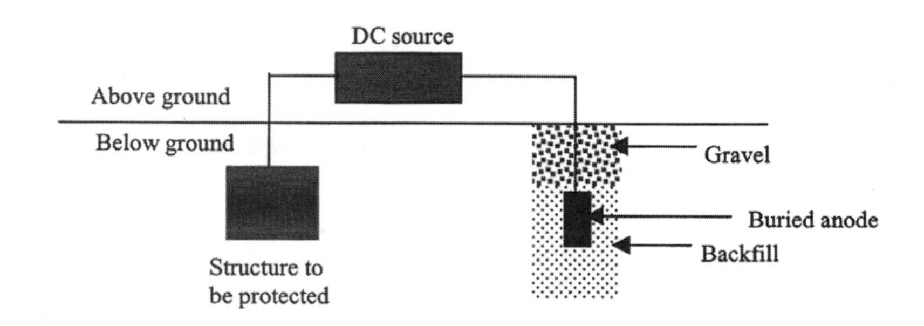

Figure 6.1.8 *Cathodic protection by the impressed current method*

Large structures can be protected by one anode. In this method, a high electrical potential can be applied between the cathode and anode (e.g. 30 V at 50 A) since the system is easily isolated. The anode material can also be chosen so that it does not corrode. Hence maintenance costs can be kept low. Typical anode materials are:

- Carbon for buried structures: A pit of coke is a cheap, neutral conducting material.
- Platinized titanium: This is more expensive but would be used in water.

The impressed current system is expensive to install because of the d.c. source. The major disadvantage of the system, however, is that there is

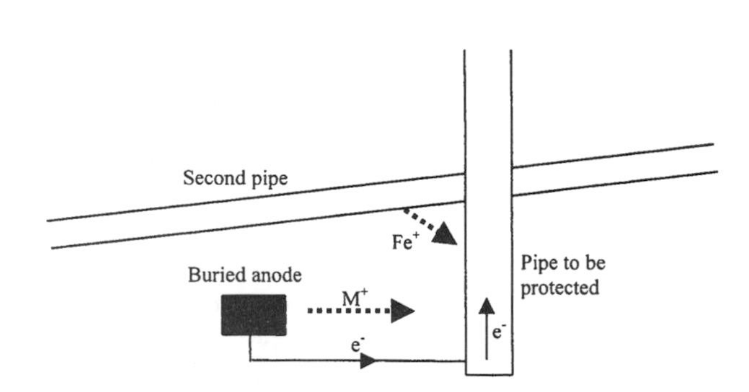

Figure 6.1.9 *Stray current corrosion*

always a danger of stray current corrosion involving other nearby metal structures. This is illustrated in Figure 6.1.9.

If there are two pipes passing very close to each other and one is being protected, an electrical potential will develop between the pipe being protected and the second pipe. The latter will become the anode and lose metal by corrosion! In cities such as London, there is legislation which controls the use of the impressed current systems because of stray current corrosion.

Galvanic (sacrificial anode) system

In this system, a metal lower in the galvanic series is connected to the structure to be protected. For example, blocks of zinc are connected to the hull of a ship. The steel body of the ship will become cathodic and the zinc will become anodic. Zinc is the most suitable anode in water whereas magnesium forms the most suitable anode in buried structures. In both cases, the anode corrodes relatively slowly and so provides protection for long periods of time without continuous monitoring. However, they must be inspected and replaced before they are completely consumed! This is a simple system, suitable for small installations. It is relatively cheap to install and to chcck. There is also little danger of stray current corrosion. Applying a coating on the steel cathode can extend the life of the anode. The power available is very limited and so on a large structure, there will have to be many anodes.

Protective coatings

Apply a protective coating, for example by painting the cathode surface. If the anode is painted, there is always the chance that a small area may be exposed when the surface is scratched, for example. The anode area is greatly reduced and severe corrosion will take place.

Organic coatings

- Oils and greases are often a simple but effective solution where there are moving parts, especially in abrasive contact with each other. The protection is temporary and so regular maintenance is needed. Examples of their use include bicycle chains, steel bearings and inside mechanical mechanisms such as locks and watches.
- Bitumen is suitable for buried pipes where they remain undisturbed, but as brittle material, it can be damaged easily. Bitumen can be reinforced with glass fibre or cloth tape or epoxy resin when applied to ship hulks.
- Plastics such as PVC, nylon and epoxy can be applied by dip coating or spraying and heating. This is expensive but useful where corrosion is severe or where damage can be catastrophic in situations such as under commercial vehicles and over gas and oil pipelines.
- Paints give a good finish on surfaces although they are not very effective in providing corrosion protection. The surface must be grease free, rust free and clean. Three or four coats of paint are

needed for effective protection. This forms a coating which is >0.125 mm thick. Since the paint consists of individual particles bonded together, water and oxygen can penetrate the paint layer along paths between paint particles. Applying several layers of paint increases the path length for the moisture, and therefore the electrical resistance to corrosion. However, the strength of the coating will reduce as the number of coats is increased.

- There are special corrosion paints:
  - Inhibitive paints: These contain pigments that act as inhibitors dissolving in the water, penetrating the paint film, e.g. red lead and chromates. For structural steelwork, a red lead primer coat could be finished with a good top coat to provide effective protection.
  - Sacrificial pigment: Zn dust can be used as pigment if the dry film contains 95% by weight of Zn, all the Zn will be in electrical contact with each other and the surface of the metal. The protection is then similar to galvanizing.
  - Leafing pigments: Iron oxide and mica are sometimes used together as a pigment. They form little flat plates that overlap each other in the paint film. The path length for moisture is increased.

Conversion coatings

The surface of the metal is chemically changed by several methods:

- Anodizing: A thick oxide film is formed on the surface of the metal to be protected by deliberately making the metal the anode in a suitable electrolyte. As the oxide layer that is formed is hard and electrically insulating, the metal is protected against further corrosion. The process is restricted to a few metals only and is most common with aluminium. The coating that is formed on the surface is porous but the pores can be sealed by dipping the components in boiling water. The coating can also be dyed before sealing to give a coloured finish. Conventional high-strength aluminium alloys are precipitation hardened with copper.
- Phosphating: The surface of the metal is converted to a phosphate by treatment in phosphoric acid with various additives to speed up the process. This is useful as a basis for paints as it helps to fix the porous paint. The coating helps to prevent rust undercutting paint on a steel surface. When metal corrodes, the water becomes alkaline as OH^- is produced. The alkali attacks paint under normal circumstances, causing it to lift. Phosphate coats are resistant to alkali solutions and so prevent the breaking of the paint layer.
- Chromating: This is achieved by dipping or spraying the component with an acid solution of chromate salts.

Metal coatings

- Electrodeposition: Surfaces must be thoroughly clean and free from grease. The component is then made the cathode in an electrolytic cell with an external DC supply. An example of this process is nickel plating.

- Hot dipping: The component to be coated is immersed in a molten bath of the coating metal. Low melting point metals such as zinc and tin are applied by this manner. Galvanizing is a well-known example of a hot dipping.
- Spraying: The coating metal is melted and atomized and then sprayed onto the component. Zinc and aluminium can be applied by spraying. The surface of the component to be sprayed must be roughened to achieve mechanical 'keying' to it. The surface is therefore first prepared by sand blasting.
- Cementation: Small components are packed into drums together with the coating metal in powder form. The drums are heated and rotated for several hours. This method can be used to achieve coatings with zinc, aluminium and chromium.

Some real solutions

(1) For steels:
- Zinc is a very important metal for the protection of steel. It corrodes more slowly than steel and therefore is suitable for neutral solutions but is less effective in acids and alkalis. Zinc offers 'gap protection' as it is anodic with respect to steel and so any exposed steel remains protected.
- Nickel–chromium alloys give corrosion protection and also good appearance. The nickel provides corrosion protection but tarnishes in air. The chromium prevents nickel tarnishing although being porous, it is less effective on its own.
- A thin coating of tin on steel provides a good non-toxic barrier. This is the material of the ubiquitous tin can. It can be soldered, provides good electrical conductivity, is ductile and looks good.

(2) Underground corrosion:
Steel, lead, copper and galvanized steel structures are used underground. Medium tensile steels, which are being used in engineering installations, are prone to corrosion. In countries such as Canada, aluminium pipes are used commonly. Corrosion control is necessary as aluminium alloys are susceptible to pitting. There is great variability in the soil condition for one solution to work universally. Soils can vary from desert conditions where more than 80% is silica sand (inorganic and dry with little corrosion) to peat bogs which are wet and almost 100% organic and often acidic. In the UK there are many areas with waterlogged clay with an almost complete absence of oxygen. As conditions are neutral and there is no oxygen, corrosion would not be expected. However, a high rate of corrosion is often found.

Types of corrosion of buried metals

- Bimetallic corrosion and oxygen concentration cells – anode and cathode areas could be separated by as much as 1 mile. This is called 'long line corrosion'.
- Stray current corrosion – electrical systems such as cathodic protection systems can cause severe corrosion of buried systems. DC and AC systems have specific problems.

- Microbial corrosion – this is the most common cause of underground corrosion in the UK. Bacteria in the soil cause chemical reactions that consume electrons and hence lead to cathodic reactions. Sulphates and water are used in the process and the heat generated further accelerates bacterial growth.
- Bacteria operate where there is an absence of oxygen and in neutral solutions. Raising the pH of the soil can kill bacteria. In the case of pipelines, bacteria can be killed by special tapes wrapped around buried pipes. These contain a heavy grease with anti-bacterial agents. Microbial attack can take place underground as well as on riverbeds and seabeds and even in aircraft fuel tanks.

Problem

Discuss how the surface of a steel water tank located in the following environments might corrode:

(1) Rural
(2) Urban
(3) Light industrial
(4) Heavy industrial
(5) Marine

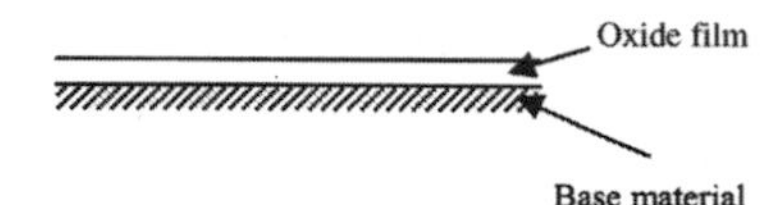

Figure 6.1.10 *Oxide film formation*

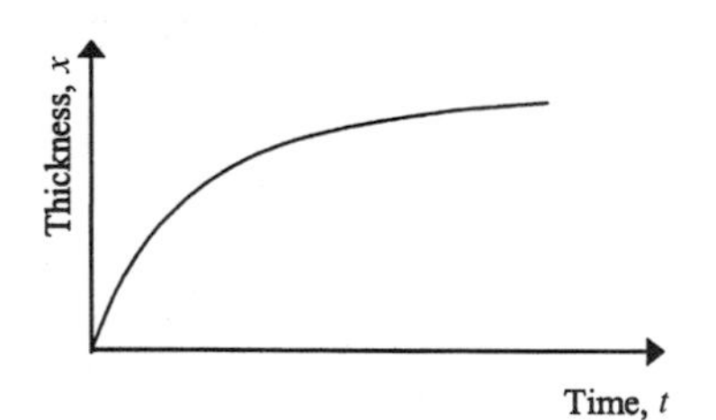

Figure 6.1.11

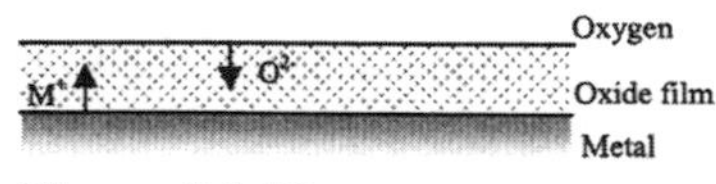

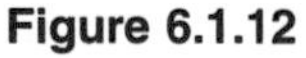

Figure 6.1.12

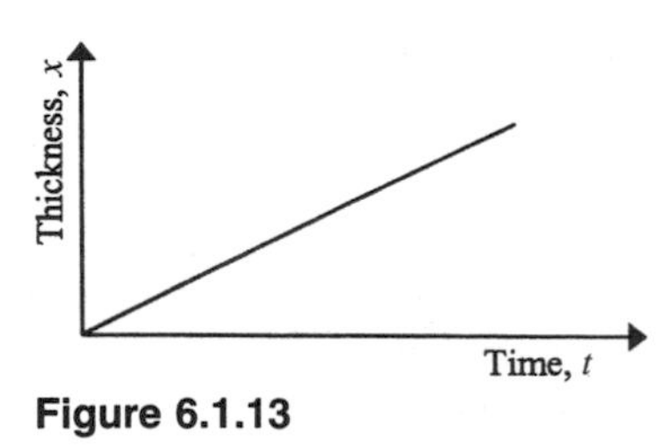

Figure 6.1.13

Dry oxidation

When the metal is exposed to high temperatures, chemical reactions can take place between the metal surface and elements in the atmosphere. This is called oxidation. Reactions take place with oxygen as well as other substances such as sulphur, carbon dioxide and nitrogen. The most common situation involves oxygen. During this process, a solid film forms on the surface of the metal (Figure 6.1.10).

The oxide film acts as a barrier slowing down the rate of oxidation. The growth of the oxide film will follow a parabolic relationship with time as illustrated in Figure 6.1.11.

In the oxide film, there are metal ions and oxygen ions. The film can only grow thicker if either metal ions, M^+, move out through the film or oxygen ions, O^{2-}, move in through the film (Figure 6.1.12). The oxide film must therefore be capable of transmitting ions and electrons.

Some metals such as chromium, silicon and aluminium form oxides which will not allow much conduction. These will provide very stable oxide films and will be resistant to oxidation. In other cases, the oxide film does not protect and the thickness increases linearly with time (Figure 6.1.13).

Possible reasons:

- The oxide film may be volatile so that it evaporates more quickly than it is deposited. This is the case with tungsten and molybdenum when heated to temperatures greater than 900°C in air. This is unfortunate as these high melting point metals are particularly useful at high temperatures.

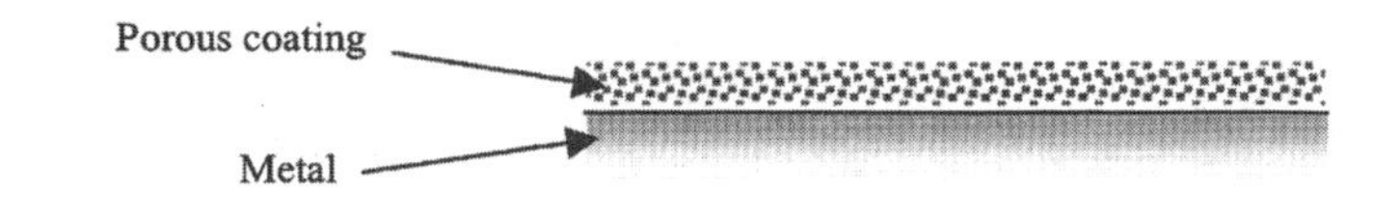

Figure 6.1.14

- The volume of oxide produced may be less than the volume of metal consumed (Figure 6.1.14). In this case, the oxide film is porous and therefore non-protective.
- The Pilling–Bedworth (PB) ratio is a useful quantity for identifying the stability of an oxide layer. This is given by the ratio,

$$\frac{\text{Volume of oxide}}{\text{Volume of metal consumed}}$$

  If this is less than 1 as is the case with magnesium, then there is usually a linear rate of oxidation.
- If the Pilling–Bedworth ratio is very high, the volume of the oxide formed is much greater than the volume of the original metal it replaces. For example, the high melting point metal niobium has a PB of 2.68. High stresses will develop in the oxide film and may cause the oxide film to crack.

Oxidation-resistant alloys

Most metals oxidize when heated in air so they can only be stable if the oxide film is protective. Some metals such as chromium form an oxide film of high electrical resistance that is protective. However, as a brittle metal, chromium is not very useful on its own. Iron and nickel are important engineering metals but they do not have particularly good oxidation resistance. The combination of chromium alloyed with iron is, however, very effective because, with the greater affinity for oxygen than iron, the chromium will oxidize preferentially. Hence, if the chromium content is sufficiently high, then the oxide will be mainly chromium oxide. Since this is an oxide film of high electrical resistance, it is protective. The following are three examples of oxide films:

Metal	*Minimum alloying content*
Iron	13% Cr
Nickel	15% Cr
Iron	10% Al

Aluminium would appear to be better than chromium at protecting iron as less is required and the electrical resistance of Al_2O_3 ($10^7\ \Omega\,cm$) is greater than that of Cr_2O_3 ($20\,\Omega\,cm$). However, the mechanical properties of Fe–Cr alloys are much better than FeAl alloys and so the former is preferred.

Temperature-resistant coatings

Steel is sometimes protected against oxidation by coating as a cheaper alternative to alloying. For example, on car exhaust systems, thin coatings of aluminium can be effective. It is of more concern that metals which have good high temperature properties, and therefore suitable for advanced technological applications such as niobium, molybdenum and tungsten, have poor oxidation resistance. Neither chromium nor aluminium can offer oxidation resistance in these cases because an oxide of the wrong metal will form. The oxide layer will also be porous and therefore non-protecting. Very thin coatings (10–100 μm thick) are more effective. These have relatively short lives but might be suitable for use where only limited exposure to high temperatures is involved such as space shuttle vehicles. Coatings of aluminium, chromium, silicon and zinc will provide stable oxides. For example, when a thin layer of silicon is applied on molybdenum a film of silica, SiO_2, forms on the silicon surface. This coating is self-repairing.

Organic materials such as rubber and PTFE can provide a few minutes' resistance to extreme temperatures (between 3000 and 4000°C). The coating can be up to 2 cm thick. This will char and burn at high temperatures and as a thermal insulator, very little heat is transmitted inwards. The method is very effective since a great deal of energy can be absorbed in the degradation process quickly. Such materials are used on space shuttles to protect their outer surface on re-entry through the earth's atmosphere.

6.2 Who is responsible for the environment?

External pressures

The last decade of the twentieth century has seen an acceleration in the push to use materials more efficiently as they become more expensive. During the 1970s, experts would have predicted that metals such as copper would have been completely mined by the year 2020 or so. Supplies of oil would have been drying up during the 1990s and therefore many polymers could no longer be manufactured by the beginning of the next millennium. These predictions have proved premature. Nonetheless, there is a severe threat to the future use of materials from the impact they are having on the environment. Much of the pollution of the environment is generated during the manufacture of products. A growing threat is now posed by the disposal of products at the end of their lives.

With mass production to meet the great worldwide demand for products, the environment is now threatened by mountains of solid rubbish and pollutants in the rivers, seas and atmosphere. Landfill sites are running out of room. In Britain, the cost of *landfilling* ordinary household waste has more than doubled from £15 per tonne in 1990 as space has become scarce. In fact, every nine months, the UK produces sufficient waste to fill Lake Windermere. A UK Government White Paper called 'Making Waste Work' (produced in 1995) has put down specific targets to tackle the problem:

- Reduce current landfill waste (240 million tonnes per year) by 60% by 2005.
- Eighty per cent of households should be within easy reach of recycling facilities.
- Forty per cent of homes with gardens should have compost heaps by 2000.

How will they make sure that these targets are reached? If targets cannot be met easily, then *incineration* will become increasingly necessary.

This is generally unpopular and could lead to the increase in emissions of harmful gases such as CO, CO_2, NO_x and dioxins. These gases have been blamed for the *'hole' in the ozone layer* above Antarctica. The level of ultraviolet radiation reaching the earth from the sun's rays will increase and lead to diseases such as skin cancer. Industrial activity and the burning of fossil fuels are also leading to *global warming* according to many experts.

Governments around the world have now begun to take these environmental threats seriously. In the United Kingdom, a new landfill tax was introduced from 1 October 1996, amounting to £7 per tonne for active waste and £2 per tonne for 'inactive' waste. (Active waste includes timber, bitumen, plaster, glass and plastics. Inactive waste includes clear, uncontaminated soil, stone, concrete, masonry free from plaster and metals.) The Association of County Councils estimated that the total cost to local authorities of the tax will be £154 million per year, adding £7 per year to the average council tax bill. This is meant to slow down the rate of landfilling and encourage better use of materials. Although recycling has become more efficient in recent years, there is a long way to go before government targets are met. More importantly, the whole debate on environmental issues has made the public including manufacturers and designers conscious of their responsibility for thc environment. The warnings about environmental effects are very serious. However, the day when currently used natural resources run out is also not very far away. The key words used by government, industry and the consumer alike are *recover*, *recycle* and *reuse*. It is therefore highly desirable for manufacturers to look at better use of materials by examining their entire *life cycle*.

(The above italicized words are part of the new vocabulary of the environmentally conscious.)

Activity 6.2.1

(1) Make a list of the reasons why attitudes have changed during the 1990s with regard to the responsible use of materials.

(2) During the life of a material, it goes through the hands of processors of the material, manufacturers of products, consumers of products, those who recover and reuse products and those who dispose of the products. Discuss how each category of person may consider the question, 'Why should I recycle?'

Material consumption

Materials available in nature have always played a key part in the progress of human civilization whether for protection against the environment, hunting for food, protection from enemies, growing food or for just getting from place to place comfortably. Many of these reasons have not changed with current products. Not only is wood and leather converted to products but plants yield materials such as cotton and cellulose. Nature also provides the raw materials for metals, polymers, ceramics, glasses and their alloys. The demand for products continues as cheaper manufacturing processes make products accessible to greater numbers of people.

The growing development of materials from their sources to modern forms is illustrated in Figure 6.2.1. At every stage of the conversion material is lost irreversibly as waste. Metals are extracted from naturally occurring minerals by the reduction process. This involves the burning of fuels and therefore consumption of chemical energy. In the case of aluminium, electrical energy is also consumed. For almost all copper and much zinc and lead, the final stage in the separation process is electrolytic.

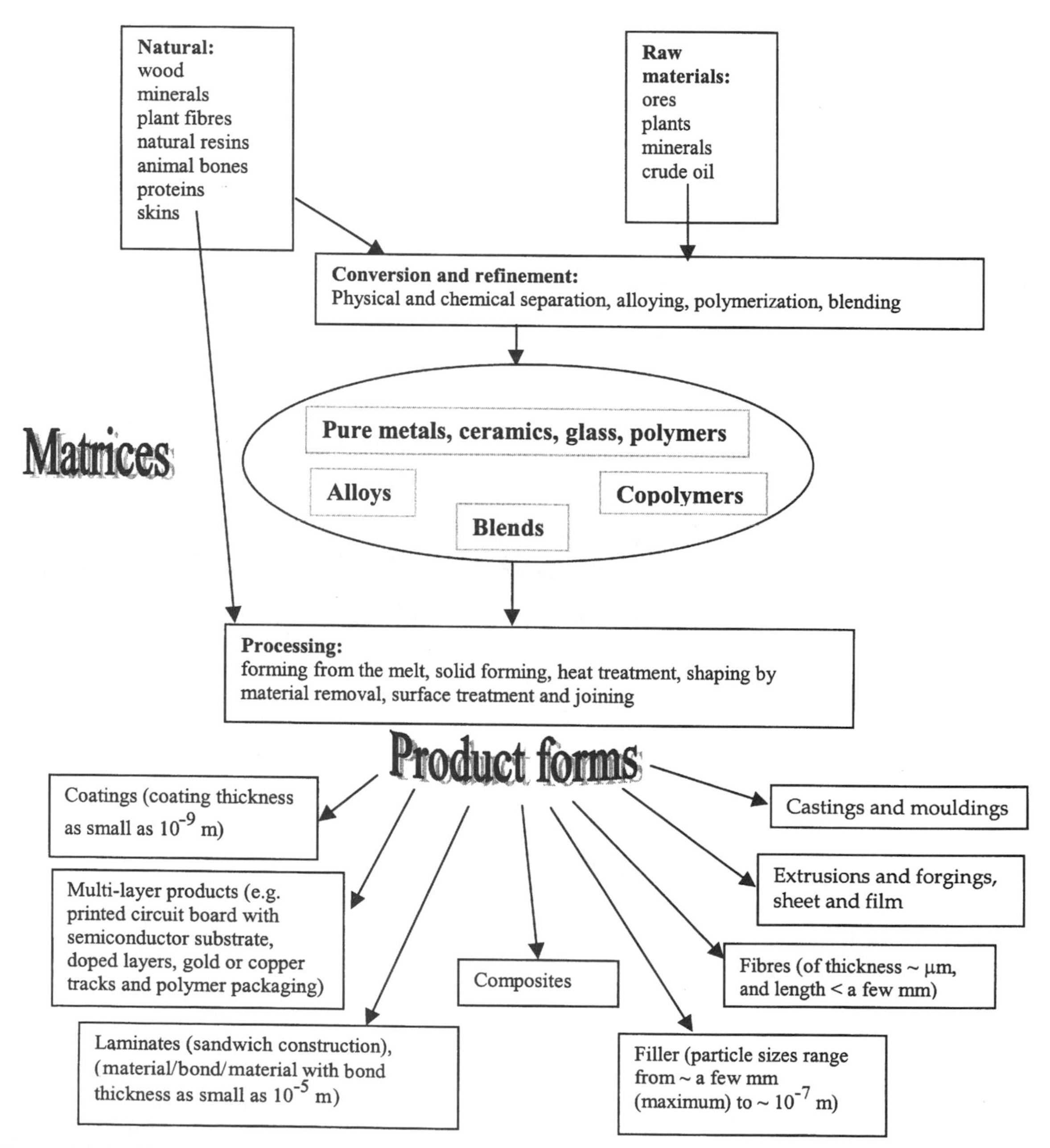

Figure 6.2.1 *The materials hierarchy*

New forms of products are complex and so recovery, recycling and reuse become increasingly difficult. However, we cannot afford to give up. We will start by considering materials from the point of view of their consumption.

Iron and steel production

The energy consumed in producing 1 tonne of crude steel is approximately 25 GJ. According to one source in the United Kingdom, 29% of this energy was wasted as waste gas and with cooling water from the production process. Seventeen per cent was lost as furnace radiation. After other losses, about 30% was usefully employed to reduce iron ore and in energy sold in by-products. By reusing 1 tonne of scrap steel, we can save 1.7 tonnes of ore, 0.68 tonnes of coke and 0.28 tonnes of limestone. However, the presence of residual elements such as copper and tin make the reuse of scrap difficult. Shredding plants that reduce cars and white goods such as refrigerators and cookers to small partly sorted fractions, go to the point of removing electric motors from the fraction by hand, so that as much copper as possible can be taken out of the scrap.

Waste products:

- Dust: Iron oxide and other contaminants (much is recycled but some is dumped).
- Alloying elements: Valuable materials such as Cr, Ni, Mn, Mo and Ti are wasted.
- Slag: Some recycled in the production of roadstone.

In 1995, 60 million tonnes of scrap steel were recycled in the USA. Nearly 2.5 million tonnes of steel were used in the manufacture of cans in 1994 in the USA of which 53% came from scrap steel. In the UK the overall recycling rate of ferrous metals in 1995 was 40%, rising to 44% in 1996.

Non-ferrous metals

The uses of non-ferrous metals are given Table 6.2.1 with some indication of relative importance. It can be seen that a small number of metals dominate the world market. Of these aluminium is the most important with copper, manganese, zinc, lead and chromium also significant. It can also be deduced from earlier data on steels that steel consumption is far greater than the total consumption of all other metals.

Aluminium

The production of aluminium begins with the mining of bauxite. The stages with some description are given in Table 6.2.2.

Use of fossil fuels in the United Kingdom makes the cost of electrical energy much greater than in countries where electricity is generated from hydroelectric sources. Scrap from secondary smelters is used for the production of foundry ingots since they are able to tolerate

Table 6.2.1 Consumption of non-ferrous metals

Metal	*Consumption by weight* (% of total)	*Uses*	*Recycling rate in 1995/1996* (% of total consumption)
Aluminium	32	Structural, transport (road, marine, air), electrical, packaging, holloware	53/44
Copper	20	Electrical, plumbing and miscellaneous for alloys (brass and bronze)	34/36
Manganese	16	Mostly used for alloying in iron/steel and for special alloys	
Zinc	12	Die castings, batteries, galvanizing, alloys (e.g. brass)	
Lead	10	Secondary batteries, plumbing, solder, cable sheathing, radiation screens	71/73
Chromium	6	Mostly used for alloying in iron/steel, for other alloys and as electrolytic coatings	
Nickel	2	Aerospace materials, alloying (e.g. stainless steel, armour plate and coinage), secondary batteries.	
Magnesium	<1	Alloying with aluminium, structural (aircraft and transport), chemical/electrochemical (e.g. corrosion protection of steels)	
Tin	<0.5	Tinplate, alloys (solder, babbit)	
Titanium	<0.2	Aerospace materials, heat exchangers, chemical plant	
Molybdenum	<0.2	Mostly alloying in iron/steel	
Cobalt	<0.2	Aerospace materials, tools	
Vanadium	<0.2	Mostly used for alloying in iron/steel	
Tungsten	<0.2	As carbides for cutting tools, electrical filaments	

(Data projected from 'Making the Most of Materials', SERC Report 1979)

impurities. Magnesium can be removed readily. Other impurities such as zinc, iron and silicon cannot be removed easily.

Copper

Copper was generally produced from sulphide ores by controlled oxidation. The mined rock is crushed to a fine powder and then separated using air bubbles and additives. Energy released as the sulphide oxidizes is used to melt the metal. The sulphur dioxide is recovered and the copper extracted and further refined. The slabs of metal produced become the anode in an electrolytic process during which copper is dissolved and subsequently deposited in almost pure form onto the cathode. This process is time consuming but yields pure copper ready for other processes. When copper is produced from oxidized ores, the rock is crushed and all of it treated with acid or ammonia to extract the copper. This process is slow and more expensive than the former.

Table 6.2.2 Energy required for aluminium smelting

Production stage	*Energy consumed* (GJ/tonne of aluminium)
Mining and shipping bauxite	This is variable, from several sources.
Refining to alumina	
Bayer process	21
Calcination	11
Electrolysis to metal	
Electric power	50
Carbon anode contribution	16
Secondary processes	
Smelting (alumina to hot metal)	212
Smelter casting	4.3
Fabrication plants (remelt cast shops)	9
Secondary smelting (foundry ingot)	14
Rolling	24
Foil (from foilstock)	18
Extrusion	25
Castings	48

Copper and its alloys such as brass and bronze are easily identified in scrap form by their colour and can be returned and recycled. The recycled material, however, deteriorates with every repeat use as the level of impurities increases. The first reuse will be in ductile brasses. This is followed by leaded brasses and then casting alloys. Finally, when very impure, the scrap is returned for refinement to pure copper.

Zinc

Mixed zinc and lead concentrates are heated together. Molten lead collects in the hearth and the zinc vaporizes. The zinc is then forced to condense in liquid lead. Further purification removes the small amounts (1.5%) of lead in the melt leaving behind the zinc.

Zinc is most often produced by electrodeposition from zinc sulphate. High purity (99.99%) is achieved although the electrolytic process is energy intensive and therefore expensive. This requires 68 GJ of energy for every tonne of zinc whereas the blast furnace route requires 18 GJ of energy per tonne of zinc while also producing 0.5 tonne of lead. The sulphur is recovered as sulphuric acid which itself has a market value. Recycling of zinc is expensive and as much of the zinc is used in galvanizing, in battery cans and in sacrificial anodes.

Lead

In the United Kingdom, lead smelting is only done in combination with zinc smelting as described above. Melting and working of lead require low levels of energy since lead has a low melting point. Process scrap is also easily returned to the melt. Recycling of battery scrap creates some handling problems associated with removal of plastic cases and of sulphuric acid.

Nickel

Nickel is extracted from nickel sulphides according to a process that is similar to that for copper sulphides. The ore is, however, mainly in the form of nickel oxides. The melting point is high and similar to that for steel. The process is therefore energy intensive. However, since the material is used in alloys for special applications, metallurgical considerations are more important than energy efficiency.

Recycling is very attractive since the metal is expensive and increasingly difficult to extract. In-house scrap generated during the production process is readily recycled. However, much nickel is used in the electroplating process and therefore forms a very small part of the scrap and cannot be recovered readily. However, since nickel is used in special steels, the process of recycling should become feasible as these materials are recovered from distinctive components.

Plastics and rubbers

Plastics and rubbers are better known to chemists and physicists as solid polymers. This is because, in practice, they always contain additions from stabilizers and plasticizers to general fillers. The processing has the following stages, each of which involves the consumption of energy.

Manufacture of the basic polymer or copolymer from monomer

The monomer can come from one of two sources:

- By extraction from vegetation in the case of natural rubber, resins and cellulosic materials. The starting materials are produced by biological conversion, involving the use of solar energy and the chemical conversion of molecules in the air and the ground. The extraction and polymerization involve further use of energy.
- By extraction from natural hydrocarbon fuels, particularly oil. Further energy is required in polymerization and manufacture of crumb or granules.

Blending and the introduction of additives

The manufacturers of products have very specific requirements. For example, the housing of a computer joystick could be made from either ABS or polypropylene. The material is chosen for toughness during handling and also because the viscosity of the melt can be controlled to allow the filling of a precise mould cavity. The final product has a well-defined surface that allows two parts to be snap-fit together as well as a textured surface to allow easy gripping. However, it also has to be rigid. For this combination, a lubricant may be added to reduce its viscosity during moulding. Of course, the colorant is another additive. In thicker sections, a general filler such as talc or wood flour may be added to reduce cost while increasing toughness. The manufacturer of the material responds to specific suggestions by the manufacturer of products in the development of commercial grades. The grades are also coded in such a way that the nature of the additions can be worked out with a little help. The introduction of each addition has energy and other environmental implications for the additive as well as the compounding into polymer grades.

Forming processes for thermoplastics, thermosets and rubbers

Plastics processing. The polymer granules are converted into the final product by processes such as injection moulding, extrusion, calendering and blow moulding. Energy could be involved in drying the polymer before processing and then in raising the temperature of the polymer to allow it to flow in the process. Energy is also expended as the melt is forced through dies or into moulds. Energy is also wasted as material is lost as flash, in runners and in spoiled product. Thermoplastic scrap can be recycled into products. However, thermosetting scrap cannot be recycled but can be ground and added as filler for new products. The data in Table 6.2.3 shows the energy used in producing the polymer and converting to plastics and rubbers. Unlike metals and ceramics, polymeric materials have an inherent energy content and therefore can be used as a fuel, for example. This is called their *calorific value*.

The energy cost of the plastic product includes the energy cost of the raw material unlike the other materials. For example, according to data from ICI, 1.41 tonnes of crude oil yields 1.38 tonnes of naphtha from which 1.05 tonnes of ethylene can be produced. The ethylene monomer can be converted into 1.0 tonne of polyethylene.

Rubber processing. The process for the manufacture of natural rubber is quite different from that for synthetic rubber. Latex extracted from the tree, *Hevea brasiliensis*, is dried and treated to produce rubber sheets. The following process is then undertaken:

- The compounding and mixing stage could involve drying the material and mechanical operations such as cutting of the bales. Further mastication of the rubber by shearing is energy intensive. The continuous working of the rubber allows a homogeneous mix to be achieved. Heat is generated and has to be removed and therefore is wasted. Scrap material at this stage can be reused.
- The compounded material is then formed by extrusion, calendering, compression moulding, transfer moulding or injection moulding or dip coating. Energy is required to drive the machinery and to reduce the viscosity of the material by heating.

Table 6.2.3 Comparison of energy input and fuel value

Material/product	*Energy content*		*Calorific value* (GJ/tonne)
	Polymer manufacture (GJ/tonne)	*Polymer conversion* (GJ/tonne)	
Polyethylene	75–109	11–12.4	39
PVC	81–102	11.5–15.4	25
SBR polymer	135	–	37
Rubber compound	90–130	10–25	30
Tyre (worn)	143	26.5	22

- In the third stage, the rubber is cured or vulcanized by chemical reaction. There is considerable energy input at this stage. Once cured, the rubber product can be removed from the mould without cooling. However, flash and runners and poor products cannot be recycled and are therefore lost unless ground and used as filler.

The energy requirements for the production of rubber and the conversion of the rubber into tyres are summarized in Tables 6.2.4 and 6.2.5 respectively.

Table 6.2.4 Energy requirements in natural rubber production

Process	*Energy* (GJ/tonne)
Crepe preparation	0.25–0.38
Crumb drying	4.2–4.3
Transport from Malaysia–UK	1.5

Table 6.2.5 Energy requirements to convert rubber products (tyres)

Process/component	*Energy* (GJ/tonne)
Milling	0.65
Extrusion	0.16
Vulcanization	0.5–3.2
Tyre ingredients	143
Tyre processing	26

Synthetic rubber is produced from polymer stock derived from oil or other sources. This involves refinement to produce the base monomer and then polymerization or copolymerization and blending. The energy involved is summarized in Table 6.2.6. Many thermoplastics can be converted into synthetic rubbers or elastomers by providing a cross-linking mechanism. In the case of thermoplastic elastomers (TPE), this link is broken at high temperatures, giving a rubber at working temperatures and a thermoplastic at processing temperatures.

Table 6.2.6 Energy used in synthetic rubber production

Energy source	*Energy requirement* (GJ/tonne)
Electricity	13.67
Refined oil products	66.57
Natural gas	75.14
TOTAL	155.38
Byproduct fuel credit	16.01
NET TOTAL	139.37

Glass

The manufacture of glass begins with the refinement of silica-based minerals containing other oxides. For example, the most common type, soda lime glass, consists of about 70% by weight silica (SiO_2) and soda (Na_2O) and lime (CaO). Alumina (Al_2O_3) is another important mineral in glass. The raw material is high quality sand. Their largest use is in the form of glass containers. This is followed by flat glass (e.g. window panes). Other important forms include fibre glass and moulded products. Most require a high degree of transparency and providing clear glass is not mixed with green or brown glass, this can be recycled. The basic ingredients including scrap glass known as cullet are heated to about 1500°C to form a melt. After conditioning, the glass is formed into products. The energy consumed is summarized in Table 6.2.7.

Table 6.2.7 Energy consumption for different glass making processes

Operation	*Energy consumption* (GJ/tonne)						
	Flat	*Container*	*Domestic*	*Fibre insulation*	*Fibre reinforcement*	*Lead crystal*	*Tubing*
Mixing	0.05	0.03		1.9	0.5		
Melting		10.02	15.7	15.2	18.6	38.7	12.2
Forehearth		0.97			10.8		
Forming	12.50	0.5	5.1	15.7	1.4		6.0
Annealing		0.5				1.56	
Processing	1.50			3.3	4.0	0.93	
Services	0.15		2.4	3.8	1.6	3.12	3.2
Overall	14.20	13.00	23.3	43.3	35.8	45.75	21.8

Table 6.2.8 Comparative data on the energy requirement for feedstock material

Materials	*Density* (kg/m^3)	*Energy requirements*			
		Feedstock (GJ/tonne)	*Conversion* (GJ/tonne)	*TOTAL* (GJ/tonne)	*By volume* (GJ/m^3)
Aluminium	2700		246	246	660
Steel billet	7800		44	44	340
Tinplate	7800		55	55	430
Copper billet	8900		53	53	460
Bottle glass	2400		20	20	50
Paper and board	800		62	62	50
Cellulose film	1450		194	194	290
PVC	1380	24	62	86	120
Polystyrene	1070	57	83	140	150
LDPE	920	49	50	99	90
HDPE	960	50	53	103	100
Polypropylene	900	52	61	113	100
Bricks				4	
Refractories				2–200	
Glass				24	
Pottery				10–250	
Cement				13	

(All figures are subject to variability and are dependent on method of calculation, material grades and environment.)

Ceramics

The production of refractories, bricks, pipes, whitewares, vitreous enamelling and high performance ceramics requires high levels of energy as the processing temperature is well above 1000°C. The refractories themselves are used to contain and conserve the energy in the production of other materials. The final products are then shipped directly to retailers involving transportation costs.

To summarize, the production of materials involves energy in all stages from the extraction of raw materials, the refinement into base materials, the conversion into commercial materials, manufacture of products and their use. At every stage additional energy is consumed in transportation. There are also costs to the environment as pollutants are released and materials are also inevitably wasted. Table 6.2.8 shows the energy consumption for some examples from each class of material.

Activity 6.2.2

Consider a ball-point pen as a product. Identify the component parts, the materials used and the processes involved in the manufacture. Identify a possible route for recovering each component part and explain the problems encountered if recycling is to be achievable.

6.3 Packaging

If there is one particular product that has led the public to look seriously at the wastefulness of materials, it must be packaging. In the summer, the street could be littered with toffee wrappers, plastic bottles and plastic food cartons. Each one has been designed to preserve food or drink and so remains inert and visible. All this has led to a serious look at why packaging is being used and what we do with it once it has reached the end of its short life.

The first stage is the definition of packaging. In short, packaging forms a category of product that is required for the effective delivery of anything from raw material to processed goods from a manufacturer or retailer to a customer or user. According to the European Parliament and Council Directive 94/62/EC on 'packaging and packaging waste', the following are functions that packaging is required to carry out:

- Containment of the product – e.g. a water or milk bottle, a toffee wrapper to keep the product from reacting with the environment, an insulating container or chamber to keep food or drugs at low temperature, a steel and concrete container for isolating nuclear waste.
- Protection – e.g. foam packing around an expensive product such as a television set to resist accidental impacts during transportation.
- Handling – e.g. cardboard boxes for the carrying of products.
- Delivery – e.g. LDPE film used to wrap up 12 bottles of lemonade for easier management of the bottles.
- Presentation – PVC film around PET bottle carrying the printed information about the contents of the bottle.

There are three types of packaging:

- Sales or primary packaging: This is any packaging that helps to attract the customer or final user to the product. If this packaging is removed, the product may be damaged.
- Grouped or secondary packaging: This is packaging that can be removed without affecting the properties of the product. This will include boxes or wrapping containing many products.
- Transport or tertiary packaging: This is packaging needed to aid handling and transport of products without causing damage to the product.

Activity 6.3.1

There are several different options for carrying drinks from milk to lemonade. As you can see from Table 6.3.1, the energy requirements for the manufacture of each type of container could be considered when making the right choice of container. In Table 6.3.2 a more detailed picture is given in terms of the size of the container. Look carefully at the data and rate the materials from best (= 5 points) to worst (= 1 point) according to the following criteria:

- Weight of containers as a proportion of the total weight transported.
- Energy consumed in making each container.
- Potential for preserving the container by reusing for any purpose.
- Potential for recycling the materials in the container including energy recovery.

Table 6.3.1 Energy used in producing containers

Production stage	*Energy requirement* (MJ/kg)		
	Glass	*Plastics*	*Tinplate*
Conversion to containers	4.4	15.6	4.4
Heating and lighting	1.8	5.0	1.8
Transport of containers	0.9	0.9	0.9
Raw materials production	13.2	101.0	44.0
TOTAL	20.3	126.5	51.1

6.4 Waste management

Waste materials from whatever source are by definition rejected after their primary use. As the twenty-first century begins, we see a world where the demand for products and their consumption is growing at a staggering rate. Manufacturers can satisfy our appetite by becoming increasingly efficient at making products. Advertisers and marketing

Table 6.3.2 Total energy consumption for 1 litre capacity container

Material	*Mass* (kg)	*Energy for manufacture of container* (MJ/kg)	*Energy equivalent of raw materials* (MJ/container)	*Energy recoverable by incineration* (MJ/container)
HDPE (milk)	0.030	1.22	1.43	1.36
PVC (milk)	0.050	3.63	1.17	0.88
PE coated paper (milk)	0.030	1.58	0.75	0.70
Steel with tin coating (tinplate)	0.13	9.11	–	–
Tinplate with 30% scrap	0.13	6.91	–	–
Aluminium-ended tinplate	0.12	10.47	–	–
Al-ended tinplate with 30% scrap	0.12	8.05	–	–
Glass	0.515	8.47	–	–

executives do all they can to persuade us to dispose of our products more and more quickly so that the turnover of goods can remain high. There is a tendency to appreciate style rather than function and to dispose rather than repair. As land for disposal becomes scarce and as waste becomes a major source of pollution and even hazard, it has inevitably become one of the major political issues of our time.

Landfill sites are filling up and the cost of landfilling has been increasing. It has therefore become increasingly important to divert as much waste as possible from the landfill site and if possible to prevent it being generated. All these factors have raised the awareness and desire for reducing the waste levels either by minimizing the amount of waste or by recycling as much of the waste as possible. The starting point for such approaches should be a study of the nature of waste. The stages and approaches to the management of solid waste are shown in Table 6.4.1

Table 6.4.1 Stages in solid waste management

Stages	*What can be done/Questions to be asked*
Reduce the amount of waste	Look at ways of reducing, for example, the use of paper, plastic carrier bags and other materials. How effective is this approach likely to be?
Collection of general mixed waste	Organize bins, dustcarts, manpower and collection schedule.
Collection of waste after separation at source	Organize special bins and vehicles, manpower and collection schedule. How many sites should there be?
Waste transfer stations	Look at accessibility of site, frequency of delivery of waste, gate fee, storage space needed and waste segregation by class.
Materials reclamation facility (MRF)	Look at storage needs at waste arrival bays, identification and separation equipment, baling and other compaction facilities and storage or separated materials ready for sale. How is rejected waste dealt with?
Disposal options	Incineration and ash disposal or landfill options – look at quantities of waste, their composition and restrictions on hazardous waste, frequency of delivery, site management and pollution control.
Reuse of materials	Look at further cleaning and separation, purification and additions, transportation cost and potential markets and price of reclaimed materials.

Once the information needed, as specified in Table 6.4.1, has been generated, targets can be set for reducing waste. Savings at each stage may be costed and a 'priority list for action' may be established.

The efficiency of the recycling process is improved if materials are separated before entering the waste stream. The quality of the material is also improved and therefore there is a greater opportunity for selling on the material. Reduction of the amount of waste at source and separation at source are important strategies that have gained little support. This requires a radical change of attitude by all consumers but when achieved, could lead to a totally different approach to waste.

Three groups are responsible for waste management in the UK. They are the Environmental Agency, waste collection authorities (such as the local borough council) and waste disposal authorities (such as the local county council or a private company). The first waste legislation was introduced into the UK in 1975.

Case study

Waste management at Sompting Materials Reclamation Facility (MRF), West Sussex

Waste arrives from general collections as well as pre-sorted waste from households at the site. Approximately 60% of the households in the county receive a blue box for putting out targeted waste. Between 70 and 80% of the households co-operate with the scheme. This pre-sorted waste is then separated at the Sompting MRF and stored for onward sale. The unsorted waste and the sorted waste without markets are landfilled. Energy is also generated from the adjacent landfill site by recovering landfill gas – a mixture of methane (65%), CO_2 and nitrogen which is subsequently incinerated for the generation of electricity on the site. The power output of the site generator is 0.02 MW.

Items separated at Sompting

Bottle glass (soda lime glass) with T_M = 800–1000°C.
Borosilicate glass (Pyrex and other dishes) with T_M > 2000°C – this is rejected.

Newspapers and magazines	–	Generally sold on although there are problems with brown cardboard, brown paper, cereal packets and 'yellow pages'.
Waste oil	–	This is recycled as long as there is no cooking oil or hydraulic oil.
Green garden waste	–	This is material with a potential for turning into compost.

Glass from the Sompting MRF is sent to Harlow and Yorkshire for cleaning before recycling. Hypodermic needles in the glass waste can lead to rejection by glass recyclers. Aluminium is separated by means of a spinning electromagnet under the conveyor carrying mixed pre-sorted waste. The eddy current effect causes the aluminium to be spun off the end of the conveyor. The steel containers stick to the belt and fall off at a different point.

Tin coated steel cans are worth recycling since the recovered tin has a market value of £4000 per tonne while the steel has a market value of £7 per tonne.

The plastics (clear PET, coloured PET, clear HDPE, coloured HDPE and clear PVC) are separated by means of a near infrared scanner. The rate of identification is three bottles per second with 95% accuracy. Once identified, the bottles are directed by a jet of air into appropriate containers. The rate of sorting has increased by eight times since the manual system was replaced by the automatic system. Prices for recycled materials continue to be extremely variable. This makes the operation uneconomical. For example, recycled clear PET has dropped from £200 per tonne to £70 per tonne in a matter weeks whereas virgin PET could cost more than £1200 per tonne. Recycled PVC and HDPE fetch approximately £90 per tonne while virgin HDPE costs about £500 per tonne.

Activity 6.4.1

Look at the items that are thrown away in your bin over a period of a week and catalogue them in the following way:

How frequently will the item be thrown out: several times a week; once a week; once a month; rarely
Is the item expensive or cheap?
Are the materials easily identified?
Are the materials easily separated?
Can the item be reused?
Can the materials be recycled currently?

Draw up a chart and determine how much of the waste can be reused or recycled.

6.5 Sorting and separation of plastics

The most difficult and expensive aspect of the process of recycling is the sorting and separation of materials from a mixed waste stream. As we have seen products use materials in combinations which are difficult to separate. Individual materials are also often difficult to identify. This is especially true of plastics and rubbers. For the process of recycling to be cost effective and truly effective, we need to avoid the mixing of materials as much as possible. Partial separation at the site where the waste is generated is therefore the best approach. Once this is done, we can apply our knowledge of the materials and the manufacturing process to the sorting and separation process.

Manual sorting

In the following section we will look at some examples of features that can be used to manually sort materials. Operators with little skill can identify materials from the products. Here are some examples:

- High density polyethylene (HDPE): Non-cylindrical milk containers with capacities in the range from 1 litre up to 6 pints are made in an off-white colour. They have a dimpled surface and built-in handle. Containers for household products such as detergents, shampoos, bleach and other cleaners and buckets are also made from HDPE.
- Low density polyethylene (LDPE): Carrier bags and other transparent and translucent bags are made from LDPE. Container lids are also made from LDPE. Supermarket carrier bags that crackle when they are handled are made from linear LDPE. This is in fact a copolymer of PE and another polyolefin.
- Other polyethylenes include medium density polyethylene (MDPE) used almost exclusively in the manufacture of plastic pipes.
- Polypropylene: Bottle tops, some cosmetic bottles and high clarity film are made from polypropylene. Milk crates and other storage containers are also made from polypropylene.
- Polyethylene terephthalate (PET): Carbonated drinks containers are made from PET. They are made by the injection blow moulding process and are distinguished by a button-shaped gate in the centre of the base. Although mostly transparent, some coloured bottles are also produced.
- Polyvinyl chloride (PVC): Mineral water containers are mainly made from PVC. They are produced by extrusion blow moulding. The base of the bottle shows markings indicating that the extruded tube is sealed by pinching along the diameter. Air is then blown into the tube to form the bottle shape. They are also transparent but can be distinguished from the PET container by the pinch mark or 'smile' on the base of the container as opposed to the 'button' on the base of the PET container. Many PVC material containers are given a faint blue tint.
- PVC and PET can be separated by using ultraviolet light. Under polarized light, PET will appear bright to operators whereas PVC will appear dark blue.

Manual sorting is time consuming and therefore expensive. Sorting is uneconomic for post-consumer waste but useful where the product is clearly identified before entering the mixed waste stream.

Sorting by density

Wet separation

The polyolefins (PE and PP) can be separated from other polymers such as PVC, PS and PET in water as the former will float while the latter will sink (density of water = 1000 kg/m^3) as shown in Table 6.5.1.

Polymer components are ground to less than a few mm in size and dropped into a tank of water. Factors such as the size and shape of the

Table 6.5.1 Densities of common polymers

Material	*Density* (kg m^{-3})
LDPE	920
MDPE	940
HDPE	960
PP	900
Water	1000
PS	1050
PET	1360
PVC (rigid)	1400
PVC (flexible)	1300

pieces of polymer play an important part in whether the polyolefins can be separated. For example, a piece of LDPE film may carry a particle of PET and therefore prevent their separation. Flakes of PET or PVC may trap air and therefore appear less dense than water. A small amount of detergent can be introduced to improve the wettability of the particles and so help separate materials as well as any dirt and glues. Expanded polystyrene will also float and contaminate the polyolefin mix. Different polyolefins may be separated from each other using mixtures of water and methanol. Similarly the density of water may be increased by the addition of salt (NaCl). This can be further complicated as the density may be affected by the presence of additions such as general fillers, plasticizers and lubricants. In the case of PET and PVC, the densities can overlap and depend on the degree of crystallinity and molecular weight distribution.

It is therefore important to obtain as much information about the materials as possible before any separation is attempted. The fluid will need to be filtered to remove contamination periodically. Finally the separated materials will need to be air dried.

Dry separation

A fluidized bed can be generated in a column of dry particles by pumping air through the column. The pressure should only be just sufficient to lift and separate particles from each other. The lighter fractions will then rise to the surface and the heavier fractions will sink to the bottom. Such a system is very effective in separating according to density discriminating between small differences in density, providing the particles have similar shapes and size. Larger particles of low density will, however, merge with smaller particles of higher density. Similarly, flakes of high density may float up into the zone with the zone of lower density.

Centrifugal sorting

The flotation method involves a balance between the force of gravity and the buoyancy effect of water. This balance can be taken to a very effective limit in the hydrocyclone, in which material is spun at high speed in a partly water-filled chamber, somewhat like a washing machine cylinder. The particles of material will sink or be carried on the surface along circular paths within a tapered cylinder. At the very high rotational velocity giving acceleration, $a > 1000\,g$, the water forms a skin on the inner walls of the cylinder. Granulates including contaminants such as dirt and glues are fed from the centre of the centrifuge. The particles are fired outwards at the water surface and in the process, the dirt is separated from the polymer. Low density polymers will float while higher density polymers will sink. The material is forced by the taper towards a collection point where separation to an accuracy of $\pm 5\,\text{kg m}^{-3}$ has been achieved in commercial systems. Particle shape and size do not have an effect on separation.

Other density-based methods

Other, more expensive methods of separation include the use of near critical and super critical liquids such as liquid carbon dioxide. Alternatively, preferential solvent absorption can be used to increase the density difference between PVC and PET before separation. These methods are valuable in the final separation of PVC from PET since the presence of even 10 parts per million of one in the other can affect the properties of the material.

Spectroscopic techniques

These techniques are extremely accurate as they provide information about a material in the form of a 'signature'. However, the process of recording and interpretation of the information can be slow. Some details of the techniques are given below.

Mid-infrared (MIR) spectroscopy

Infrared waves are projected onto a polymer surface. Some of the wavelengths, characteristic of the polymer structure, are absorbed and the waves returning from the surface of the polymer will show gaps in the frequency range. This picture, called the absorption spectra, is unique for individual bonds in the polymer chain. By analysing the particular absorbed frequencies, the components of the polymer can be identified. MIR works in the frequency range from 4000 to $700\,\text{cm}^{-1}$.

Note: measurements are given in terms of wave number $= 1/\lambda$, where λ is the wavelength of the waves. Normally frequency, f, is given by $f = v/\lambda$ where v is the velocity of the wave. Since the velocity of all electromagnetic waves is constant ($v = 3 \times 10^8\,\text{m/s}$) and very large, it is ignored. Therefore,

$$4000\,\text{cm}^{-1} \text{ is the same as, } \left(\frac{4000}{10^{-2}\,\text{m}}\right) \times 3 \times 10^8\,\text{m/s} = 1.2 \times 10^{14}\,\text{Hz}$$

This technique is effective with polymers containing fillers such as carbon black. The surface needs to be smooth and free of coatings and other contaminants that can mask the surface.

Near-infrared (NIR) spectroscopy

This method is more suitable for transparent and translucent plastics components such as bottles. NIR works in the frequency range from 14 300 to 4000 cm^{-1}. At these higher frequencies, harmonics and more complex vibrations within molecules, and therefore small differences in composition, can be picked up. It is ideally suited to the sorting of mixed wastes where speed of sorting is important and conditions can be difficult. Dark plastics cannot be identified. The technique is, however, relatively cheap and maintenance free.

Other spectroscopic techniques

Among other techniques, polarized light can be used to distinguish between PET and PVC. Under crossed polarized light, small differences in the crystallinity of each cause PET to glow whereas PVC remains invisible.

Absorption in the ultraviolet frequency range can also lead to differences that can help distinguish polymers.

X-ray fluorescence has proved useful in identifying small amounts of PVC in PET streams by picking up the X-ray signature of the chlorine atom. It has been very effective for sorting PVC in mixed PVC/PET streams as well.

6.6 Factors that control recyclability

Before any material or product is considered recyclable some hard questions need to be answered.

- Is there a market for the recycled material?
  Who will buy the materials and what level of quality is required?
- How reliable is the source of material?
  How much waste can be procured?
  How easily can it be identified, sorted and cleaned?
  For a recycling operation to be viable, there will be minimum requirements for volumes of materials and potential for sorting.
- Who will be responsible for collecting the waste and how much pre-sorting can be done?
  Will homeowners be provided with special bins or will they be required to take waste to communal bins? Will paper, plastics and glass be pre-sorted?
- How much do the various stages in the operation cost?
  - collection
  - identification and sorting
  - cleaning and further processing such as crushing and baling
  - transportation
  - pre-processing such as granulation and blending.
- How does the cost of recyclate compare with that of virgin material? How variable is the price of virgin material and recyclate?
- How much competition is there from suppliers of virgin material?
- How are physical properties affected by recycling and how significant is the effect on product performance?
- How does the consumer perceive the product if it is made from recyclate rather than virgin material?

Activity 6.6.1

Design for recycling

The products listed below are commonly found in the home or workshop. Consider how well suited they are for recycling, and write a short summary under the following headings:

(1) Brief description identifying the product and its operating environment.
(2) Identify the functions of the product from a designer's point of view.
(3) Identify materials used in the manufacture of all components. If the material is unclear how would you obtain this information? What are the potential problems with identification?
(4) How easily can components be separated by material? What are the problems, for example, due to the joining methods employed? How can they be overcome?
(5) What design modifications need to be made in order to aid recycling?

Products

- Torque wrench
- AC/DC adapter with cable for a mobile phone
- 3-pin electrical plug
- Loudspeaker unit
- Plastic box of individually wrapped chocolates
- Floppy disk
- Whiteboard marker pen
- Detergent bottle and top
- Computer mouse

Further reading

Trethaway, K. R. & Chamberlain, J. *Corrosion for Science and Engineering*. Longman, 1995
Tchobanopoulos, G., Theisen, H., Vigil, S. A. *Integrated Solid Waste Management*. McGraw-Hill, 1993
Wentz, C. A. *Hazardous Waste Management*. McGraw-Hill, 1995
Scheirs, J. *Polymer Recycling*. Wiley, 1998

7 Optimizing design

Summary

A product is made up from a combination of internally made and/or bought-in components. It has to be sold for a price that leaves a margin for profit and reinvestment after covering all the direct and indirect costs involved. This chapter looks at the factors which improve the attractiveness of the offering of the basic product itself.

The main problem in design is that once the initial design is complete, there is little opportunity for changing the design, and a higher cost is involved in doing so as shown in Figure 7.1.1.

Section 7.1 gives a means of measuring the performance of a process using factors which impact on the customer including the concept of value analysis. Section 7.2 details the techniques of robust design and experimental design devised by Dr Taguchi. Section 7.3 considers reliability concepts and the reduction of failure modes. Section 7.4 looks at factors to consider when designing for assembly.

Objectives

By the end of this chapter, you should:

- be aware of the customer core measures that have to be considered when choosing a process, and how to apply value analysis to balance the cost of producing a product's benefits against the value attached to these benefits by the customer (Section 7.1);
- understand the importance of the design input in minimizing the quality loss to the customer and how to minimize the number of experiments required to guide the best selection of varying input factors (Section 7.2);
- understand the concept of reliability and be able to draw up fault trees and failure mode and effects analysis to minimize faults arising and/or their effect (Section 7.3);
- understand the rationale behind standardization and modularization, and be able to apply the concept of design for assembly (Section 7.4).

7.1 Measures of the performance of a process

When selecting a process, the prime consideration will be the characteristics of the resultant product. However, there are other considerations which may also determine which actual process will be best to use.

When customers come to make a purchase they will have criteria in mind in addition to the product's function. They will also take into account the availability, delivery speed, quality and reliability, price and flexibility in design and volume. For similar products made by different organizations, these additional aspects will be the customers' order winning criteria. The processes selected play a key role in an organization's ability to satisfy these criteria, therefore these aspects must also be considered in the process selection.

The customers' order winning criteria

Selecting a process is a mixture of decision making about competing pressures of time, quality and cost. Installing process plant is a major investment which has led organizations to specialize on only a few key core processes and use other organizations' expertise to produce the less critical processes and components.

The fact that a component may be made elsewhere does not negate the designers' need to consider how the process selected impacts on what the customer requires. On the contrary, it means that designers are freed from the constraint of making use of internally available processes to consider other processes and can select the best.

Ability to produce component shape and features

It may seem obvious to start with a process's ability to produce an item, but the ease of producing the shape and features required determines all aspects of a process's impact on the order winning criteria.

Knowing what the processes are capable of doing not only helps to decide on which process to select, but also should determine the detail design of the component to make best use of that process. Without this detailed knowledge, the design process will not maximize the process's inherent advantages while minimizing any potential problems. This is why process and manufacturing engineers now play a major role in concurrent engineering teams.

There are many materials and associated processes which can produce a functional component. The selection matrix, shown in Figure 7.1.2, gives a typical, but not exclusive, linkage between materials, processes and the annual production quantity.

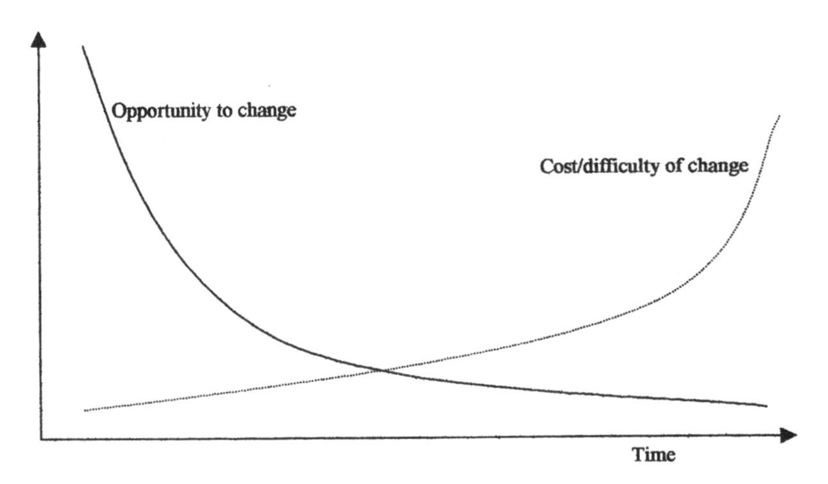

Figure 7.1.1 *Problems in changing design details as time progresses (from Swift and Booker, 1998,* Process Selection*)*

Quantity \ Material	Iron	Carbon Steel	Tool/alloy Steel	Stainless Steel	Copper & Alloys	Aluminium & Alloys	Magnesium & Alloys	Zinc & Alloys	Tin & Alloys	Lead & Alloys	Nickel & Alloys	Titanium & Alloys	Thermo-plastics	Thermo-sets	Composite	Ceramics	Reactive metals	Precious Metals
VERY SMALL 1 TO 100	15,16,17 M	15,17 36 M 41,45,46	11,17 36 M 41,45,46	17 36 M 41,45,46	15,17 36 M 41	15,17 36 M 41,45	16,17 36 M 41,45	11,17 36 M 45	11,17 36 M 45	11 36 M 45	15,17 36 M 41,45,46	11,16 M 41,45,46	23,25		26	46		45
SMALL 100 TO 1000	12,13 15,16,17 M 43,44 43,44	12,13, 15 17 M 36 41,43 44,45	11,17 36 M 41,43,44 45,46	12,17 36 M 41,43 44,45	12,13,15 17,18 33,36 M 41,43,44	12,13,15 18,18 36 M 43,44,45	13,16 17,18 36 M 45	11,13 17,18 36 M 45	11,13 17,18 36 M 45	11,13,18 36 M 45	12,13 15,17 36 M 41,43 44,45	11,16 M 43,44 45,46	22,23,25	22	22,26	46		45
SMALL/ MEDIUM 1000 TO 10 000	12,13, 15,16,17 M 42	12,13 15,17 31,32.37 M 42 43,44,45	12,17 31,32,37 A 42,43 44,45	12,17 31,32,37 A 42,43 44,45	12,13,14 15,18 31,32,37 A 42,43,44	12,13,14 15,18 31,32,37 A 43,44,45	13,14 16,18 31 A 45	13,14,18 32 A 45	13,14 32	13,14 32	12,13 15,17 31,32,37 A 42 43,44,45	31,37 A 42,43 44,45	21,22 23,24	22	22	42		45
MEDIUM/ HIGH 10 000 TO 100 000	12,13 37 A	31,32,33 37,38 A 45	32,33,38 A 42	31,32 33,37 38 A	12,14 31,32,33 37,38 A	12,13,14 31,32,33 37,38 A 45	13,14 31,38 A	14 32,38 A	14 32,38	14 32,33,38 A	33,37,38 A 42,45	37,38 A 42,45	21,23 24,27	21,22,27	22	37		33
HIGH 100 000+	12,13 37	31,32 33,38 A			12 32,33 37,38 A	12,13,14 32,33,38 A	13,14 31,38 A	14 32		14 32			21,24,27	21,22,27		37		
ALL QUANTITIES	11	11,16 34,35	16	11,16 34,35	11,16 34,35 45	11,16 34,35	11 34,35	11 34,35			11,16 34,35	34,35				45	16	16

PROCESSESS

M = Manual machining
A = Automatic machining

11 = Sand casting
12 = Shell moulding
13 = Gravity die casting
14 = Pressure die casting
15 = Centrifugal casting
16 = Investment casting
17 = Ceramic mould casting
18 = Plaster mould casting

21 = Injection moulding
22 = Compression moulding
23 = Vacuum forming
24 = Blow moulding
25 = Rotational moulding
26 = Contact moulding
27 = Continuous extrusion

31 = Closed die/upset forging
32 = Cold forming
33 = Cold heading
34 = Sheet metal shearing
35 = Sheet metal forming
36 = Spinning
37 = Powder metallurgy
38 = Continuous extrusion

41 = Electrical discharge machining
42 = Electrochemical machining
43 = Electronic beam machining
44 = Laser beam machining
45 = Chemical machining
46 = Ultrasonic machining

Figure 7.1.2 *Process selection matrix (from Swift & Booker's Process Selection, 1998 – from design to manufacture)*

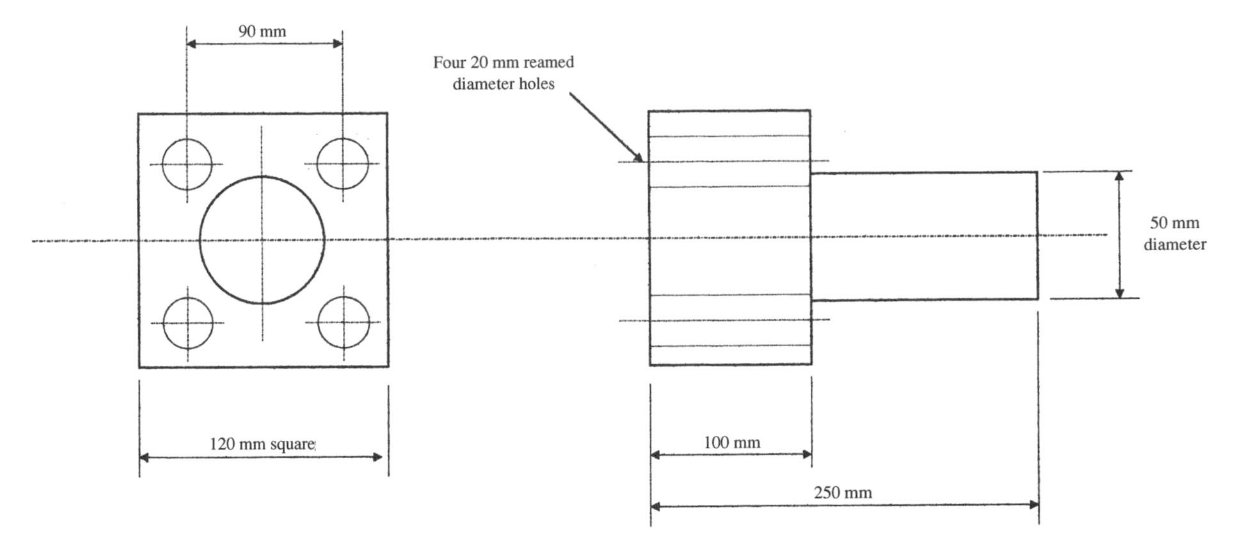

Figure 7.1.3 *Component – material: mild steel*

Activity 7.1.1

Consider the component shown in Figure 7.1.3. Identify five distinctly different processes which can be used to make it.

In addition to shape generation, designers should also examine processes such as testing and inspection, heat treatment, surface finish, protective coating and packaging.

Time to process an order

The time that a component spends actually being processed is normally just a small portion of its lead time, i.e. the overall time a component takes to move through a factory. We therefore need to examine each available process to maximize the positive impact on the time related order winning criteria.

The time taken for a component to move through the factory is based on a mixture of the following:

- The actual processing time per individual part, including any associated tool movement.
- Tool life and associated tool changing time.
- The time to load and unload a component from the process.
- The time to set up the process for an individual unit or a batch.
- Time spent waiting for processing.
- Time spent moving between operations.

The first two of these are very much dictated by the process selected, the plant used and its operation. The total time taken follows on from these and involves other factors such as the degree of automation, the organization and management of the plant, etc. The detail of working out the actual lead time is not included in this text because this is very dependent upon the plant installed, but is left to a later textbook.

Quality

Quality has achieved major importance in supplier selection. Many customers are demanding that all components they receive will be error free. Quality has different aspects.

Capability. The capability of a process is its basic ability to consistently produce within a set tolerance band with suitable surface finish. Figures 7.1.4 and 7.1.5 give an indication of the capability of a range of basic processes to produce surface finishes and tolerances. 'Interpretation of quality loss' (page 273) discusses the importance of ensuring tolerances achieved minimize the quality loss to the customer.

Conformity. No process will produce exactly the same component time after time. There will be variations, and errors, but these must fall within acceptable levels. The process selected, especially the plant used, will

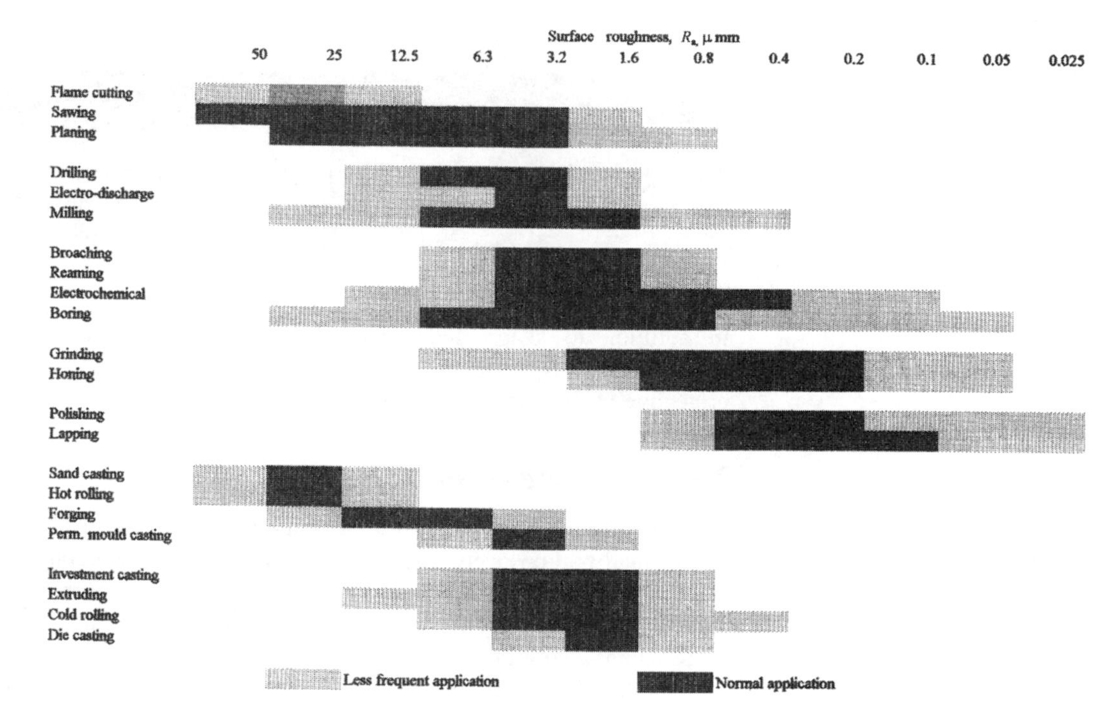

Figure 7.1.4 *Surface roughness achievable from different processes (from Koshal, D. 1993* Manufacturing Engineer's Reference Book*)*

Figure 7.1.5 *Tolerances achievable from different processes*

PROCESS	TOLERANCES ± mm
Sand casting	0.03 – 1.60
Forging	0.80 – 1.50
Die-casting – plastic	0.40
– zinc	0.05 – 0.20
Presswork	0.10 – 1.40
Planing	0.10 – 0.30
Drilling	0.10 – 0.18
Reaming	0.02 – 0.04
Milling	0.08 – 0.12
Turning	0.10 – 0.12
Turning – precision	0.05
Broaching	0.02 – 0.04
Honing	0.010 – 0.016
Grinding	0.007 – 0.016
Lapping	0.002 – 0.010

determine the natural variability of the output. Measurement and control will determine how closely its output adheres to the specification limits.

Reliability. This is a product's ability to continue to deliver its function over time. This is discussed in Section 7.3. The process could directly affect the material characteristics of the components going into a product and may also have inherent problems which must be guarded against.

It may appear that to achieve high quality is costly in time and money. Most companies find, however, that it is cost effective by reducing inspection, scrap and reworking costs and lost time. High quality standards further enable them to achieve higher prices, and therefore higher profits, for their products.

Cost

Costs needs to be calculated under different conditions and using different plant, tooling and material handling. Figure 7.1.6 (from Swift and Booker's *Process Selection* (1998)) demonstrates the effect of batch size on level of investment in automation – the higher batch size justifies the spending of money on process plant and associated equipment to reduce operational time and processing cost.

The true determination of the actual costs is often constrained by an organization's accountancy system – especially the method of attributing overheads. When considering alternative processes, only the true marginal costs should be used, i.e. exclude general overheads but include specific ones relating directly to the process. These should be

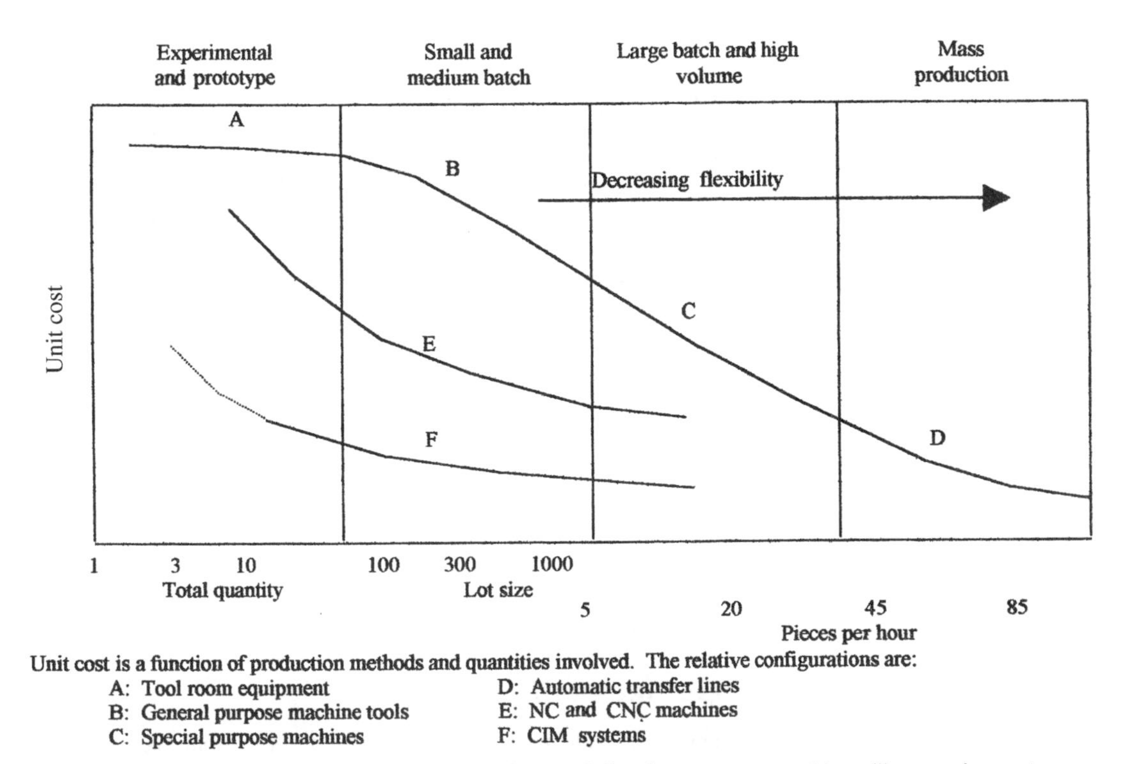

Figure 7.1.6 *Effect of batch size on investment in specialized processes and handling equipment*

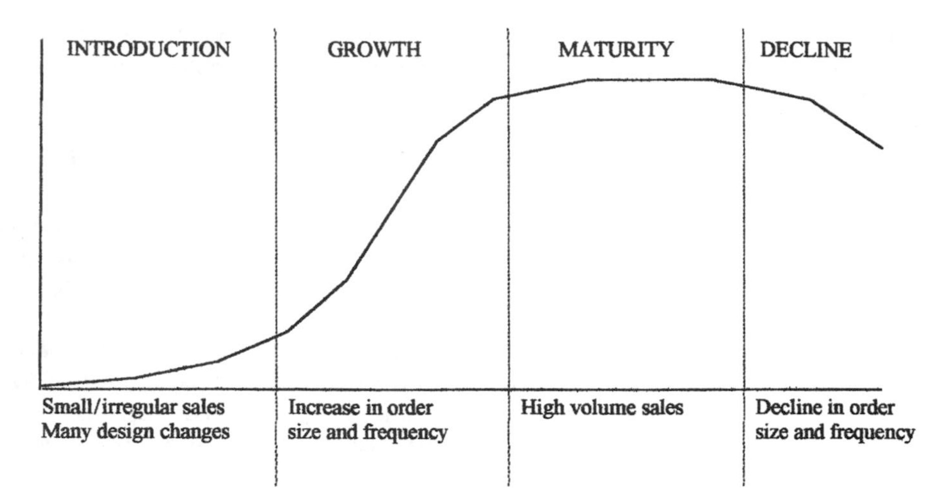

Figure 7.1.7 *Change in order pattern at different stages of the product life cycle*

used with care especially when making decisions on matters such as *make or buy.*

Costs are determined by a number of factors set at the design stage. This is where the processes that can be used are determined. 'Processing cost drivers' (below) discusses the cost factors based on a cost calculation system developed by Swift and Booker in *Process Selection.* Cost comparisons of different materials and processes should be carried out during the design phase before the particular design is finalized.

Flexibility

A design and its associated processes need to be flexible in several ways.

Volume flexibility is the ability to produce different quantities efficiently. Part of this flexibility is due to the varying needs and constraints laid down by the product's life cycle at various stages (see Figure 7.1.7). The other aspect is that quite often the actual demand is unknown at the design stage, but still must be assumed when determining the process and hence its influence on the design.
Product flexibility is the process's ability to handle variations on the product's features. Many products sold are variations of a main product and the selected processes must be easily set up or adjusted to produce any of the possible variations that may be called for.

Processing cost drivers

The actual cost for producing a component is made up from a number of factors. This section discusses the main factors in design which will determine the cost of any process selected. It can be used to determine which process will result in the best overall cost to meet the design specification.

Input material

The material selected is the starting point. Each input material has a cost made up of two components – the basic cost of the material itself and

Table 7.1.1 Relative material costs

Material		*Cost relative to mild steel by volume*
Mild steel		1.00
Cast iron		1.15
Alloy steel		3.80
Stainless steel		5.00
Aluminium alloy		2.00
Zinc alloy		3.00
Copper alloy		6.30
Thermoplastic	Nylon, PMMA	1.60
	Others	0.44
Thermosets		0.85

the required volume of the form it is supplied in, e.g. ingot, bar, plate, etc.

Table 7.1.1 gives an indication of the relative costs of material against that of mild steel. Basic costs can vary considerably over time, therefore these need confirmation at the time of an actual design.

Looking again across the matrix in Figure 7.1.2, it can be seen that many processes are more suitable to certain materials than others. Each process will involve an amount of waste, which will be valued differently depending on its recyclability and the market availability of that waste material. The prices of recycle materials are even more volatile than the basic material prices.

In material removal processes, the cost of the volume (as shown in Figure 7.1.8) of the starting material is often used. This is because the price received for waste swarf is normally very low in comparison, but this depends on what processes are available within the company, e.g. cutting fluid recovery and swarf cleaning.

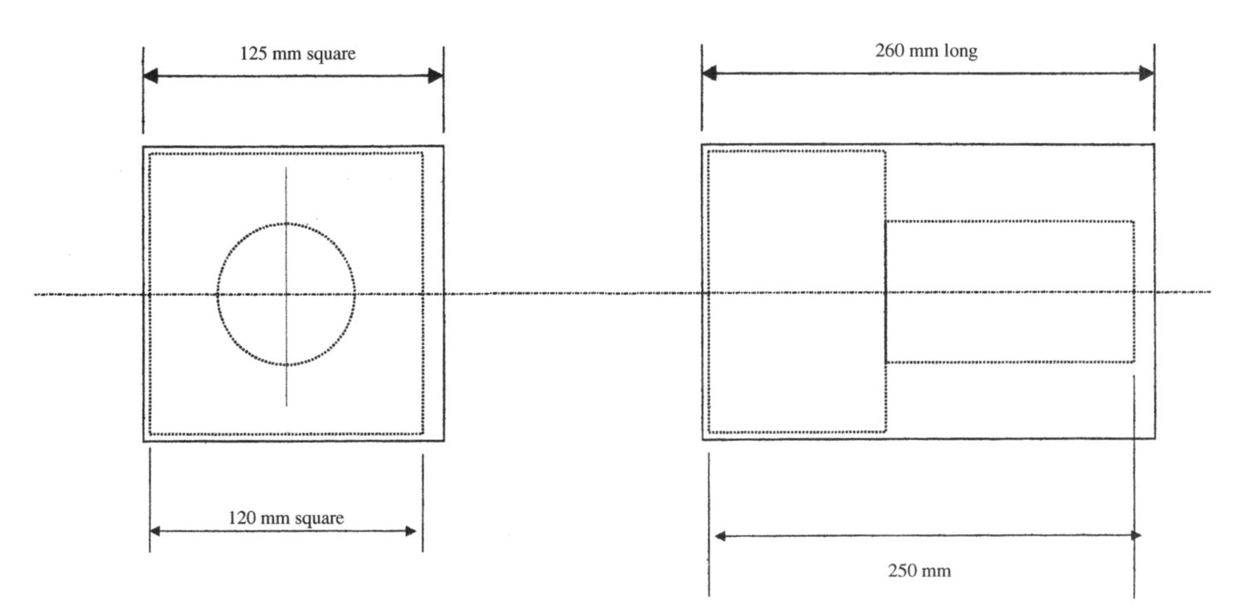

Figure 7.1.8 *Starting size for machining component as per Figure 7.1.3 – material: mild steel. Dotted line showed component and full line shows required starting size including saw cut allowance*

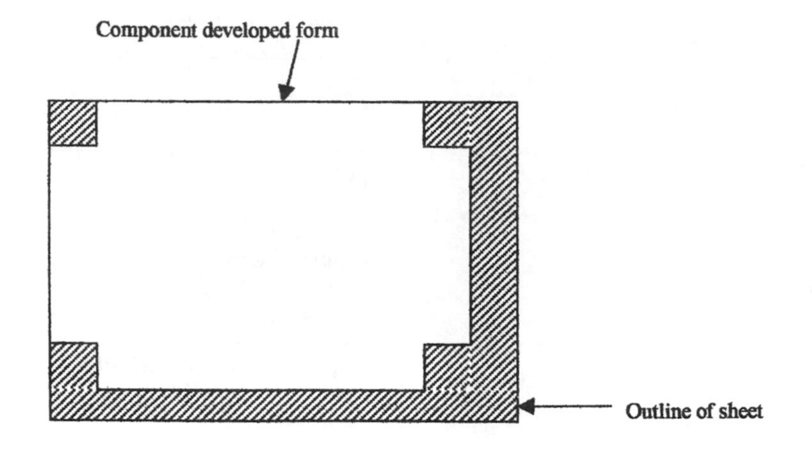

Figure 7.1.9 *Developed form of sheet metal component. The cross-hatched area shows portion of sheet unused by component*

In sheet metal working the total area used, as in Figure 7.1.9, should be used – with any recoverable money for the offcut material set against it, but only if it can be recovered or used in another product.

The other processes also have varying amounts (1%–20%) of waste produced which must be considered as an addition to the finished volume.

Shape and features

The basic shape and features determine what processes can be used. The more complex the shape, the higher the cost of the process and/or of the tooling involved.

Swift & Booker in *Process Selection* give a good, rough guide to the effect of these factors on the manufacturing cost for a range of processes. Their values are not sufficient to determine the full cost involved of a finished design, but they are a useful guidance at the initial design stage to see the effect of different shapes and features. Each design requires a careful calculation to determine the actual real cost involved.

Tolerance and finish

The tolerances and surface finish required do severely constrain the processes which can be used directly to achieve the product shape as shown in Figures 7.1.4 and 7.1.5. Extra processes may have to be included just to achieve a design limit set, adding to the time and cost involved. Therefore in design it is important to ensure that no excess finish over and above that required for functionally is selected.

Tooling

The selection of a process is often determined by the quantity involved over a period of time. This is partially based on the need to produce a set amount per time period. This is also because each process has associated with it the capital cost of the plant and the cost of operating it – including any associated tooling. The process itself does determine the basic tooling required, but the quantity determines what tooling can be economically purchased as it has to be recovered from the product sales. There are many choices of tool material available to achieve a required suitable life – and each of these has associated manufacturing costs.

Building up a product cost

From the previous sections, it can be seen that there are many considerations to go into the costing calculation. The full details of a product cost build-up can be seen in Chapter 5 of the accompanying textbook in this series – *Business Skills for Engineers and Technologists*.

Value analysis

Value analysis (VA) and value engineering (VE) are in fact the same technique applied at different stages in a product life cycle. VA is applied to existing products and VE during the design stage. Both have the same aims:

- Increase the value to the customer for the same, or lower, cost.
- Produce the same, or greater, value for the customer at a lower cost.

In actual fact what the technique does is to determine if all the attributes which a product carries have the same relative cost attached to achieving them.

Value to the customer

Any product, or service, must have a value to a customer so that they will exchange money for it. This value is made up of a combination of what in marketing terms is termed the product's attributes. These are discussed in full in *Business Skills for Engineers and Technologists*, the companion textbook in this series, but in brief they are:

- Tangible benefits: The main needs met by the product.
- Intangible benefits: Meeting other physiological needs.
- Feature attributes: Appealing to the senses of sight, touch and smell.
- Signal attributes: Adding to the prestige of the product.
- Product range and depth: To widen the potential customer base.
- Packaging: A mixture of protection and adding to the other attributes.

All of these attributes are there because they have some value to our customers. These attributes have to be built into the product to make it appeal to the customer, and this involves an associated cost in doing so. That cost should be in correlation with the resultant value. Value analysis is the technique of judging this correlation.

The first stage is to identify the main attributes that a product has – starting with its primary function and breaking this down into constituent secondary functions. It is best to describe these using just a verb and a noun, but sometimes that can be unnecessary:

A ball point pen has the basic function: 'mark paper'.
A table lamp has the basic function: 'provide light'.
A restaurant has the basic function: 'provide food'.

The support functions are the other functions that a customer expects in addition to the basic function. The ball point pen could have additional functions of:

No leakage.
Free flowing ink.
Protect writing tip.
Ease to hold and manipulate.
Looks attractive.

When looking across a range of items, there can be other attributes desired by customers. For the ball point pen this may include:

Range of colours of inks.
Range of width of line.

It is important therefore to decide at the beginning what exactly is to be analysed – a range of products, an individual product, a subassembly or even an individual component. It is always better to analyse at as high a level as possible, but sometimes that can create a very complex analysis and it does need to be broken down. Care should be taken that the interconnectivity between parts is closely considered.

It is also possible to continue the process down to quite detailed analysis when the need arises, i.e. the constituents of the ink in the case of the ball point pen.

Each of the attributes needs to be given a rating reflecting its value to the customer. One way of doing this is to carry out a paired comparison to determine the relative value ranking using the customer viewpoint.

Example 7.1.1

Paired comparison

Let us take, as an example, a simple ball point pen. The analysis will be at the individual pen level.

The basic procedure is to draw up a matrix, as in Figure 7.1.10, with each attribute heading both a row and a column.

You then move along the row comparing that row's attribute against each of the column's attributes. Award (place in the intersection box) one of the following numbers:

- 2 (two) if the row attribute is considered of a higher value to the customer.
- 1 (one) if both jobs are judged equal value.
- 0 (zero) if the column attribute is judged of higher value.

Note each attribute is compared against each of the others on two occasions – once using the row position and the other

	Free flow	Easy hold	Protect point	No leaks	Can refill	Hold pocket	Looks good	TOTAL
Free flow	X	2	2	2	2	2	2	12
Easy hold	0	X	1	0	2	2	2	7
Protect point	0	1	X	0	2	2	2	7
No leaks	0	2	2	X	2	2	2	10
Can refill	0	0	0	0	X	2	0	2
Hold pocket	0	0	0	0	0	X	0	0
Looks good	0	0	0	0	2	2	X	4
							Check total	42

Figure 7.1.10 *Paired comparison of attributes of a ball point pen*

using the column position. These two comparisons must agree (i.e. the two box contents add up to two).

Once all the attributes have been compared add up the scores per row to give a total relative value for that attribute. Check the total of all the attributes' values summed by comparing it to the multiple of (attributes × (attributes −1)).

Rank the attributes in order of their total value.

The final run of the paired comparison is as shown in Table 7.1.2. The highest score was gained by the free flow attribute – the lowest by the hold pocket attribute. This gives us the first part of the comparison – the relative ranking of each attribute to the customer. This is only a relative value, but one that should feel correct. It can be converted to a more definite value, but this may be difficult and the points gained serve fairly well.

Table 7.1.2

Function	*Points gained*	*Ranking*
Free flow	12	1
No leaks	10	2
Easy hold	7	3
Protect point	7	3
Looks good	4	5
Can refill	2	6
Hold pocket	0	7

We now move onto the determination of what cost is involved in producing these attributes.

Cost of producing an attribute

Each product is made from raw material and components with some operations that are applied to it. The marginal costs of these should be identified and a matrix produced with each significant part/assembly forming one axis, i.e. in the first column. In the second column goes the total cost of that part/assembly. The other axis is formed by the attributes each heading a column – as in Figure 7.1.11.

The cost per part/assembly is then apportioned under each of the attributes, i.e. how much of its cost can be attributed to producing that attribute. Once all the cost has been attributed then the apportioned cost

Part	Cost per 1000 units in pence	Cost apportioned to function						
		Free flow	No leaks	Easy hold	Protect point	Looks good	Can refill	Hold pocket
Ink	400	100% = 400						
Point assembly	700	60% = 420	10% = 70			10% = 70	20% = 140	
Assembly collar	125						100% = 125	
Inner ink tube	50	50% = 25	50% = 25					
Outer barrel	400			50% = 200		25% = 100	25% = 100	
End plug	75		100% = 75					
Cap	250				70% = 175	10% = 25		20% = 50
Total	2000							
Cost per function		845	170	200	175	195	365	50
% of total cost		42%	8.5%	10%	8.8%	9.8%	18%	2.5%
Cost rank		1	6	3	5	4	2	7
Value rank		1	2	3	3	5	6	7

Figure 7.1.11 *Cost of parts apportioned to functions*

per attribute is summed. These sums are then ranked in decreasing order of magnitude, i.e. the highest being given the highest rank.

You may be surprised when doing this apportioning that you have some costs left over which have not been allocated to an attribute. This indicates either of two happenings:

- You have unnecessary costs, that can be easily eliminated, or
- You have missed out some important attributes (more likely). The attributes, which are often missed out, relate to product protection or customer safety, e.g. outer covers. If this happens – simply add some extra columns with these attributes.

If a value ranking is high in relation to the cost, you may not be putting enough value into the other attributes and presenting a mixed package to your customer. There could be the opportunity to use a similar technique to increase the other values offered.

Where the cost is high in relation to the value, then that cost must be reduced by finding an alternative means of achieving the same value. The customer pays for value – not cost. Finding these alternatives is the same process as that involved in the initial stages of the design of a new product (see Chapter 1).

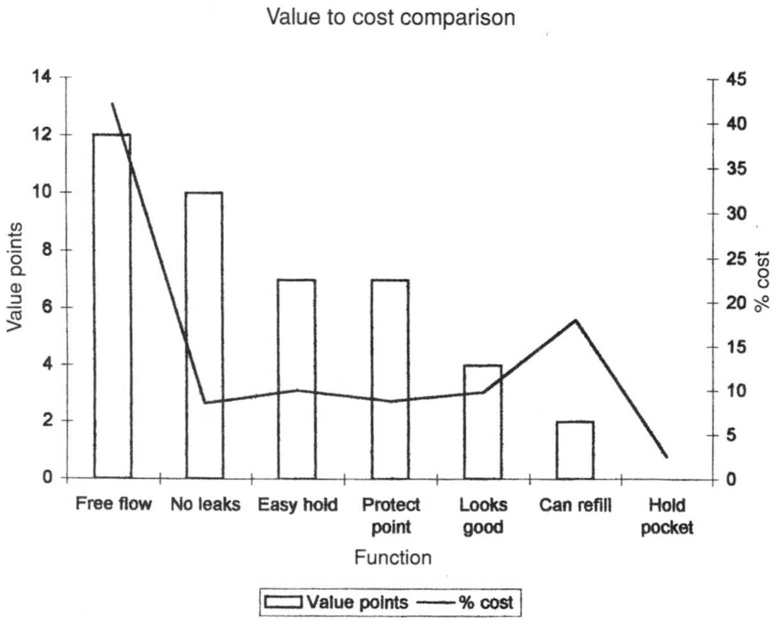

Figure 7.1.12 *Comparison of function value and apportioned cost*

It remains now to compare the two rankings of value and cost against each attribute in Figure 7.1.10. One way is to produce a combination graph using the points awarded under the paired comparison against the percentage of cost as in Figure 7.1.12. Where there is a major mismatch, there is a resultant opportunity.

In the ball point pen example, the mismatches which stand out are:

- The function 'No leaks' is high in customer value, but low in cost.
- The function 'Can refill' is high in cost, but low in customer value.

There are several opportunities presented here. The main one could be to eliminate the last two functions on the customer value list, i.e. 'Can refill' and 'Hold pocket', hence saving some 20% of the cost without substantially reducing the value to the customer.

As the individual pen is part of a range of colours and ball point widths then a substantial saving could be made while still offering a product that the customer values.

Problems 7.1.1

(1) Give several advantages and disadvantages of the different processes you selected in page 265 to make the component in Figure 7.1.3.

(2) Take a simple product – like an office stapler. Estimate the total processing time to make the parts and then assemble them, including the packing operation. How long would you guess the lead time within a factory is in comparison to the actual processing time for this product?

(3) Look at a common product – say ball point pens. How many different varieties of pens of exactly the same outer shape are there? How many parts are common between the varieties?

(4) Examine an old design of a common household product against a modern design. Compare the number of parts and determine if the new design represents better value at a lower cost for the customer.

(5) Compare the packaging cost on two cosmetic products that have a similar function but substantially different prices. Can you justify the difference in cost?

7.2 The Taguchi approach to design

Dr Taguchi's definition of quality:

> 'The quality of a product is the (minimum) loss imparted by the product to society from the time the product is shipped.'
>
> (*Quality by Design*, N. Belavendram, Prentice-Hall)

Dr Genichi Taguchi started out as a statistician, before turning to quality aspects, first with the development of the statistics involved in statistical quality control (SQC). In the 1950s he championed in Japan a different

Extracted from BS 4500 : 1969

BRITISH S

SELECTED ISO F

Clearance fits

Diagram to scale for 25 mm. diameter

+ H11 H9 H9 H8 H7 0 g6 f7 e9 d10 c11 −

Nominal sizes		Tolerance		Tolerance		Tolerance		Tolerance		Tolerance	
Over	To	H11	c11	H9	d10	H9	e9	H8	f7	H7	g6
mm	mm	0·001 mm	0·001 mm	0·001 mm	0·001 mm	0·001 mm	0·001 mm	0·001 mm	0·001 mm	0·001 mm	0·001 mm
—	3	+ 60 0	− 60 − 120	+ 25 0	− 20 − 60	+ 25 0	− 14 − 39	+ 14 0	− 6 − 16	+ 10 0	− 2 − 8
3	6	+ 75 0	− 70 − 145	+ 30 0	− 30 − 78	+ 30 0	− 20 − 50	+ 18 0	− 10 − 22	+ 12 0	− 4 − 12
6	10	+ 90 0	− 80 − 170	+ 36 0	− 40 − 98	+ 36 0	− 25 − 61	+ 22 0	− 13 − 28	+ 15 0	− 5 − 14
10	18	+ 110 0	− 95 − 205	+ 43 0	− 50 − 120	+ 43 0	− 32 − 75	+ 27 0	− 16 − 34	+ 18 0	− 6 − 17
18	30	+ 130 0	− 110 − 240	+ 52 0	− 65 − 149	+ 52 0	− 40 − 92	+ 33 0	− 20 − 41	+ 21 0	− 7 − 20
30	40	+ 160 0	− 120 − 280	+ 62 0	− 80 − 180	+ 62 0	− 50 − 112	+ 39 0	− 25 − 50	+ 25 0	− 9 − 25
40	50	+ 160 0	− 130 − 290								
50	65	+ 190 0	− 140 − 330	+ 74 0	− 100 − 220	+ 74 0	− 60 − 134	+ 46 0	− 30 − 60	+ 30 0	− 10 − 29
65	80	+ 190 0	− 150 − 340								
80	100	+ 220 0	− 170 − 390	+ 87 0	− 120 − 260	+ 87 0	− 72 − 159	+ 54 0	− 36 − 71	+ 35 0	− 12 − 34
100	120	+ 220 0	− 180 − 400								
120	140	+ 250 0	− 200 − 450	+ 100 0	− 145 − 305	+ 100 0	− 84 − 185	+ 63 0	− 43 − 83	+ 40 0	− 14 − 39
140	160	+ 250 0	− 210 − 460								
160	180	+ 250 0	− 230 − 480								
180	200	+ 290 0	− 240 − 530	+ 115 0	− 170 − 355	+ 115 0	− 100 − 215	+ 72 0	− 50 − 96	+ 46 0	− 15 − 44
200	225	+ 290 0	260 550								
225	250	+ 290 0	− 280 − 570								
250	280	+ 320 0	− 300 620	+ 130 0	− 190 − 400	+ 130 0	− 110 − 240	+ 81 0	− 56 − 108	+ 52 0	− 17 − 49
280	315	+ 320 0	330 − 650								
315	355	+ 360 0	360 − 720	+ 140 0	− 210 − 440	+ 140 0	− 125 − 265	+ 89 0	− 62 − 119	+ 57 0	− 18 − 54
355	400	+ 360 0	400 760								
400	450	+ 400 0	− 440 840	+ 155 0	− 230 − 480	+ 155 0	− 135 − 290	+ 97 0	− 68 − 131	+ 63 0	− 20 − 60
450	500	+ 400 0	480 880								

Figure 7.2.1 *ISO fits – hole basis. (Extracts from BS4500:1969. Reproduced with permission of BSI under licence number 2001/SK0152. Complete British Standards can be obtained from BSI Customer Services (Tel. +44 (0)20 8996 9001))*

TANDARD

ITS—HOLE BASIS

Data Sheet
4500A
Issue 1. February 1970
Confirmed August 1985

		Transition fits				Interference fits					
H7 / h6		H7 k6		H7 n6		H7 p6		H7 s6		Holes / Shafts	
Tolerance		Tolerance		Tolerance		Tolerance		Tolerance		Nominal sizes	
H7	h6	H7	k6	H7	n6	H7	p6	H7	s6	Over	To
0·001 mm	0·001 mm	0·001 mm	0·001 mm	0·001 mm	0·001 mm	0·001 mm	0·001 mm	0·001 mm	0·001 mm	mm	mm
+10 0	−6 0	+10 0	+6 +0	+10 0	+10 +4	+10 0	+12 +6	+10 0	+20 +14	—	3
+12 0	−8 0	+12 0	+9 +1	+12 0	+16 +8	+12 0	+20 +12	+12 0	+27 +19	3	6
+15 0	−9 0	+15 0	+10 +1	+15 0	+19 +10	+15 0	+24 +15	+15 0	+32 +23	6	10
+18 0	−11 0	+18 0	+12 +1	+18 0	+23 +12	+18 0	+29 +18	+18 0	+39 +28	10	18
+21 0	−13 0	+21 0	+15 +2	+21 0	+28 +15	+21 0	+35 +22	+21 0	+48 +35	18	30
+25 0	−16 0	+25 0	+18 +2	+25 0	+33 +17	+25 0	+42 +26	+25 0	+59 +43	30	40
										40	50
+30 0	−19 0	+30 0	+21 +2	+30 0	+39 +20	+30 0	+51 +32	+30 0	+72 +53	50	65
								+30 0	+78 +59	65	80
+35 0	−22 0	+35 0	+25 +3	+35 0	+45 +23	+35 0	+59 +37	+35 0	+93 +71	80	100
								+35 0	+101 +79	100	120
+40 0	−25 0	+40 0	+28 +3	+40 0	+52 +27	+40 0	+68 +43	+40 0	+117 +92	120	140
								+40 0	+125 +100	140	160
								+40 0	+133 +108	160	180
+46 0	−29 0	+46 0	+33 +4	+46 0	+60 +31	+46 0	+79 +50	+46 0	+151 +122	180	200
								+46 0	+159 +130	200	225
								+46 0	+169 +140	225	250
+52 0	−32 0	+52 0	+36 +4	+52 0	+66 +34	+52 0	+88 +56	+52 0	+190 +158	250	280
								+52 0	+202 +170	280	315
+57 0	−36 0	+57 0	+40 +4	+57 0	+73 +37	+57 0	+98 +62	+57 0	+226 +190	315	355
								+57 0	+244 +208	355	400
+63 0	−40 0	+63 0	+45 +5	+63 0	+80 +40	+63 0	+108 +68	+63 0	+272 +232	400	450
								+63 0	+292 +252	450	500

BRITISH STANDARDS INSTITUTION, 2 Park Street, London, [illegible]A 2BS
SBN: 580 05766 6

way of achieving improved quality in products by moving the responsibility for conformity and reliability to the design of product and processes where the true causes lay. Before this the manufacturing function had the sole responsibility of achieving high conforming and reliable products. He was amongst the first to realize that some 80% of the final quality level is achieved at the design process.

This section looks at Taguchi's concept of quality loss, then his principles of robust design before reaching his concept of how to design experiments to minimize the number of iterations required.

Interpretation of quality loss

Dr Taguchi's interpretation of the loss due to the products' variation was substantially different from that recognized by others previously in the application of tolerances.

Limits and fits – historic tolerancing and interchangeability of parts

Until the late nineteenth century, each product was individually made. This meant that at the assembly stage, each component had to be individually matched to other mating parts – often by making final alterations. This led to problems in servicing as replacement parts again had to be modified to fit into the product.

In the 1790s, Eli Whitney received an urgent order for 10 000 rifles from the US Government. He developed the concept of interchangeability by setting a narrow range of size for each component to be made to. This ensured that any component could be used at the assembly stage, and also afterwards in the field as a matching replacement.

This gave rise to the design concept of limits and fits to meet the operating function of an assembly as specified in ISO 286, an extract of which is shown in Figure 7.2.1.

The need to set a range rather than specifying the ideal size is due to it being impractical to repeatedly make a component to an exact size by any manufacturing process. This is due to natural variations in the set-up, the process involved, the machine and tool used and the

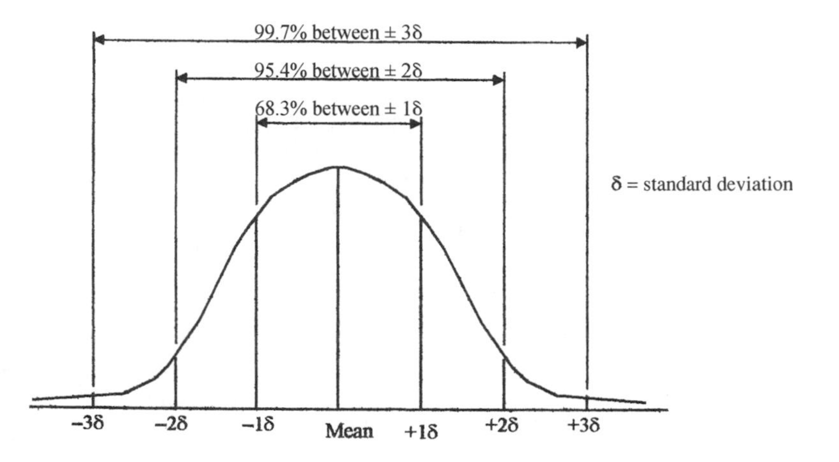

Figure 7.2.2 *Normal distribution curve showing percentage bands across normal distribution*

characteristics of the material involved. The natural variation from a process produces a range of sizes forming the normal bell-shaped curve (see Figure 7.2.2).

The accepted practice therefore was to select a tolerance band around the nominal size within which a component was considered acceptable for use for the designated function.

- If a component is produced within the set tolerance band it should be considered as good quality.
- If the component is outside the tolerance band it is considered as poor quality and therefore unsuitable for use.

Therefore a small difference in size is sufficient to determine the difference between a satisfactory and an unsatisfactory component. The different regions are shown in Figure 7.2.3.

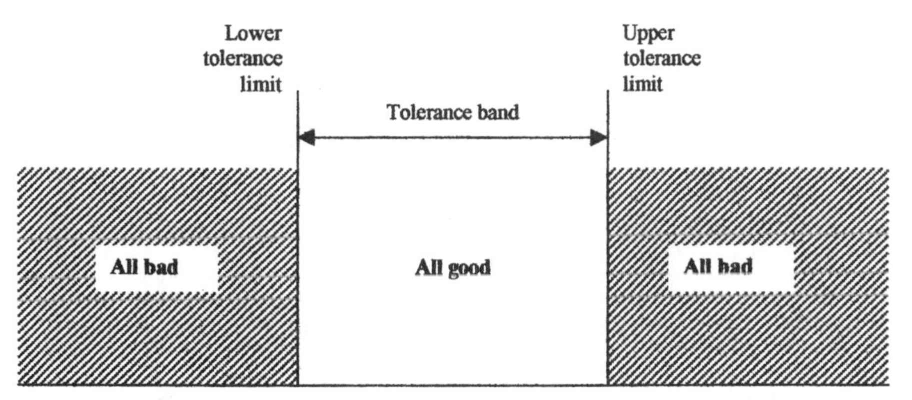

Figure 7.2.3 *Normal classification of good and bad components according to traditional limits and fits*

Having developed a working set of limits and fits over a number of years, this has been considered a satisfactory basis for producing high quality functional products for many years.

Quality loss function (QLF)

A particular event demonstrates why we should question the concept of using acceptable tolerances as the basis for producing the best functioning of a design.

In the 1970s, a television set was produced in two factories of the Sony Corporation using identical equipment and processes. One factory was in Japan, the other in the USA. Both plants operated to the same design and tolerance specification with very few sets leaving the plants outside that laid down specification.

The sets according to the theory of interchangeability should therefore appear identical to the customer. Customers, however, could often discern a distinct difference, which they valued, between some sets made in Japan and some made in the USA – even without knowing where they were made, i.e. no preconceptions were operating. These particular Japanese-made sets were, on average, considered superior in colour definition to their US cousins.

As the sets were identical in design and made to the same specification on similar processes, what was the difference between them in practice?

The difference came down to one particular subassembly, a voltage regulator which controlled the voltage to the colour density to a tolerance of 115 ± 20 volts. Both the Japanese and the US factory held to this tolerance. The difference was in how they selected the target they operated to:

- The US plant selected as a target that they would be within the tolerance band, and if the subassembly was within this band no further action was taken.
- The Japanese plant selected the nominal voltage as their target and continually attempted to come as close as possible to that.

The result of this was that both plants were operating to the same outer limts, but within these limits the actual spread of results being obtained was substantially different (see Figure 7.2.4). Many more Japanese sets were being dispatched with the regulating voltage close to the nominal one.

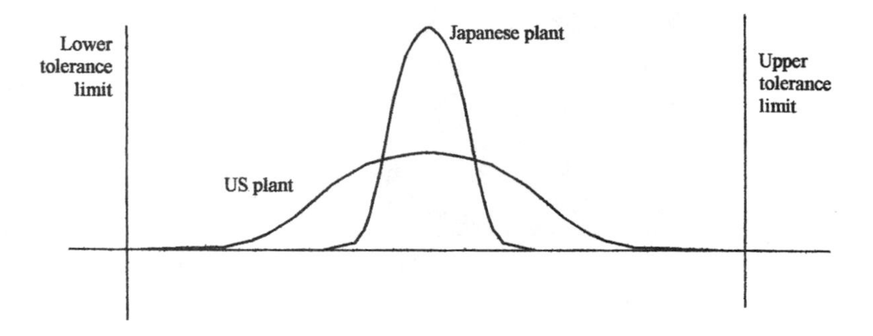

Figure 7.2.4 *Comparison of range of voltages in units coming from both plants*

On further investigation, both processes were found to be capable of producing a similar dispersion of voltages. However, the US plant by concentrating at the tolerance limits often operated with a wider range of values, or with the mean being produced on any particular batch different from the ideal nominal value.

Dr Taguchi came to the conclusion that any departure from the ideal design results in a functional loss, or as he puts it, a quality loss. This concept of a quality loss is a logical one, as we shall see.

If according to a design there is an ideal measurement for a component, any difference from that measurement in the actual product means that the ideal design function is not being met. Let us consider some examples:

A shaft rotating in a bearing

The ideal is that the shaft radial movement is tightly constrained, yet allows freedom for the shaft to turn.

- The system of limits and fits is such that this is the minimum clearance that will be achieved in an assembly, because any less and the shaft will not be free to rotate.

- Setting limits on the mating components means that in practice all clearances are actually greater than the ideal.
- The shaft is therefore not correctly restrained radially – this means losses in the functional operation (vibration and noise) and a lower life.

An electrical component

The component has an ideal operating voltage envelope.

- Failure to hold to this ideal during the product operation means a deterioration in controllable operating characteristics and a loss in life.
- Conversely if the component has a range of possible operations under a set voltage, then the whole circuit could be subjected to suboptimal conditions.

Figure 7.2.5 shows the difference between the two concepts:

- Working to a set of tolerance limits means only two distinctly contrasting states can be achieved: completely satisfactory (i.e. inside the tolerance band) or completely unsatisfactory (i.e. outside the tolerance band).
- The QLF concept shows only one completely satisfactory condition – exactly on the target measurement. As the actual size moves away from this target, it becomes less and less valuable, i.e. it incurs a greater loss.

There are several types of targets:

- A set nominal measurement, e.g. a shaft diameter: Any difference, up or down, results in a lesser value than the ideal would, i.e. a loss occurs as Figure 7.2.6(a).

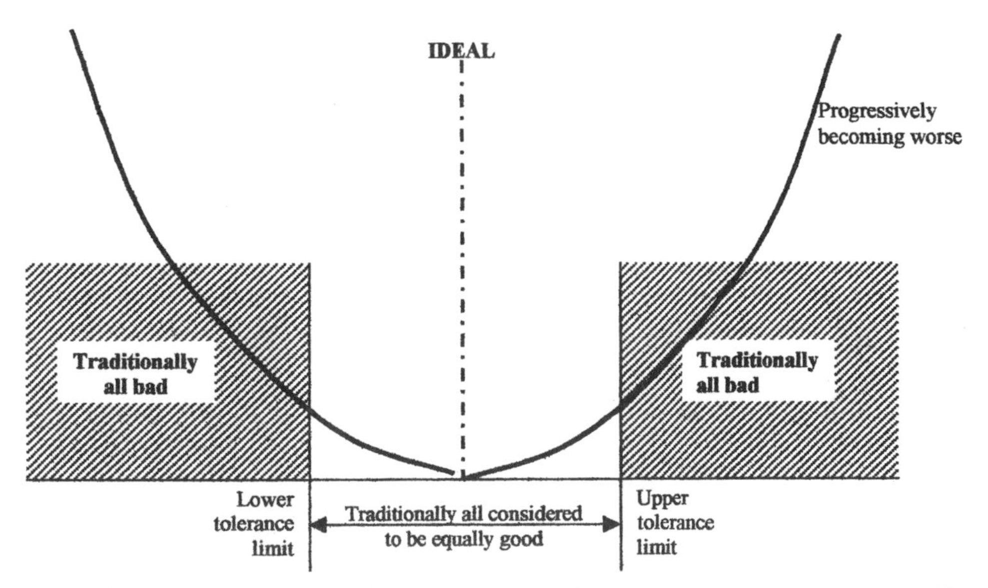

Figure 7.2.5 *Difference between Taguchi's quality loss function curve and the traditional classification of a size in relation to the ideal*

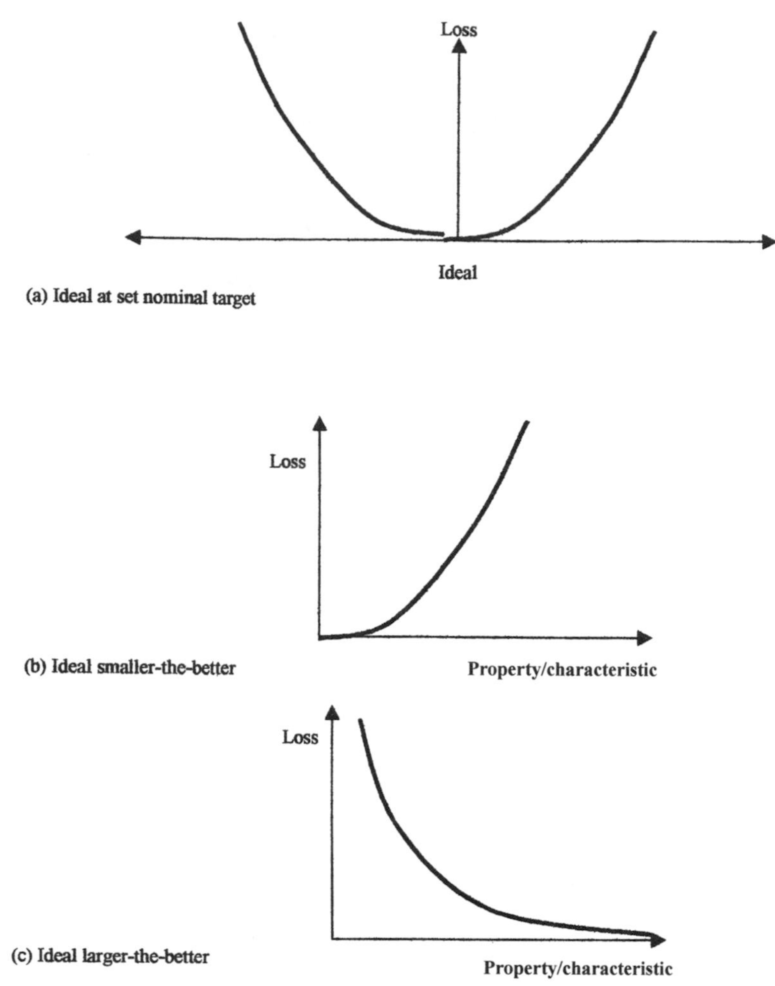

Figure 7.2.6 *The different ideal targets*

- The smaller-the-better, e.g. weight: In theory the best value is zero, therefore the higher the actual weight, the greater the loss as in Figure 7.2.6(b).
- The larger-the-better, e.g. strength: In theory the best value is infinity, therefore any strength less than this means a loss as Figure 7.2.6(c).

Of course in practice the latter two targets, i.e. the smaller-the-better and the larger-the-better, may be unobtainable. In many situations they are design aims which conflict with each other and need to be balanced one against the other. This does not mean that they are not ideals to be aimed towards.

Dr Taguchi's formulae for calculating the quality loss

The quality loss, L, is related to the difference from the ideal target. The loss Dr Taguchi determined formed an exponential shape as the difference increased – so he uses the square of the difference to calculate the loss.

When the design criteria is to a nominal-the-best basis then we can find the loss based on:

For a single item, $L(y) = k(y - m)^2$
where
- y = the actual measurement of the variable
- m = the ideal target for the variable
- $(y - m)$ = the variance
- $L(y)$ = the monetary loss when variable = y
- k = the cost coefficient

Group of items: $L(y) = k(\sigma^2 + (\mu - m)^2)$
where
- σ = the group standard deviation, its variability
- μ = the central mean of the group, i.e. the bias

When the design criteria is smaller-the-better, i.e. an ideal target of zero, the variance, y–m, reduces to y as m is zero:

Single item: $L(y) = k(y)^2$

Group of items: $L(y) = k(\sigma^2 + \mu^2)$

When the design criteria is larger-the-better, i.e. an ideal target of infinity, we use the square of the inverse of the variance:

Single item: $L(y) = k(1 \div y)^2$

Group of items: $L(y) = k/\mu(1 + (3\sigma^2 \div \mu^2))$
where μ = The average measurement

Examples

(1) Nominal-the-best
A computer monitor has a voltage specification for the contrast circuit of 110 ± 10 volts. These limits were set at the points when 50% of customers viewed the set as poor quality and would bring the set back for repair or replacement. The average cost for such an event was put at £100.

Therefore we have for a single item,

$L(y)$ = £100

$y - m$ = 10 volts
$\therefore (y - m)^2$ = 100 volt^2

$L(y) = k(y - m)^2$ for a nominal-the-best target
$\therefore$ £100 = $k \times 100$
$\therefore k$ = £1.00 per volt^2

From this we can work out the quality loss on a set for any voltage.

For example, a voltage of 115 volts is 5 volts away from the nominal 110 volts. This would give a loss of $(5^2) \times$ £1.00, i.e. £25. As this is £75 less than the initial limit loss we could afford to spend up to this £75 to ensure all assemblies were within a reduced specification of ±5 volts.

Sometimes it can be more profitable overall to increase the number of operations if the customer values the increase in quality. If it is possible to tighten the output specification, perhaps with an extra operation to adjust a characteristic, then the QLF can be used to determine the cost effectiveness of introducing this additional work. This involves using the QLF equation to find the breakeven point and hence reset the specification limits.

(2) Smaller-the-better

An aerospace component has an estimated loss = £500 for a weight of 10 kg.

Therefore we have,

$L(y)$ = £500

y = 10 kg
$\therefore y^2$ = 100 kg^2

$L(y) = k(y)^2$ for a small-the-better target
$\therefore$ £500 = $k \times 100$
$\therefore k$ = £5 per kg^2

Using a lighter material with a weight of 5 kg would reduce the loss to $(5^2) \times$ £5, i.e. £125. We would therefore have a saving of £375 to offset against any extra cost of the lighter material.

(3) Larger-the-better

On a material, the average loss that occurs with a tensile strength of 600 MN m^2 is £10.

Therefore we have,

$L(y)$ = £10

y = 600 MN m^2
$\therefore 1 \div y = 1 \div 600$
$\therefore (1 \div y)^2 = 1 \div 360\,000$

$L(y) = k(1 \div y)^2$ for a larger-the-better target
$\therefore$ £10 = $k \times (1 \div 360\,000)$
$\therefore k$ = £3 600 000 per (1 ÷ tensile strength in MN $m^2)^2$

A cheaper material with a tensile strength of only 400 MN m^2 would give a new quality loss of £22.50. Therefore the £12.50 increase in the quality loss due to the strength reduction should be set against the savings from using the cheaper material.

Taguchi's robust design process

One of the key roles of design is to reduce economically any variation of a product's function in use. This is done by identifying the various factors that affect the quality characteristics. If we can eliminate, modify or change these, we minimize the variations which affect the proper functioning of the product.

The factors that cause variations Taguchi called noise factors. These factors can be broken down into three types:

- External noise, e.g. arising from the operating environment and the conditions of use by the customer.
- Internal noise, e.g. deterioration over time in the component or manufacturing process.
- Item-to-item noise, e.g. variations which arise in the component or manufacturing process.

The quality role of activities

Product design. Good design produces a noise-resistant product through developing the best design, using the best combination of components and product operating conditions. It controls noise factors by keeping them within narrow tolerances using high grade components.

Process design. Achieves conformity by selecting the best technology and setting the optimum process operating conditions. It controls noise factors by keeping operating conditions within a narrow range.

Manufacturing Achieves conformity by process diagnosis and adjustment that enable predictions to be made. It controls by constant measuring and taking corrective action to keep processes in correct operation.

Customer service. Maintains reliability by careful instruction and training of operators and maintenance personnel. Design aids by design for maintainability and incorporating on-line diagnosis.

Principle of robust design

The quality loss function has three possible components if we look at the formula for a group of items with a set nominal target:

$$L(y) = k\,(\sigma^2 + (\mu - m)^2)$$

- The cost coefficient, i.e. k.
- The bias, i.e. μ or $(y - m)^2$.
- The variability, i.e. σ^2.

The cost coefficient is determined by the customer reaction and resultant costs involved in rectification. The bias can normally be reduced by careful adjustment of the process to centre on a target output. Reducing the variance is normally more difficult, but it can be achieved to some degree by:

- Screening out bad products before they are brought into use which means expensive inspection and scrap or reworking costs. This is expensive and wastes what we have already put into making these items.
- Discovering and eliminating causes as they occur (which should always be ongoing anyway).
- Narrowing the tolerance applied, but this usually involves extra cost during the process or in purchasing more durable components.
- Applying robust design.

All noise factors cannot be entirely eliminated, hence each product will differ from the ideal. Robust design attempts to make the product less sensitive to noise factors and hence reduce losses. The characteristics can be of three types:

- A variable, i.e. an objective measurement along a sliding scale.
- An attribute, i.e. a subjective measurement such as degree of sheen.
- Digital, i.e. straight good or bad.

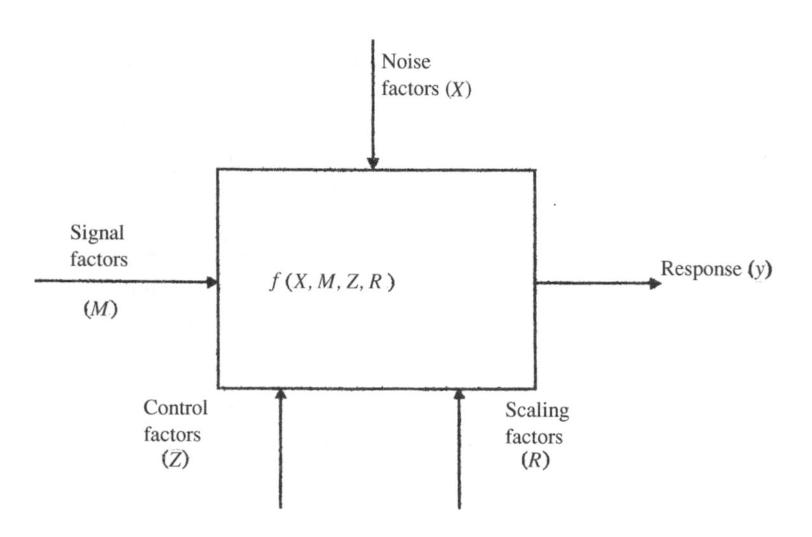

Figure 7.2.7 *Factors influencing the performance of a product or process*

The factors which influence the characteristics can be classified as in Figure 7.2.7 into:

- Noise factors, X – usually difficult, expensive or impossible to control.
- Control factors, Z – values controlled by the designer.
- Signal factors, M – settings by the user.
- Scaling factors, R – used to change the mean level of a characteristic.

We wish to maximize the predictable parts (M, Z and R) whilst minimizing the unpredictable part (X), i.e. optimize the signal/noise ratio for the different types of targets. We do this by careful selection of different parameters which can require a considerable trial and error process. We achieve significant results with a careful design of these combinations by experiments.

Dr Taguchi – design of experiments

A product, or process, has often many different possible input variables each with different settings. With these it is often difficult to determine the best combination to optimize the overall effect. Each of the many possible combinations can be carried out and the results analysed but this would mean a substantial number of experiments.

Dr Taguchi's approach to designing experiments indicates the significant variables with a minimum of experiments. The steps to be followed are as per the flow chart in Figure 7.2.8.

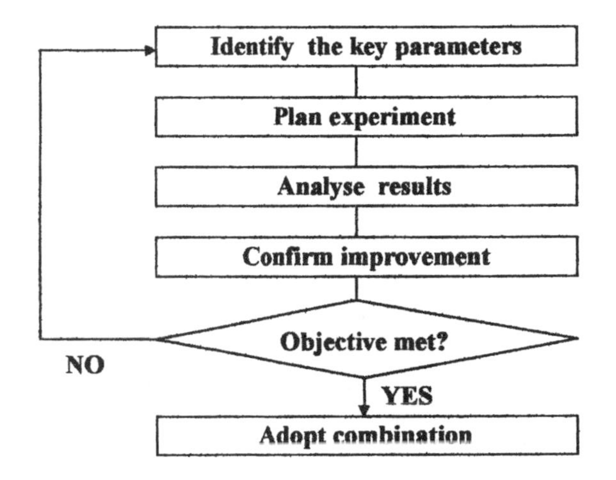

Figure 7.2.8 *Flow chart of a series of experiments to maximize robust design factors*

Identify the key factors

In order to clearly identify the key factors, a brainstorming session should be held to collect all the possible variables. These are often recorded in a cause/effect diagram as in Figure 7.2.9.

It is necessary to separate these into the factors which can be controlled readily and those that are more expensive or difficult to control, i.e. noise factors, such as ambient temperature, humidity, background sound, etc.

Design the experiment

Having decided on the controllable key factors, the next step is decide on the range of settings to be used. Each factor should have a minimum

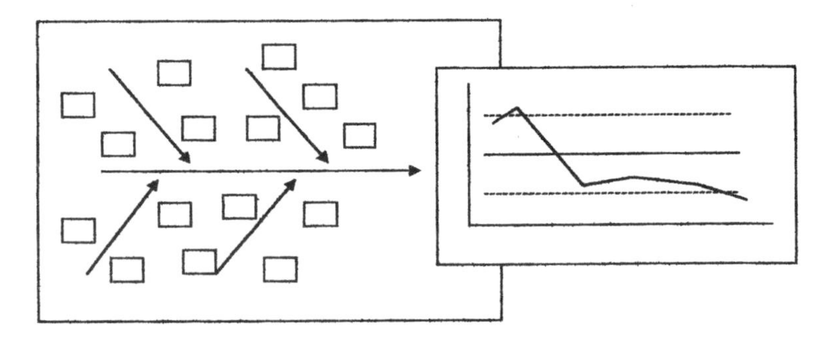

Figure 7.2.9 *Cause and effect diagram coupled to measurement of response*

of two settings but three settings gives a better indication of the actual effect of each setting.

An orthogonal array is then selected depending on the number of factors and maximum number of settings (see table and appendix). Each factor is allocated one column of the array and the varied settings of that factor are called for by numbers in that column.

Array name	*Factors*	*Settings*	*Trials*
$L_4\ (2^3)$	3	2	4
$L_8\ (2^7)$	7	2	8
$L_9\ (3^4)$	4	3	9
$L_{12}(2^{11})$	11	2	12
$L_{16}\ (2^{15})$	15	2	16
$L_{16}\ (4^5)$	5	4	16
$L_{18}\ (2^1 \times 3^7)$	{1	2}	18
	{7	3}	
$L_{27}(3^{13})$	13	3	27

The rows form a unique combination of all the factors' settings. Tests are carried out with the factors set as per each row and the result recorded. By recording other noise factors occurring during the tests, these also can be recorded at the time against the specific row under test for correlation.

Analyse results

In the analysis stage, all the test results are averaged to produce a grand average.

The results against each setting of each factor are then also averaged.

These setting averages are then compared to the grand average to determine which setting produces an average more favourable (see example for graphing technique). That setting is then selected as the most favourable for that factor. Note that the results may indicate a setting outwith the best in these trials, which would indicate further trials with different settings being made gradually zooming in on a true best.

It is possible to predict the overall best result by summing the variance from the grand average of each significant factor's favourable setting. A further confirmation experiment must be carried out using each factor set at its best to gauge the actual success of the correlation.

Parameter design

Choice of factors

- Levels – a minimum of two. The more levels used, the more experiments needed.
- Range of factor levels: A wide range is better as more effect is shown.
- Wider range may not show up detailed effect, e.g. less linearity.
- Feasibility of levels, i.e. is it possible to run the combinations?

- Sliding scales: Known effects of combinations of factors can be reduced by sliding scales:

Temperature		Low	High
Speed	Low	160°C	170°C
	High	200°C	220°C

- Safety of plant and people.

Effect of factor

- Mean mainly, i.e. an adjustment factor.
- Variance mainly.
- Mean and variance both significantly – use carefully.
- No significant effect – set at level of cost or convenience.

Orthogonal array

- Equal proportions of experiments.
- Equal proportions of remaining factor levels.
- Equal proportions of combinations of factor levels.

Example 7.2.1

A cable is to be formed by using a plastic tube with plug ends glued on as in Figure 7.2.10. A problem has been experienced with the plug being pulled out.

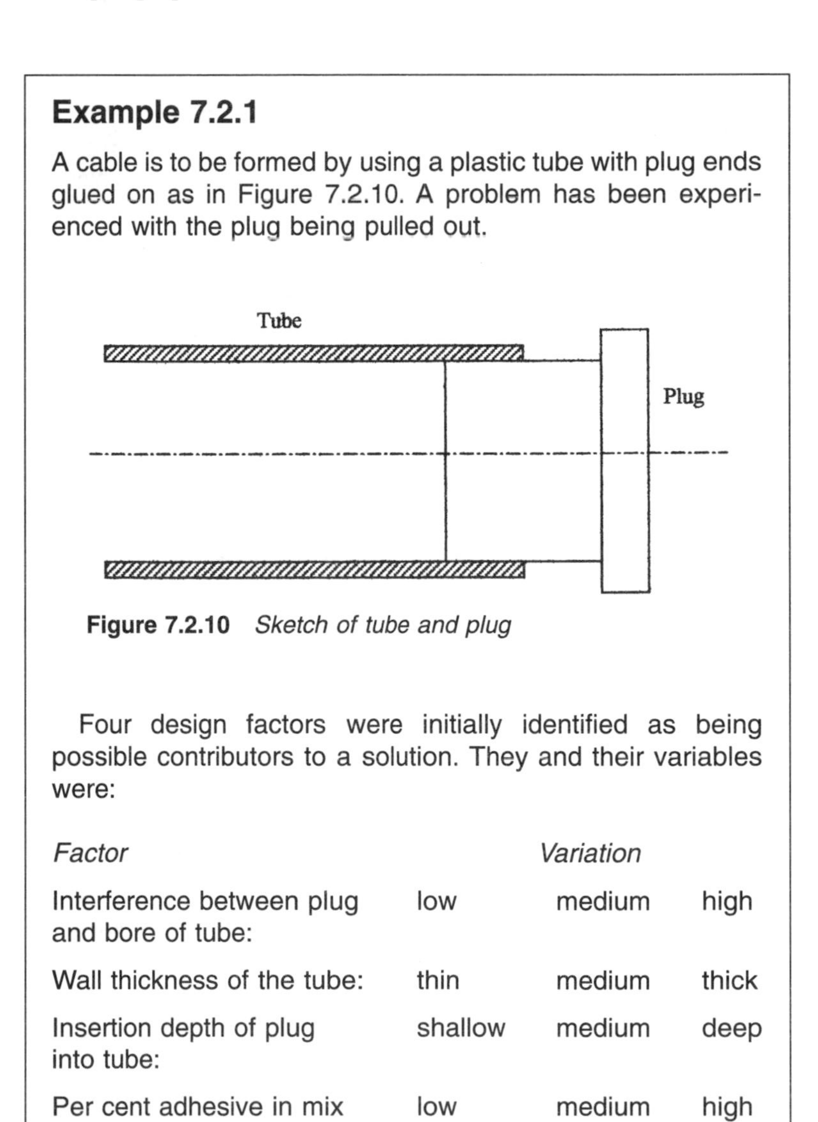

Figure 7.2.10 *Sketch of tube and plug*

Four design factors were initially identified as being possible contributors to a solution. They and their variables were:

Factor	*Variation*		
Interference between plug and bore of tube:	low	medium	high
Wall thickness of the tube:	thin	medium	thick
Insertion depth of plug into tube:	shallow	medium	deep
Per cent adhesive in mix of binder/hardener:	low	medium	high

This gave four factors with three settings each, and called for an L9 orthogonal array, i.e. nine tests to be made, namely:

	Parameter/setting			
Row	A	B	C	D
1	1	1	1	1
2	1	2	2	2
3	1	3	3	3
4	2	1	2	3
5	2	2	3	1
6	2	3	1	2
7	3	1	3	2
8	3	2	1	3
9	3	3	2	1

The results of the tests were:

Experiment	*Interference*	*Wall thickness*	*Insertion*	*Adhesive content*	*Pull-off force* (kg)
1	low	thin	shallow	low	19.1
2	low	medium	medium	medium	21.9
3	low	thick	deep	high	20.4
4	medium	thin	medium	high	24.7
5	medium	medium	deep	low	25.3
6	medium	thick	shallow	medium	24.7
7	high	thin	deep	medium	21.6
8	high	medium	shallow	high	24.4
9	high	thick	medium	low	28.6

This gave a grand average for the pull-off force of 23.41 kg with a standard deviation of 2.75 kg.

We can now produce for each factor's setting the average and the variance of the associated experiments from the grand average.

Factor	*Setting*	*Experiment*	*Average*	*Variance*
Interference	low	1, 2 & 3	20.47	–2.94
	medium	4, 5 & 6	24.90	1.49
	high	7, 8 & 9	24.87	1.46
Wall thickness	thin	1, 4 & 7	21.80	–1.61
	medium	2, 5 & 8	23.87	0.46
	thick	3, 6 & 9	24.57	1.16
Insertion	shallow	1, 6 & 8	22.73	–0.68
	medium	2, 4 & 9	25.07	1.66
	deep	3, 5 & 7	22.43	–0.98
Adhesive content	low	1, 5 & 9	24.33	0.92
	medium	2, 6 & 7	22.73	–0.68
	high	3, 4 & 8	23.17	–0.24

These can also be shown graphically, as in Figure 7.2.11.

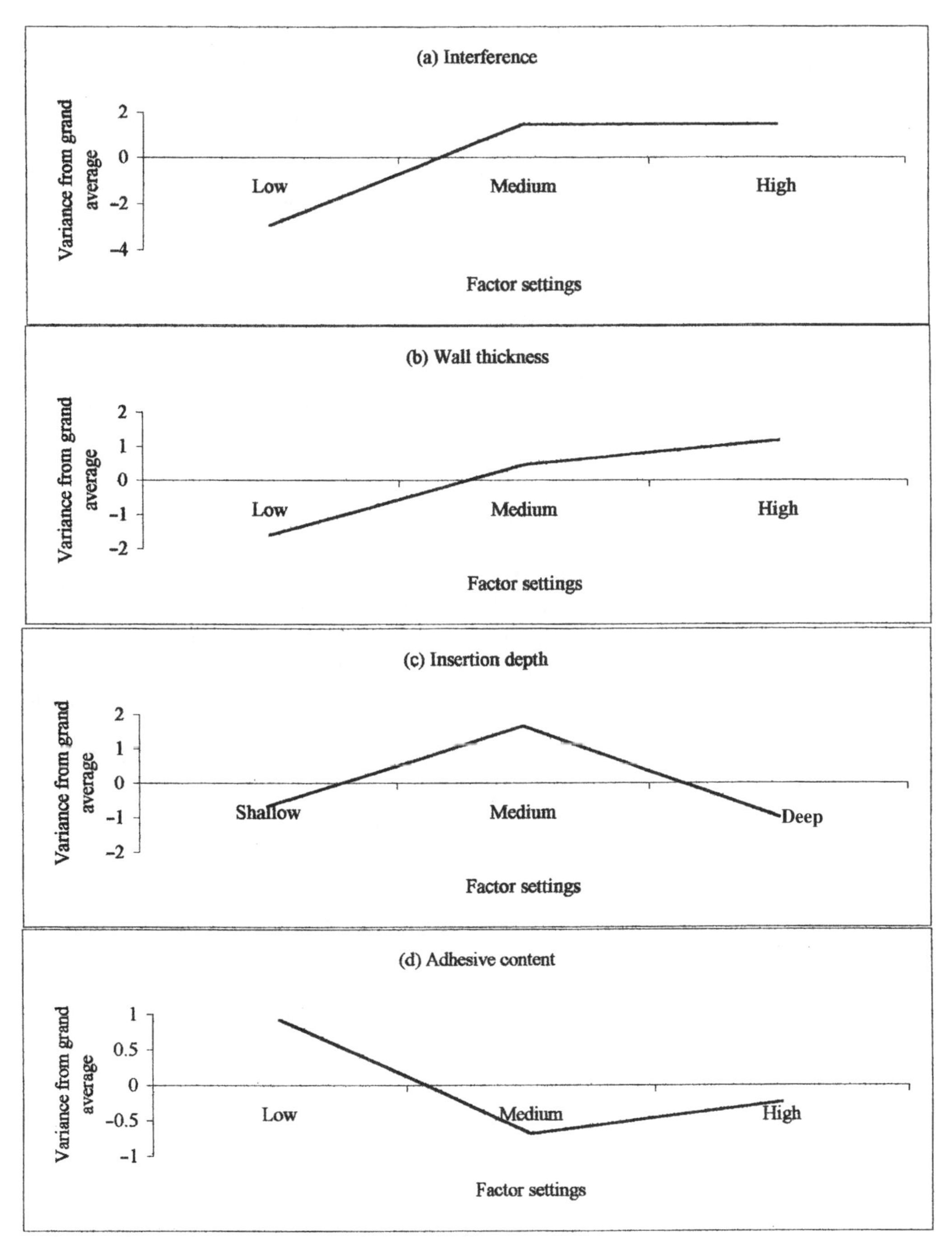

Figure 7.2.11 *Graphs of results from experiments*

From these results the best individual setting for each factor is:

Medium interference	+1.49	effect
Thick wall	+1.16	effect
Medium insertion	+1.66	effect
Low adhesive content	+0.92	effect

Total effect indicated is the sum of these, i.e. +5.23 kg, which gives a predicted combination pull-off force required of 28.64 kg. This is higher than any of the actual test results, and can be tested for by producing to this specification and comparing the actual force required.

It is normal practice in these type of experiments only to use approximately the best 50% of the results to allow for experimental error. In this case that would be to use only:

Medium interference	+1.49	effect
Medium insertion	+1.66	effect

Which would give a predicted increase of 3.15 kg, but we can see that this is bettered, therefore we would use all factor best settings as they appear significant.

However, if we further examine the graphs in Figure 7.2.11, we can see from their shapes that the maximum values may even be higher than those indicated from the experiments:

- Interference appears to have a maximum between the medium and high settings.
- Wall thickness appears to have a maximum with an even larger wall thickness than used in the experiments.
- Insertion depth appears to have a maximum around the medium value.
- Adhesive content appears to have a maximum with a lower content of adhesive than used in the experiment.

By taking intermediate settings from these actually taken on each of the factors, we may be able to improve on the anticipated best combination. A further series of investigatory experiments using these new factor settings should be taken to check if this is so.

Once we have reached what appears to be the best combination of factor settings, we would have to carry out a series of experiments to confirm them.

In addition, we should always be re-examining our assumptions regarding what factors are most influential.

Orthogonal arrays

Degrees of freedom

This is the number of independent comparisons that are made. For a factor, the degrees of freedom is one less than the number of levels. In an orthogonal array the degrees of freedom is one less than the number of experiments because one degree is taken up by the overall mean.

Interactions

It is possible to see if a pair of factors have a combined effect different from each factor's individual effect. This could happen in medical trials with two medicines where we want to know the effect of each applied in isolation and what happens when we apply them both at the same time.

One way is to use an $L_4(2^2)$ array:

Experiment	A	B	Column A × B
1	1	1	1
2	1	2	2
3	2	1	2
4	2	2	1

The first two columns contain the two main factors each set at two levels (say used and not used) which are compared as normal by finding the average of results at each level. The third column shows the combined effect and gives four results:

A1 × B1; A1 × B2; A2 × B1; A2 × B2

We could restate these as:

- A not used/B not used;
- A not used/B used;
- A used/B not used;
- A used/B used.

These are then analysed in the response table as normal.

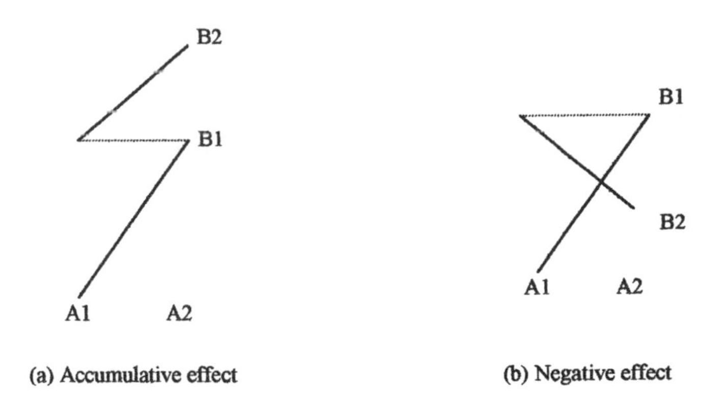

Figure 7.2.12 *Combination effects of two factors on end result*

In the response graph all four interactions would be shown as in Figure 7.2.12. These show additive and negative (crossed) combinations. The effect of A × B is:

A × B = AB – A – B.

Normally interactions are not examined in robust design. Where it is felt that such interactions do exist and are significant then we can allow for this by using the interaction table. This allows for possible interaction by allocating set columns for these:

	With column						
Column	**1**	**2**	**3**	**4**	**5**	**6**	**7**
1		3	2	5	4	7	6
2			1	6	7	4	5
3				7	6	5	4
4					1	2	3
5						3	2
6							1

Thus we can see that the interaction of columns 1 and 2 would be recorded in column 3. Interaction tables are available in Taguchi and Konishi's book *Orthogonal Arrays & Linear Graphs: Tools for Quality Engineering*, 1987, ASI Press. This also describes the arrays graphically so that interactions can be specifically designed.

Basically in a linear graph the interaction between two columns is found by allocating two of the dotes to these and using the column number on the joining line to record their interaction settings. There are a complete range of Linear graphs in Taguchi and Konishi's book.

Problems 7.2.1

(1) Why do you think that the Japanese produced closer to the nominal size than the USA plant if both plants had identical specifications?
(2) What do you think happens when you have two conflicting ideals of maximum strength and minimum weight?
(3) What do you think are the causes of noise in Figure 7.2.7?
(4) In the example of the tube and plug, can you think of other factors that also may be important?
(5) Can you think of a used/not used combination in your own field that you may wish to examine?

7.3 Improving reliability

The full study of reliability involves the use of complex statistical formulae which are not covered in this textbook. This section introduces reliability in a mainly non-mathematical way to enable designers to evaluate their designs to determine faults and therefore devise means to:

- Eliminate their occurrences.
- Reduce their frequency.
- Reduce their effect.
- Recognize they are happening, or are about to happen.

We justify this limited approach because:

- It is difficult to gather accurate data on components' reliability performance, especially where the application is new.
- Manufacturers are constantly changing their product and production methods which means that historical data may not match present components' performance.
- Many products have a multitude of components which makes calculating the combined reliability a complex mathematical calculation.

This does not denigrate the full study of reliability, but because of the complexity, it is a specialist area which is outwith the normal training of a designer.

As designers we are interested in satisfying the customers' need to have an operational product for the life they expect; we should therefore try to produce our products to set reliability criteria.

Reliability concepts

If a product meets all its design functional criteria at the production stage it will enter service with a customer. However, sooner, or later, it will not be able to continue to fully meet that criteria. In other words, it will have failed either partially or fully.

The study of reliability centres around estimating how long in time the product will remain reliable, i.e. continue to meet its functional specification.

Products come in two categories, with respect to how it can be affected by a failure to perform:

- Non-repairable: When the unit fails to function, it must be discarded and replaced.
- Repairable: A repair should be able to restore the product to full functionality after it fails.

We have several conditions whose probability we would like to know:

- $R(t)$, the reliability function, is the probability that the unit will remain functioning at a time t.
- $F(t)$, the probability that the unit will cease to function at a time t.

Over time the probability of failure normally rises, but at any one time t:

$$R(t) + F(t) = 1$$

Measures of reliability

We are interested in being able to predict three main aspects:

- The time when an item will fail.
- How many will fail within a set period.
- How much available operational time will a repairable unit have.

To determine the first, we need to have information from many products on how long they operate before they fail, i.e. T, the time until failure occurs. A graph of the results of non-repairable failures, uptime would appear as in Figure 7.3.1. Later we shall be considering the difference with repairable items.

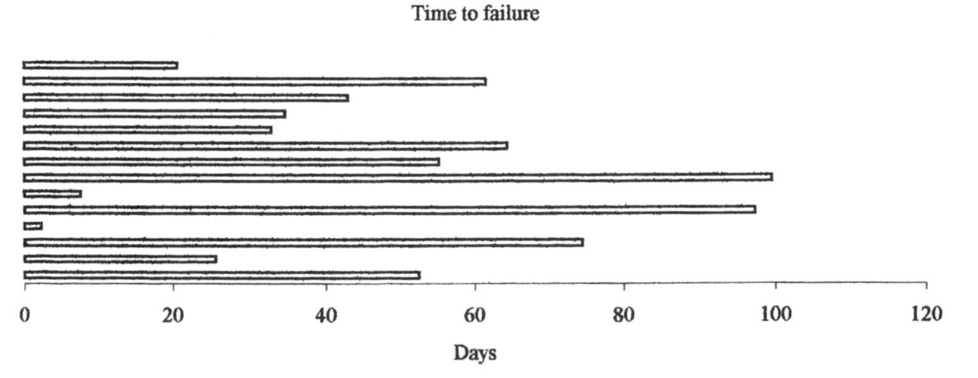

Figure 7.3.1 *Graph of time-to-failure of a non-repairable item*

From this information we can establish an average, or mean, operational time:

$$\text{Mean time to failure, } MTTF = \frac{\text{total of time to failure}}{\text{number of units failed}}$$

$$= \frac{\sum_{i=1}^{i=N} T_i}{N}$$

We can also express $MTTF$ for an infinite number of units as:

$$MTTF = \int_0^{\infty} R(t)\,\mathrm{d}t$$

where $R(t)$ = the reliability function.

Knowing $MTTF$, we can find the mean failure rate, λ, which is the inverse of $MTTF$. In the case of non-repairable units:

$$\text{Mean failure rate, } \lambda = \frac{\text{number of units failed}}{\text{total of time to failure}}$$

$$= \frac{1}{MTTF}$$

$$= \frac{N}{\sum_{i=1}^{i=N} T_i}$$

Example 7.3.1

If we have noted the hours of operation until failure of a series of ten light bulbs, we may arrive at the following lives in hours:

210, 230, 190, 260, 240, 220, 240, 200, 180, 230

$$\text{Therefore the } MTTF = \frac{\Sigma\text{ (bulb lives)}}{\text{number of bulbs}} = \frac{2200}{10} = 220 \text{ hours}$$

This is the arithmetic average figure – note that the actual lives ranged from 180 hours to 260 hours, with only one bulb having the same actual life as the $MTTF$ value.

The mean failure rate, λ, is the inverse of the $MTTF$ value, i.e. 0.0045 per hour.

Note that the mean failure rate only really becomes useful when we are considering a number of units where each unit is replaced when it fails.

Where an item is repairable the graph of the time changes to that in Figure 7.3.2. The times T_D shown in this graph represent the downtimes, i.e. when the unit is not functional due to faults. This may be due only

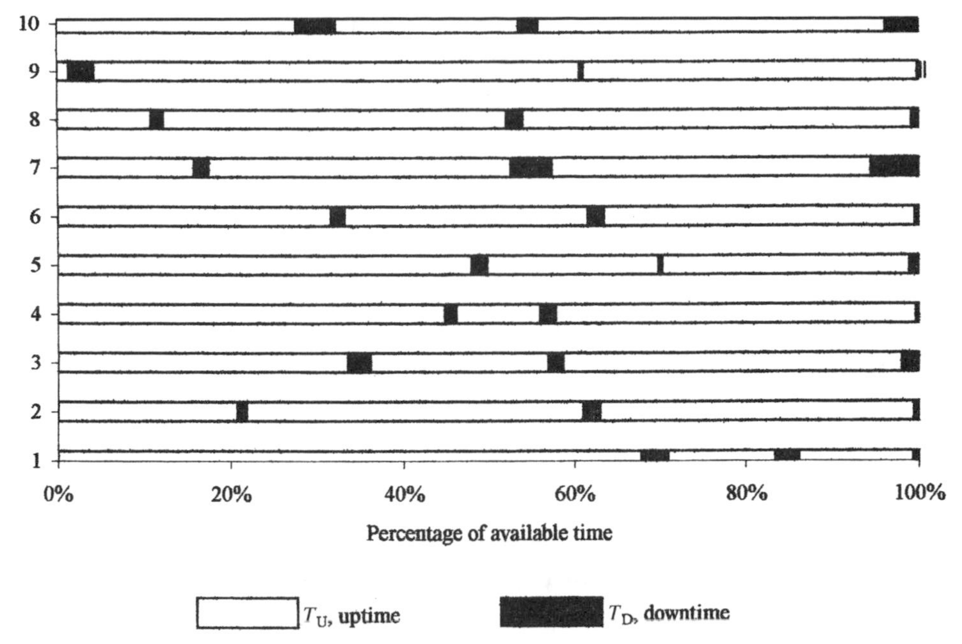

Figure 7.3.2 *Graph of available operational uptime and associated downtime on repairs*

to the repair time, but in practice would probably include a period of time waiting to be repaired.

The total downtime therefore cannot be fully allowed for by a designer as some of this time will be outside the design remit. The designer can only factor in a time to carry out the actual repair. The repair time is the time we assume as T_D in the following formulae:

$$\text{Mean down time, } MDT = \frac{\text{total downtime}}{\text{number of failures}}$$

$$= \frac{1}{N_F} \sum_{j=1}^{j=N} T_{Dj}$$

The total downtime, N_F MDT, in this case being the sum of the individual downtimes, i.e. $= \sum_{j=1}^{j=N} T_{Dj}$.

The total uptime can be found by subtracting the total downtime from the total available time, NT.

$$\text{Total uptime} = NT - N_F\, MDT$$

Knowing this we can calculate the mean time between failures:

$$\text{Mean time between failure, } MTBF = \frac{\text{total uptime}}{\text{number of failures}}$$

$$= \frac{NT - N_F\, MDT}{N_F}$$

This time we use the *MTBF* to calculate the mean failure rate, λ:

$$\text{Mean failure rate, } \lambda = \frac{\text{number of failures}}{\text{total uptime}}$$

$$= \frac{N_F}{NT - N_F\, MDT}$$

$$\text{And availability, } A = \frac{100 \times \text{total uptime}}{\text{total uptime} + \text{total downtime}}\ \%$$

$$= \frac{100 \times MTBF}{MTBF + MDT}\ \%$$

Example 7.3.2

Consider a process plant working 40 hours per week. In a 46 week year (allowing for plant shutdown for holidays, etc.) total possible working time is 1840 hours. During the year the plant has 20 breakdowns which gave a total downtime of 30 hours. We can then calculate the reliability statistics:

NT = 1840 hrs

$$\text{Mean downtime, } MDT = \frac{\text{total downtime}}{\text{number of failures}} = \frac{30}{20} = 1.5\,\text{hrs}$$

$$\text{Total uptime} = NT - N_F\, MDT = 1840 - 30 = 1810\,\text{hrs}$$

$$\text{Mean time between failure, } MTBF = \frac{\text{total uptime}}{\text{number of failures}}$$

$$= \frac{1810}{20}$$

$$= 90.5\,\text{hrs}$$

$$\text{Mean failure rate, } \lambda = \frac{\text{number of failures}}{\text{total uptime}} = \frac{20}{1810} = 0.011 \text{ per hour}$$

$$\text{Availability, } A = \frac{100 \times \text{total uptime}}{\text{total uptime} + \text{total downtime}}\ \%$$

$$= \frac{100 \times 1810}{1810 + 30}\ \%$$

$$= 98.4\%$$

The other criterion we are interested in is the instantaneous failure rate, or the hazard at any particular time. To find this we need to calculate:

- The probability that an item survives to a particular time.
- The probability that the item will then fail in a very short time after that time.

The instantaneous failure rate is the difference between these. The mathematics involved are outwith the scope of this textbook, but we will use the concept to show common patterns of failure.

Patterns of failure

The most common pattern of failure rate is the bathtub curve as shown in Figure 7.3.3. This consists of three distinct phases.

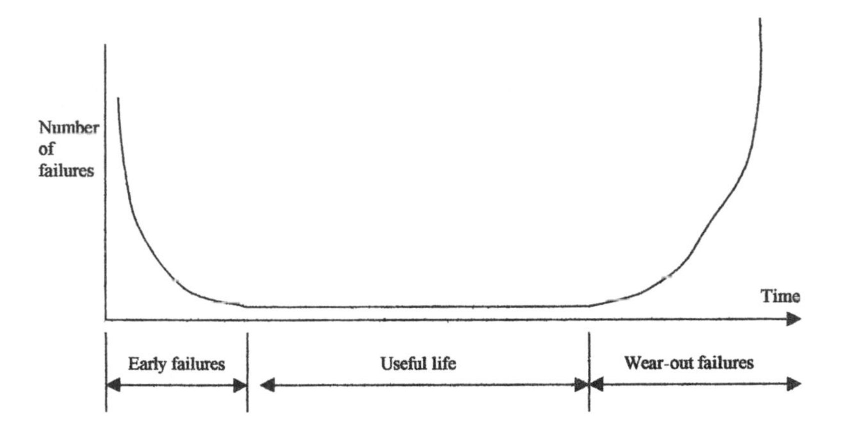

Figure 7.3.3 *Bathtub pattern of failures*

Early failure. This may be due to some poor quality components or manufacturing faults which have not shown up before leaving the plant. It may also be a function of unfamiliarity with the operation of the product. As time progresses the number of these early failures falls until the next phase is reached. This region is often allowed for with a burn-in period during which the early failures occur and are replaced.

Useful life. The failure rate settles into a low steady failure rate in this region.

Wear out. Eventually the components come to the end of their useful life and the failure rate starts to increase again.

Although the bathtub curve gives a useful indication for a product hazard rate, the life of individual components tends to follow a range of patterns. Although we are not examining these in detail in this text, we will demonstrate the pattern of failure of the two most common:

- Constant failure and exponential function: This is characterized as shown in Figure 7.3.4. Here we have a high early failure phase followed by a slowly decaying rate with no wear-out phase apparent. The slowly decaying phase could be approximated as a constant failure rate. The reliability function $R(t) = \exp(-\lambda t), t > 0$.

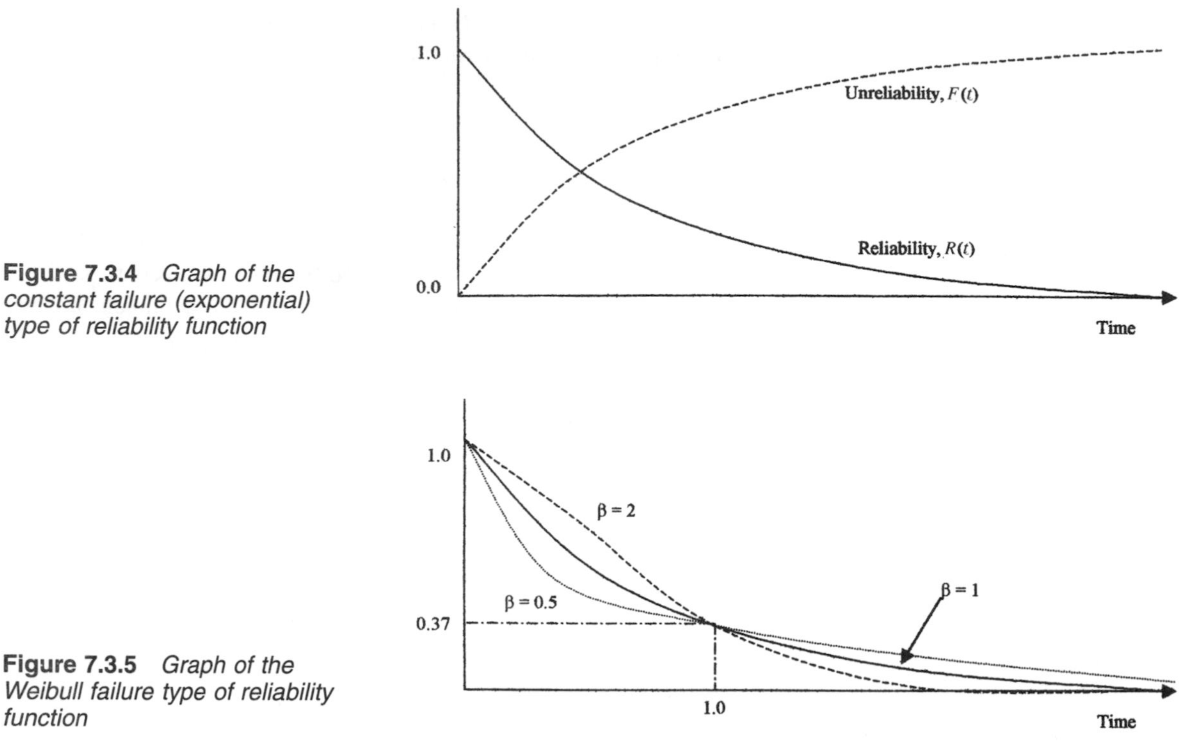

Figure 7.3.4 *Graph of the constant failure (exponential) type of reliability function*

Figure 7.3.5 *Graph of the Weibull failure type of reliability function*

- The Weibull failure rate is the other common form as shown in Figure 7.3.5. Here a shape parameter β is used to denote the way in which failures occur. Where $\beta = 1$, failure rate is constant. $\beta > 1$ signifies increasing failure rate and $\beta < 1$ decreasing failure rate. In this case $R(t) = \exp(-(t/\alpha)^{\beta})$, $t > 0$.

Failures – types and causes

To attempt to list all possible failures would be impossible. The following is a typical list of faults that could happen to a product:

Cracked/fractured	Contaminated	Overheating
Distorted	Intermittent	Burned
Undersize	Open circuit	Overloaded
Oversize	Short circuit	Omitted
Fails to open	Premature	Incorrectly assembled
Fails to close	Delayed operation	Scored
Internal leakage	Binding/sticking	Noisy
External leakage	Loose	Arcing
Fails to start	Incorrect adjustment	False response
Fails to stop	Seized/jamming	Using extra power
Corroded	Worn	

Sometimes these faults will have serious consequences, e.g. serious injury or complete product functional failure, and at times they will not have a significant effect. At other times several faults on their own may be insignificant, but if they all happen at the same time their cumulative effect could be serious. We therefore need to closely examine all designs

Table 7.3.1 Some sources of failure

Type	*Examples*
Specification	Omissions and Errors
Design	Mistakes in application and calculations; poor drawings
Manufacture	Omitted or erroneous actions; poor plant condition; poor training; equipment failure; poor procedures; poor instructions/drawings
Installation	Omitted or erroneous actions; poor procedures; poor instructions/drawings
Operation	Omitted or erroneous actions; poor plant condition; poor training; equipment failure; poor procedures
Maintenance	Omitted or erroneous actions; poor procedures; poor instructions/drawings
Environment	Range of temperature, humidity, vibration, corrosion not allowed for
Accidents	Fire and explosion in other plant
Acts of God	Earthquake, flood

and systems to spot where problems can arise and if possible alleviate their effect. Table 7.3.1 demonstrates parts of the design – production cycle and what failures can arise in them:

Many faults appear to arise through human error, but often the root cause lies not at the operator/worker level but elsewhere in the system, i.e. a lack of a proper system or poor training.

Common human errors are:

- Inadequate information: Many times through an omission of knowledge of procedures or consequences.
- Poor communications – written or spoken: Again often omissions can occur.
- Inadequate design: This can apply to both equipment and procedures.
- Lapses of attention: Sometimes this is due to distractions but can also be due to overconfidence.
- Mistakes: Normally a genuine error in matching action to an event. Can be due to lack of experience or training.
- Misperceptions: Can be due to inexperience or a preconceived notion.
- Deliberate: Fairly limited in occurrence but must be allowed for.

Therefore the designer not only needs to examine a product under ideal conditions but needs to fully investigate what can go wrong in that product's use.

Sources of information on failures

Manufacturers' data sheets

Many manufacturers, especially in the electronics supply field, publish reliability data sheets as shown in Figure 7.3.6. This particular example

MEAN TIME BETWEEN FAILURES.
Calculation of *MTBF* using the table of factors below.

"N"	60%	90%	95%	99%
0	1.0800	0.4320	0.3380	0.2180
1	0.5000	0.2580	0.2140	0.1520
2	0.3250	0.1900	0.1610	0.1200
3	0.2400	0.1500	0.1300	0.1000
4	0.1910	0.1270	0.1100	0.0870
5	0.1600	0.1100	0.0960	0.0764
6	0.1375	0.0960	0.0850	0.0695
7	0.1200	0.0860	0.0760	0.0630
8	0.1070	0.0770	0.0700	0.0578
9	0.0960	0.0710	0.0640	0.0535
10	0.0875	0.0655	0.0590	0.0500
11	0.0800	0.0605	0.0555	0.0468
12	0.0745	0.0565	0.0520	0.0442
13	0.0690	0.0530	0.0490	0.0415
14	0.0640	0.0500	0.0460	0.0398
15	0.0600	0.0475	0.0435	0.0376
16	0.0570	0.0450	0.0415	0.0360
17	0.0540	0.0430	0.0395	0.0344
18	0.0510	0.0405	0.0378	0.0330
19	0.0480	0.0387	0.0361	0.0315
20	0.0460	0.0370	0.0346	0.0304

This information is calculated using a 60% confidence level.

CTR period covered	Product series	Test submitted – load	Units tested	Total test hrs.	Rejects (>20%)	*MTBF* figure	Failure/rate (%) per 1000 hrs
Aug88-Feb00	H8	1000 Hrs @ 70°C	920	920,000	0	993600	0.101
Aug88-Feb00	H4	1000 Hrs @ 70°C	919	919,000	0	992520	0.101
Aug88-Feb00	H2	1000 Hrs @ 70°C	920	920,000	0	993600	0.101

NOTE:
To obtain the total test hours required for a given *MTBF* or failure rate per cent, 1000 Hrs at a given confidence level, divide the *MTBF* figure by the factor quoted for a given confidence level at 'n' number of failures.

The table of the factors gives the lower limit so the user may expect the *MTBF* to be equal or better than the result calculated.

Total test hours = no. of units tested x test hours
MTBF = Total test hours x factor
Failure 1/*MTBF* per hour
Failure rate % per 1000 Hrs = failure rate x 10^5
1 FIT 1 Failure in 10^9 hours
FIT = Failure Rate x 10^9

Figure 7.3.6 *(Reproduced courtesy of Meggitt Electronic Components Ltd)*

shows the *MTBF* of a Holco axial leaded precision resistor from Meggitt Electronic Components Ltd.

Service records

An organization's own service record contains a substantial amount of real information about faults in the field and remedial actions taken as shown in Figure 7.3.7, taken from BS 5760. An important omission from this particular sheet is time – there are no dates.

Pareto analysis

It is important that effort in quality improvement investigations be targeted to achieve the most beneficial results. Pareto charts can effectively show up the relative frequency, or effect such as lost time or high costs, of particular types of faults occurring.

If faults are collated by cause, the results can be charted in a histogram (see Figure 7.3.8) with the heaviest occurrences at one end

BSI Handbook 22: 1983
BS 5760 *(continued)*

EXAMPLE

Average number of instruments under guarantee = 51
Total number of failures = 116

Fault area	Total reported	Failure rate	Total replaced	Comments
Drainage problems	12	0.235		
Burner tube	10	0.196	6	
Cloud chamber	9	0.176	5	4 complete, 1 chamber bung
Realign burner	9	0.176	2	2 replaced
Realign lamps	7	0.137		
Fuel inlet elbow	6	0.118	6	
Sample capillary	6	0.118	3	
Hollow cathode lamps	6	0.118	6	Sb. cd. - 4 unspecified
Nebulizer	5	0.098	5	4 St. 1 inert.
Log printed circuit board	4	0.078	4	
Align entrance				
Optics	4	0.078		
Meter	4	0.078	3	
Compressor (SP93)		0.078	3	2 units, 1 pressure guage
Wiring faults	3	0.059		
Burner door hinges	3	0.059	3	
Photomultiplier	2	0.039	2	
Filter disc adjust	2	0.039		
Lamp holders	2	0.039	2	
Mirror attachments	2	0.039	1	
Wavelength calibr.	2	0.039		
Burner ht. scale	2	0.039	2	Adhesive
Needle valves	2	0.039	1	
Flowmeter	1	0.020	1	
Spoilers (cl. ch)	1	0.020	1	
Gland nuts	1	0.020	1	Cracked (overnight?)
Prism actuator	1	0.020		
Electronic chassis	1	0.020	1	
Printed circuit board connector	1	0.020	1	
SW.1 (main)	1	0.020	1	
SW.7 (mode)	1	0.020	1	
HT. smoothing	1	0.020	1	Capacitor added
RV6, 40K	1	0.020	1	

40 instruments needed services under guarantee calls
No. of instruments with 1 call = 28
2 calls = 7
3 calls = 3
4 calls = 1
5 calls = 1

No. of instruments having 1 failure = 11
2 failures = 10
3 failures = 10
4 failures = 4
5 failures = 2
6 failures = 1
7 failures = 1
16 failures = 1

No. of faulty installations = 11
No. of failures on installation = 23
Total service under guarantee calls involving specific failures = 60
Total service under guarantee calls = 72

Figure 7.3.7 *(Extracts from BS 5760: Part 3 from the BSI Handbook 22: 1983. Reproduced with permission of BSI under licence number 2001/SK0152. Complete British Standards can be obtained from BSI Customer Services (Tel. +44 (0)20 8996 9001))*

under a variety of different criteria. These are normally converted into percentages of all faults for a fuller effect. This enables us to separate the important few from the trivial many when prioritizing our design effort.

This analysis often shows up that only a few occurrences account for the bulk of the happenings. This is highlighted by plotting the important characteristic under both the occurrence against each item and also in a sorted manner showing the cumulative effect of the most occurring items. Can be used to direct effort in:

- Quality: Assigning frequency against faults, or causes, it highlights the items where to concentrate design and manufacturing engineering and management. Elimination, or reduction, of these will have the greatest overall effect.
- Maintenance: Identifying the most frequent or longest downtime faults can lead in the short term to effective planned maintenance

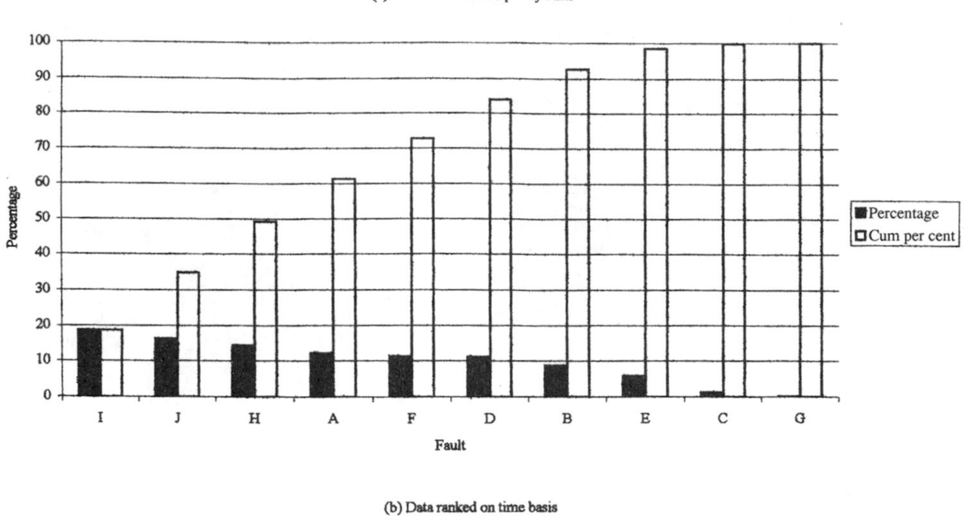

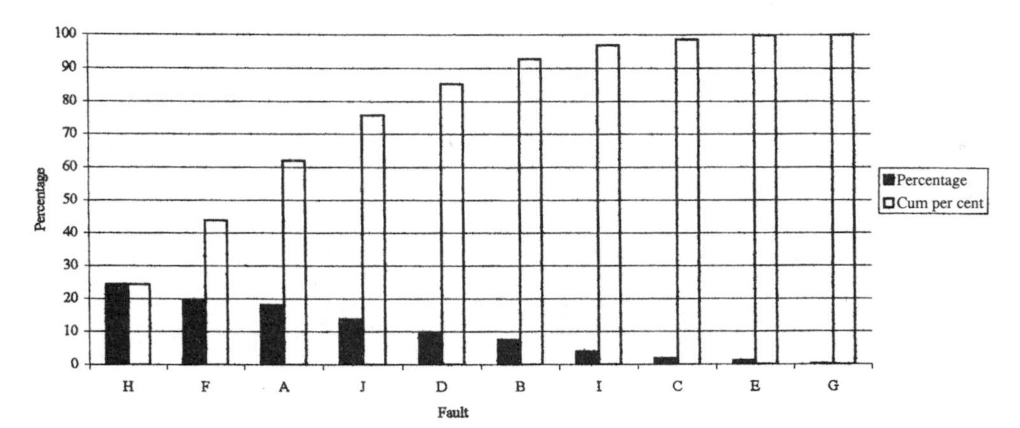

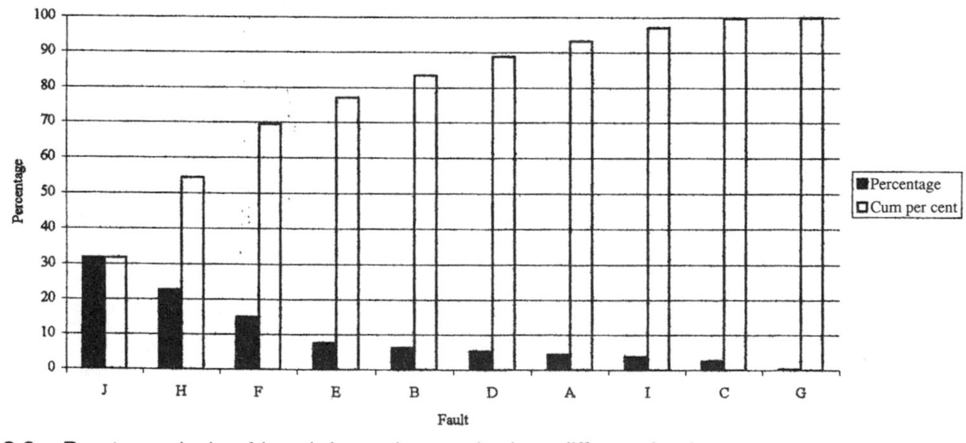

Figure 7.3.8 *Pareto analysis of breakdown data ranked on different basis*

and in the longer term to a redesign of plant and equipment to reduce the fault and hence reduce maintenance and increase availability.

- Spare parts: By identifying the high usage items it enables the designer to highlight potential problem areas.

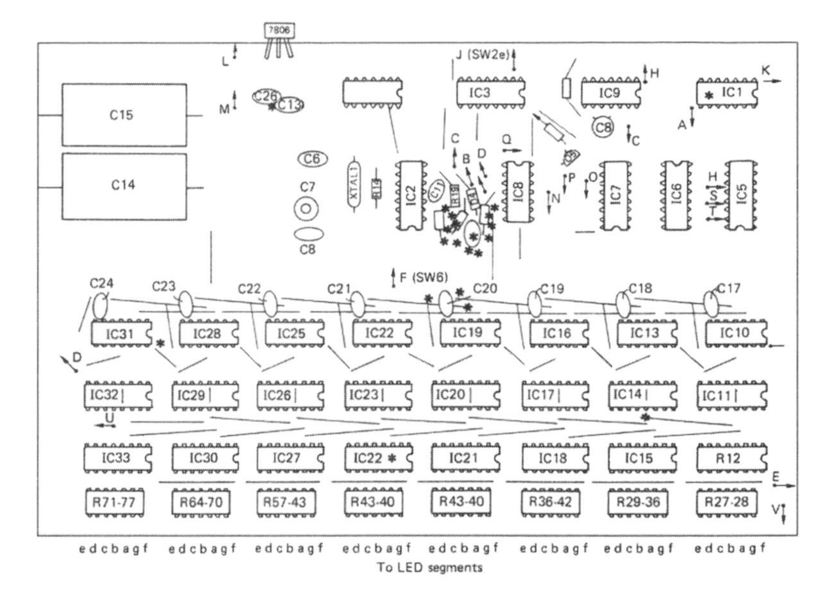

Figure 7.3.9

Measles charts

In complex assemblies, it is often useful to mark a sketch of the assembly with the location of any fault which arises. A clustering of marks, as in Figure 7.3.9, readily identifies the location of recurring problems, e.g. a problem in wave soldering can be identified through clustering of solder faults in one area.

This technique is particularly useful for identifying the location of repeated failures and hence leads to targeting improvements in the process itself or in the design. It can also be used in pressure circuits and cast/moulding operations to show common areas where porosity has occurred.

Experience: brainstorming and fishbone diagrams

As there are many possible causes for variations in quality, it is often useful to carry out a brainstorming session to identify possible causes. During these sessions it is useful to produce a fishbone, or Ishikawa, diagram to show all possible causes as in Figure 7.2.9 (page 287).

The effect being investigated is shown at the end of a horizontal arrow. Potential causes are then shown as labelled arrows entering this horizontal arrow. Each main arrow may have other arrows entering it as possible sub-causes. The sub-causes may in turn be due to a number of possible sub-causes.

The process continues, perhaps using a series of fishbones, until all possible causes, however unlikely, are listed. Once this is done, each cause is considered for further detailed investigation.

Failure reduction design techniques

Fault tree analysis (FTA)

FTA is the logical analysis of the chain of lower level events which have to occur to cause another event to happen. In failure analysis it is used to trace back to find the root causes that allow systems or components to fail.

FTA can be used to:

- Identify the different failure causes to enable solutions to be found to mitigate them.
- Show where combinations of events have to occur to cause failure.
- Identify weaknesses in a design to give opportunity to increase robustness.
- Check the effectiveness of fail-safe or emergency systems.

It is based on Boolean algebra logic, as in circuit design, and employs a set of symbols as in Figure 7.3.10. The AND and OR gates show how

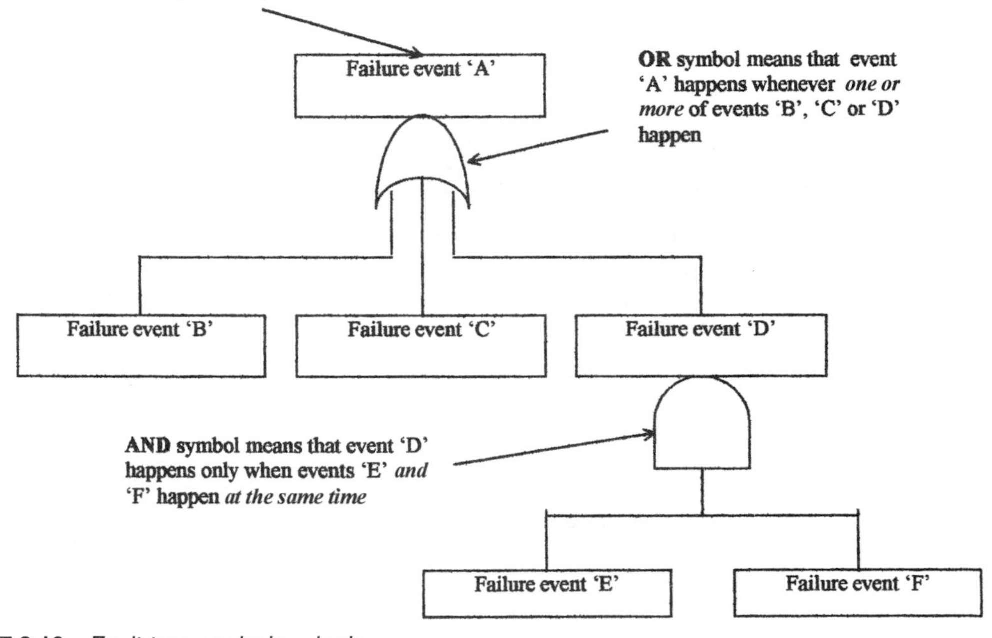

Figure 7.3.10 *Fault tree analysis – logic*

a combination of lower level events will cause the effect above to happen:

- The AND means that all the lower events must occur before the higher effect transpires.
- The OR means if any one of the lower events occurs then the higher effect will transpire.

Using the example shown in Figure 7.3.11 of a VCR, failing to record a particular programme, the process stages are:

- Identify the system failure effect, e.g. desired programme not recorded.
- Write a clear description of that effect in a box.

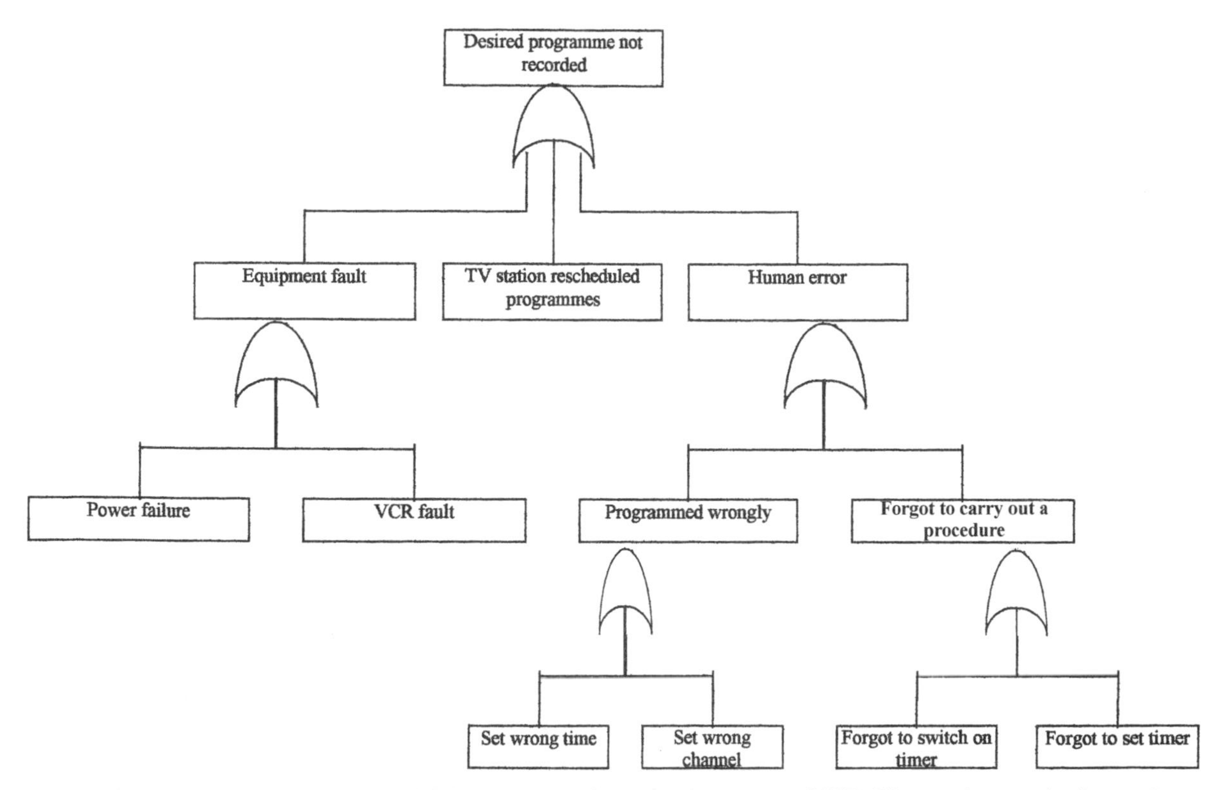

Figure 7.3.11 *Fault tree diagram of failure to record a programme on a VCR. The equipment fault events are not proceeded with as most common fault lies in human error sphere*

- List causal failures which if they occur could lead to that failure. Try to give as many causes as possible, e.g. failure in timer mechanism; power failure; human error; etc.
- Divide, if possible, the list of causal failures into groups for possible sub-causal failure identification.
- Draw the logic diagram showing the different levels of causal failures and how they can lead to the final system failure.
- Decide on the lower level of cut-off for this particular diagram. The lower level causal failures may appear in different final system FTAs.

Failure trees can often be used as troubleshooting procedures to show common faults and their possible causes.

Failure mode effects analysis (FMEA) and failure mode effects criticality analysis (FMECA)

FMEA is a design procedure in which we investigate the consequences that the potential failure modes of a component or subsystem may have. It starts from the opposite end to the FTA, which starts with an overall system failure. It can be applied to a product or a process.

Figure 7.3.12 shows an FMEA for an electric motor where each component and subsystem is subjected to the following questions, the answers being listed in the named column:

- The part under examination.
- What its exact function is.

FMEA

Indenture level:	Design by:	Prepared by:
Sheet no.:	Item:	Approved by:
Mission phase:	Issue:	Date:

Item ref	Item description-function	Failure entry code	Failure mode	Possible failure causes	Symptom detected by	Local effect	Effect on unit output	Compensating provision against failure	Severity class	Failure rate (F/Mhr)	Data source	Recommendations and actions taken
1.1.1	motor stator	1111	open circuit	winding fracture	low speed roughness	low power	trip	single phase protection temperature trip	4			
		1112	open circuit	connection fracture	low speed roughness	low power	trip	single phase protection temperature trip	3			
		1113	insulation breakdown	persistant high temp. manufacturing defect	protection system	overload	no output	annual inspection temperature trip	4			
		1114	thermistor open circuit	ageing connection fracture	protection system	none	no output	fitted spare	3			recommend consideration spare connected through to outside casing
		1115	thermistor short circuit	failure thermistor	protection system	reduced trip margin	no output if load high	fitted spare temperature trip	3			recommend consideration spare connected through to outside casing
1.1.2	motor cooling system	1121	inadequate cooling	blockage low diff. pressure	high temperature stator detected by thermistor	winding excessive temperature	motor excessive temperature	temperature trip stator	2			
		1122	leakage to atmosphere	piping connection	motor temperature	motor inadequate cooling	motor excessive temperature	temperature trip 2 hourly check	2			
		1123	leakage from atmosphere	piping connection	low output	air in system	none	2 hourly check	2			
1.1.3	motor bearing	1131	seal external leakage	wear bearing failure	low level lub oil sump	loss of lub oil	none unless leak severe	daily check	3			

Figure 7.3.12 *Example of an FMEA worksheet. (Extracts from BS 5760: Part 5 1991. Reproduced with permission of BSI under licence number 2001/SK0152. Complete British Standards can be obtained from BSI Customer Services (Tel. +44 (0)20 8996 9001))*

- The potential failures modes, i.e. how it can fail, are identified.
- The possible causes are then listed, i.e. what made it fail.
- How the fault is detected.
- What the effect of the fault is:
  - On the subsystem.
  - On the entire system.
- What protection there is against the effect of the fault.

Other columns can list any pertinent data re reliability. The final column records what recommended design, or operating, measures need to be carried out.

The corrective measures are then fully examined and designed to reduce the net effect by reducing the possibility of the cause arising, reducing the consequences or improving the detection. Actions may include redesign, additional parts/subsystems or adding warning or reaction systems.

Failure mode effects and criticality analysis (FMECA) is basically the same as FMEA with the addition of ranking the effect by the product of the criticality, i.e. the severity of the effect, the probability of it occurring and the ease of detecting it beforehand.

The addition to FMEA of the criticality analysis is judged under three column headings on a scale of 1–10 as per the example of the Ford in-line fuel injection pump shown in Figure 7.3.13:

p = probability of occurrence — 1 = almost never; 10 = almost certain

d = difficulty of detection — 1 = highly obvious; 10 = no prior warning

c = criticality of effect — 1 = negligible effect; 10 = system failure

Overall critical number $= p \times d \times c$

The higher the critical number, the greater the priority given to solving the cause. In the example shown in Figure 7.3.13, the highest risk number is 160 against the erosion in the bore of the delivery valve holder. It should attract the greatest input even though it is not deemed to have a high criticality. The high scores of 5 (probability of occurrence) and 10 (difficulty of detection) make it a priority.

The wear in the delivery valve has a much higher criticality of 9, but the other aspects are lower, hence the lower overall risk number of 108.

A step deeper in the FMECA methodology is to allocate a percentage probability of a particular failure mode occurring and, following that occurrence, the probability of each of the different effects happening.

Increasing reliability

It may appear obvious to state that we need to address unreliability. If a reliable system is what the customer needs, we have to approach that as best we can. Improvements in the design will probably involve extra time and money, so we need to ensure that we use these in the most effective way.

Product – I.C. engine	Subsystem – in line F.I. pump	P Probability of occurrence	F M E C A No. PR001
Process –	Location –	C Severity of failure	Date
System –	User –	D Difficulty of detection	Engineer
		RN Risk Number = $P \times C \times D$	

(1) Part No. Component	(2) Function	(3) Failure mode	(4) Effect of failure	(5) Cause of failure	(6) *P*	(7) *C*	(8) *D*	(9) *R.N.*	(10) How can failure be eliminated or reduced
Cams and follower rollers	Plunger drive	Fatigue pitting	Loss of power irregularity (or complete functional failure in extreme cases)	Overload (e.g. nozzle blockage, dirty or no oil, poor surface finish)	2	3	10	60	Supplier to determine failure definition. Limit operating pressure
Centre bearing bridge	Support camshaft centre bearing Ties pump sides at centre	Cracked across centre of bridge	Mainly cosmetic Slight oil seepage		1	2	10	20	Problem exists on present die cast pumps. Supplier likely to shell mould on our 8 cyl. pump therefore more material on bridge. To be assessed when samples are tested
Delivery valve holder		Erosion in bore	Nozzle blockage, loss of performance	High flow rates Cavitation	5	3	10	150	To be assessed when samples are tested
Delivery valve	Unloading	Consistency U/L wear	Engine performance pump phase and delivery balance	Manufacture, adjustment poor filter/water	3	9	4	108	M/C test, LP systems detail
Delivery valve spring	Controls unloading	Fatigue/erosion/ wear (omission)	Loss of engine performance, smoke	Hydraulic duty	4	4	4	64	Specification detail
Delivery stop peg	Control delivery lift	Wear crushing (omission)	As above	As above	2	4	3	24	Injection rates
Plunger element		'F' slot polishing	Seizure or sticking leading to functional failure	Particulates produced by erosion of pump body in transfer gallery or erosion shield	1	10	10	100	Slots now rolled into element prior to hardening and grinding. To be assessed when samples are tested

Figure 7.3.13 *Failure mode effects and criticality analysis (FMECA). (Extracts from BS 5760: Part 3 from the BSI Handbook 22: 1983. Reproduced with permission of BSI under licence number 2001/SK0152. Complete British Standards can be obtained from BSI Customer Services (Tel. +44 (0)20 8996 9001))*

If we take the three factors that are used in FMECA to calculate the risk number, they can identify the separate areas in which we can make improvements which would result in a lower risk number. Concentrate on the factor that has the highest number.

Reduce the probability of occurring

Use proven components and designs

Perhaps an obvious solution but it does need stressing. Most problems arise from new applications or components which we are unfamiliar with. Subject any new parts to extensive life tests.

Increase factor of safety

When we carry out the design, it is better to use components which can cope with more severe conditions than the design specification. A natural way to achieve this is to increase the factor of safety used in the design calculations.

Reduce criticality

Redundancy

In the aerospace industry, it is common to have a back-up system available for many functions. Even on the instrumentation side, there can be two or three simultaneous circuits relaying information and a computer control system can carry out majority voting to determine true results.

There are different ways of arranging this back-up:

- Active: Here the redundant system is kept fully operational at all times. Examples include twin recording systems.
- Passive: Here the redundant system is held in reserve and only comes into use when the primary system fails. The control system to switch in the back-up unit requires careful examination to ensure it operates correctly when required to.
  Sometimes the back-up unit can deteriorate if not in use – e.g. some batteries used as a back-up to mains supply will deteriorate, if not periodically drained and recharged, and human back-ups can suffer from boredom which reduces their effectiveness and reaction time.
- Semi-passive: Here the back-up is kept switched on, but at a reduced load so that it will quickly come on line. Radio and television transmitters are examples of this.

Increase detectability

Add warning devices

These monitor particular devices and information can be passed to continuous readouts, as in the temperature of a car engine, or can sound alarms as in many fire detection systems. The cost and extra complexity of these warning devices need to be carefully justified against any increased benefit they deliver.

Condition monitoring systems can be installed to continually monitor key subsystems and give advance warning of potential failure. These are especially important on capital goods and now can be linked to the telephone system for remote analysis.

Easy replacement

When all other avenues are less successful and the design still has to contain a degree of unreliability, you can ease the operation by making the unreliable system easy to access and repair/replace.

Remember that if you add another system to improve detection or ease replacement you are adding a further opportunity for something to go wrong.

Problems 7.3.1

(1) Why would the armed forces be particularly interested in the reliability of a product it purchases, e.g. a rifle?
(2) What are the needs that a student has in reliability terms from a PC?
(3) If your product demonstrated the bathtub pattern of failure, what would you recommend to reduce the number of early failures being experienced by your customers?
(4) Think about a recent design you have been involved with. How many of the list of failures on page 300 could apply? Are there any other failures that may occur which are not on the list?
(5) Think of a common problem – say being late for a meeting. Draw a fault tree diagram to demonstrate possible reasons. Convert this into a failure mode and effect matrix and develop means to eliminate the causes, or minimize their effect.

7.4 Increasing product effectiveness

The design sets the basic parameters which decide how easy it is for an organization to make products profitably. This means not just meeting the customers' functional needs but also the other characteristics desired by the customer in a simple and cost-effective manner. This is achieved through variety control, i.e. standardization and modular design, and by designing products for ease of assembly.

Variety control

Although producing a variety of products increases the opportunity to appeal to a wide pool of customers, it also imposes a burden on an organization's effectiveness. This results in extra cost and reduced efficiency through:

- Dissipating the advertising and selling effort over more products.
- Requiring a greater after-sales cost through increased training and extra spare part holding.
- Requiring more design effort and extending the time involved in design.
- Preventing designers attaining a deep knowledge of the important design aspects.
- Making reliability more difficult to achieve.

- Increasing the stock of raw materials, components and finished goods.
- Increasing the number of purchase orders.
- Increasing the number of set-ups and special tooling required during manufacture.
- Reducing the opportunity for increased automation.
- Reducing the run sizes.
- Increasing the production control activities.
- Making achieving quick deliveries more difficult.
- Specials cost more to procure than standard items.

All of these add up to a need to control *excess* variety both in end products and the parts which go into them. However, this must be done in a way that does not stifle creativity and opportunities for advances in design.

Standardization

The first obvious solution is to use recognized standard parts wherever possible in a design. Standardization enables interchangeability of parts between products as well as eliminating many of the problems listed above. Using standard parts enables the company itself either to concentrate on the design and manufacturing of these parts, or to purchase them from specialists.

Standardization does not mean you have to use only industry stock items from a wide variety of sources, this could limit the uniqueness of your design. It does mean that you should attempt to use the same part in as many products as possible. This will probably involve redesign to make one part capable of fulfilling several different functions.

This may mean that a part can then do more than is required of it in some products. This may appear an extra cost, but the extra material cost is normally offset by the reduction in the support costs.

Good examples of this are:

- The electrical junction box used in household circuits. This has several press-out sections in different locations to enable wiring to enter from different points.
- Use of standard packing cases with different internal support to hold a variety of different shaped products.
- Dual voltage electrical goods where the operating voltage is set by the customer using a simple switch.
- Ball point pens where the body and internal parts are the same, with only the ink and end caps being different. Even then the end caps use the same tooling with a slightly different blend of raw materials.
- Using a small range of bolts to cover all fastening needs.

A few companies have attempted to mix freedom of choice and using standard items by imposing an extra percentage onto the costing of non-standard parts. This means that the use has to be economically sound to overcome the extra cost imposed.

Standardization can also be pursued even when the requirements mean that different parts are different sizes. A range of different sized parts can be made to the same design. This saves design effort and ensures common experience and training of the manufacturing, assembly and even after-sales servicing operations.

It is often more effective to modify a standard part than to design and make a part uniquely for a slightly different function.

Modular design

The second solution lies in modular design. Modular design means building up unique products using a standardized range of main sub-assemblies and parts. This is common in the automobile and electronics industries. This effect is sometimes referred to as the mushroom technique of starting with ordering standard parts, combining them into unique assemblies and then blending these assemblies into a range of end products as in Figure 7.4.1.

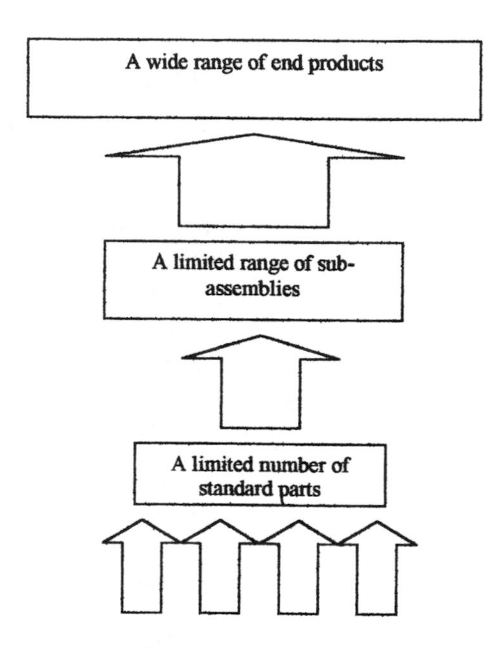

Figure 7.4.1 *The mushroom effect of using modular design*

This simplifies long-range planning by aggregating forecasts into family groups which are easier to forecast than the individual variations. It reduces the lead times from customer order to delivery as it is only at the last moment that a blend of modules needs to be made to form a unique end product.

If we consider the range of top level variations which are available for a particular model from an automobile manufacturer, these would run into hundreds made up from all the possible mixes of:

- Body shape: Saloon, hatchback, estate, two door, four door, etc.
- Interior trim: Material, colour, dashboard layouts, instruments, etc.
- Exterior trim: Paint finishes, colours, trim, etc.
- Engines: Diesel, petrol, size, etc.

If each vehicle awaited the customer order before parts were even ordered, the time before delivery would stretch into weeks. By ordering many of the parts beforehand in family groupings and only bringing

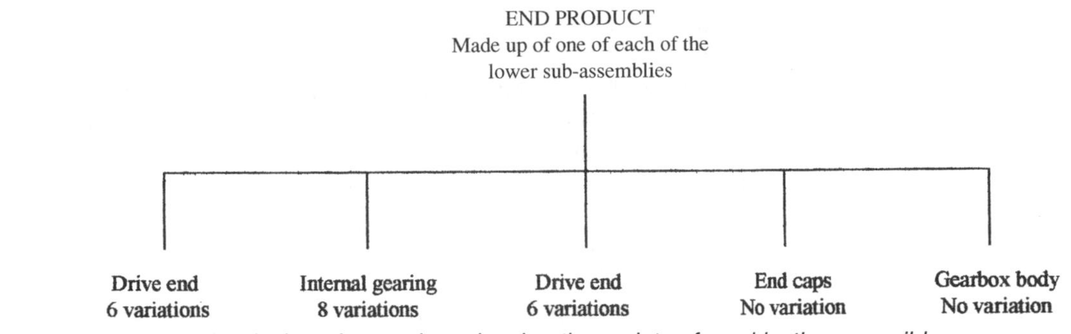

Figure 7.4.2 *Modular design of a gearbox showing the variety of combinations possible*

them together on the final assembly line, the actual requirement can be produced within days.

Modularity can also be applied at a part level. A simple gearbox may have variations at three points – each end and the reduction ratio (internal arrangement of gears within a common body) as in Figure 7.4.2. If we said that there were the following varieties:

- Six variations in drive end.
- Eight variations in gearing.
- Six variations in drive end.
- Common end caps.
- Common gearbox body.
- This gives total possible variations of $6 \times 8 \times 6 = 288$.

If the gearboxes were required as spare stock, where delivery time is very important, this could mean the company holding a stock of at least 288 gearboxes – more if they wished to have more than one of each available.

However, if the gearbox was designed for last-minute assembly, then only a stock of one of each part, i.e. a total of 20 items, would be required. This would allow any variation to be assembled and despatched.

Therefore by holding a small number of each of the variations at the lower level, it is easy to assemble these to produce any required combination of end product. This is particularly useful when the demand for the end product is small but a large number of possible variations may be asked for.

This ability to quickly react to a customer's urgent delivery need will turn into a strong competitive advantage in dealing with customers.

Designing for assembly (DFA)

To produce an assembly often requires a great degree of logistics to ensure all of the different parts are available in the correct quantity at the correct time. It can be a fairly skilled operation which is labour intensive or requires complex automation. The process itself can also be the source of defects, such as missing/wrong parts, wrong way round assembled or errors in the fixing process, e.g. soldering.

It is important therefore that all designers involved in the design of assembly follow an effective design for assembly procedure to:

- Reduce the number of parts – saving parts cost and reducing logistics.
- Reduce the possibility of items being assembled wrongly.
- Make handling of parts easy – giving opportunities to apply low cost automation.
- Make the task of assembly quick and easy, therefore at a low cost.
- Make assembly definite hence reducing defects.

A design for assembly procedure needs to address all these aims and the following system developed by Lucas Engineering and Systems Ltd does that through three iterative stages of analysis as shown in Figure 7.4.3:

- Functional analysis to eliminate or combine parts.
- Handling analysis of how parts are transported and oriented.
- Fitting Analysis of how parts are gripped and inserted into the assembly.

It is important that all parts including labels are examined under this process. Where a large number of components is involved, it may be more efficient to examine major sub-assemblies separately, but care

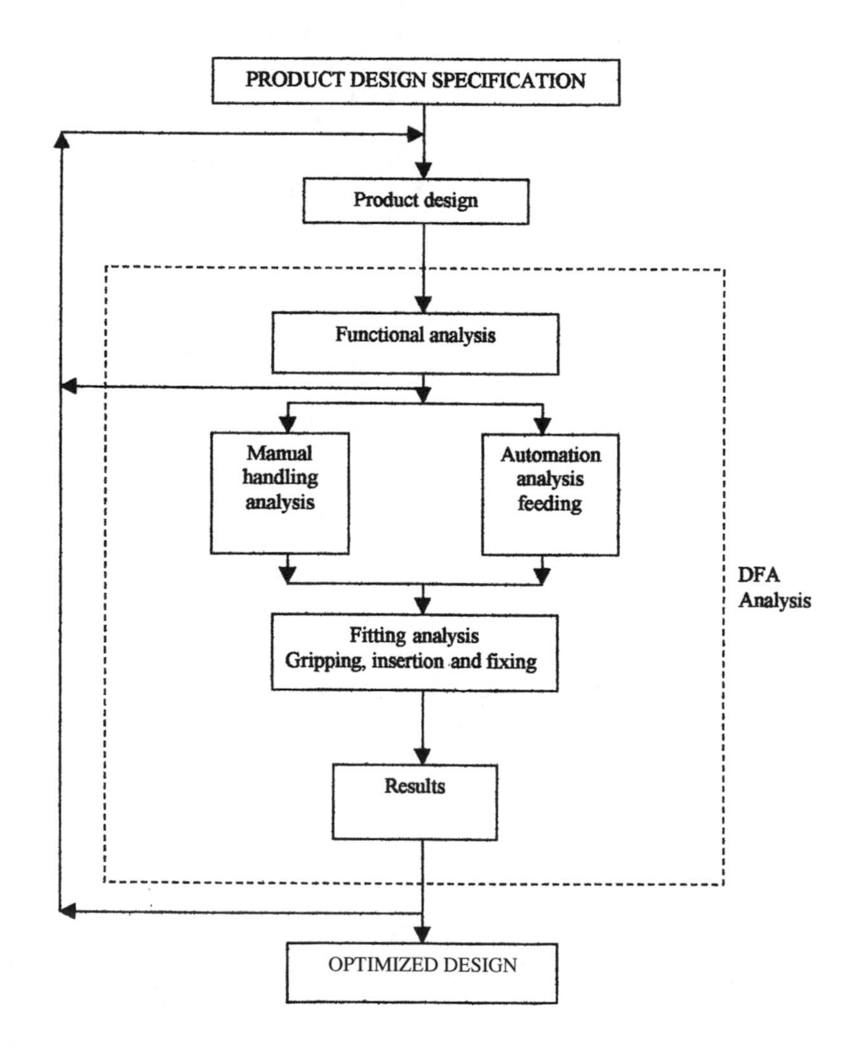

Figure 7.4.3 *Stages in design for assembly procedure*

should be taken that an opportunity to eliminate or combine parts is not missed by doing so.

Functional analysis

The objective of this stage is to separate parts into those which are necessary for the function of a product and those which are not critical. The essential parts are designated as 'A' parts, the others as 'B' parts.

Every part goes through a series of up to five questions as shown in Figure 7.4.4, to determine if it should be classified as an 'A' or a 'B' part. These questions relate to each part's need for movement, different material and servicing – especially in relation to the other parts in an assembly. If there is a need for any one of these, the part will be designated as an 'A' part. If there is no need for any of these, the part will be a 'B' one.

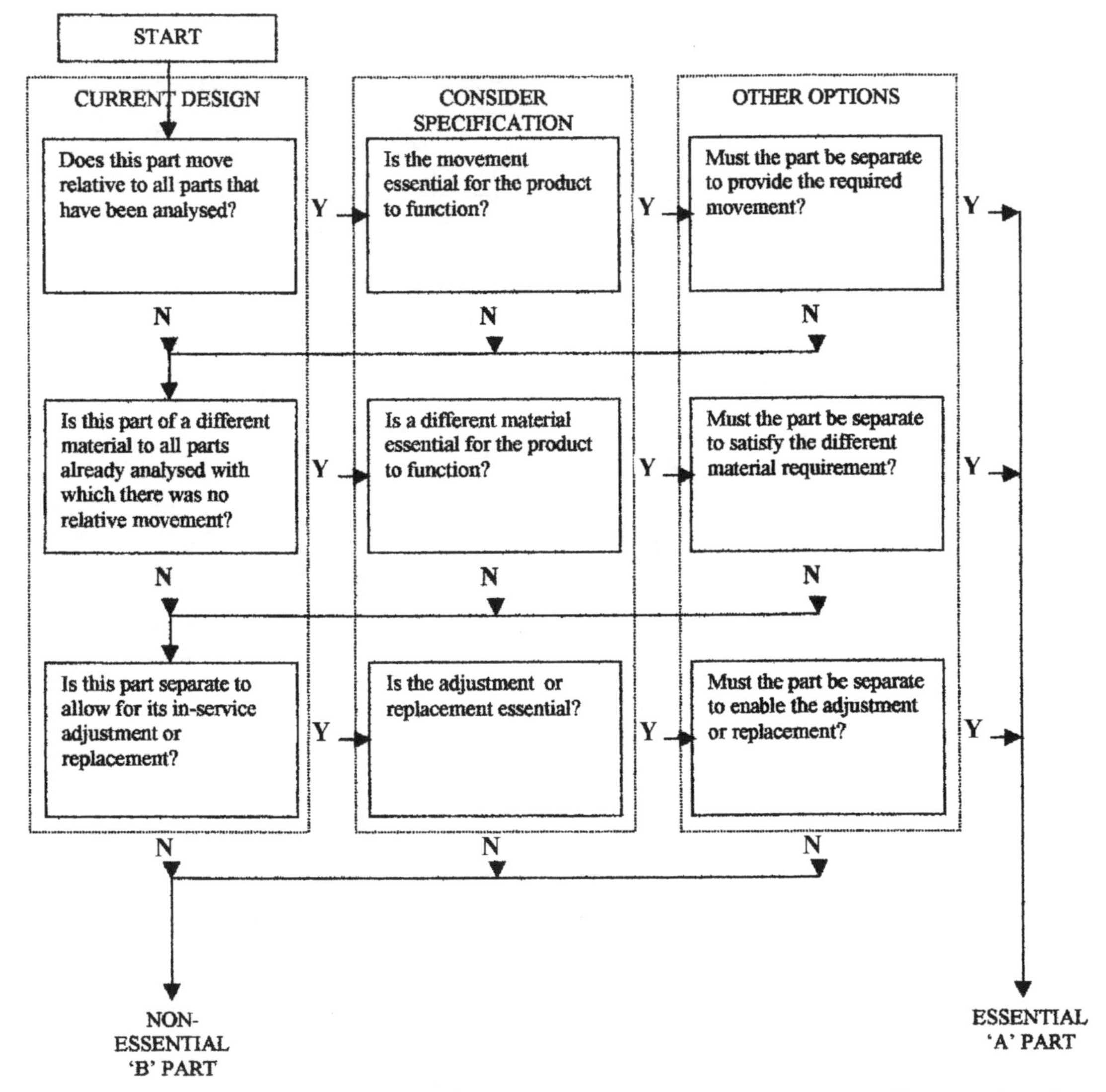

Figure 7.4.4 *Functional analysis. To start, pick one component that is obviously an 'A' part and analyse other parts in relation to this*

The parts which are labelled as 'A' are accepted under this procedure although it may be possible to redesign them more effectively.

The parts labelled as 'B' parts give the greatest opportunity for redesign to eliminate them or combine them with other 'A' or 'B' parts without affecting the product's functions. The minimum target design efficiency is to at least 60% of the total parts rated as 'A' parts. Having no 'B' parts will therefore achieve the ideal of 100% in design efficiency.

Only once the assembly is accepted as being within the design efficiency limits, do we move onto the next stages in analysing the remaining parts.

Handling analysis

A part can be examined under one of two routes – either under manual handling or under automatic feeding, depending on the facilities to be used. These routes are similar in areas examined, but differ slightly due to the fact that the human hand is a much more flexible and versatile tool than most automatic feeders can be and hence slightly different aspects are considered with different degrees of severity resulting.

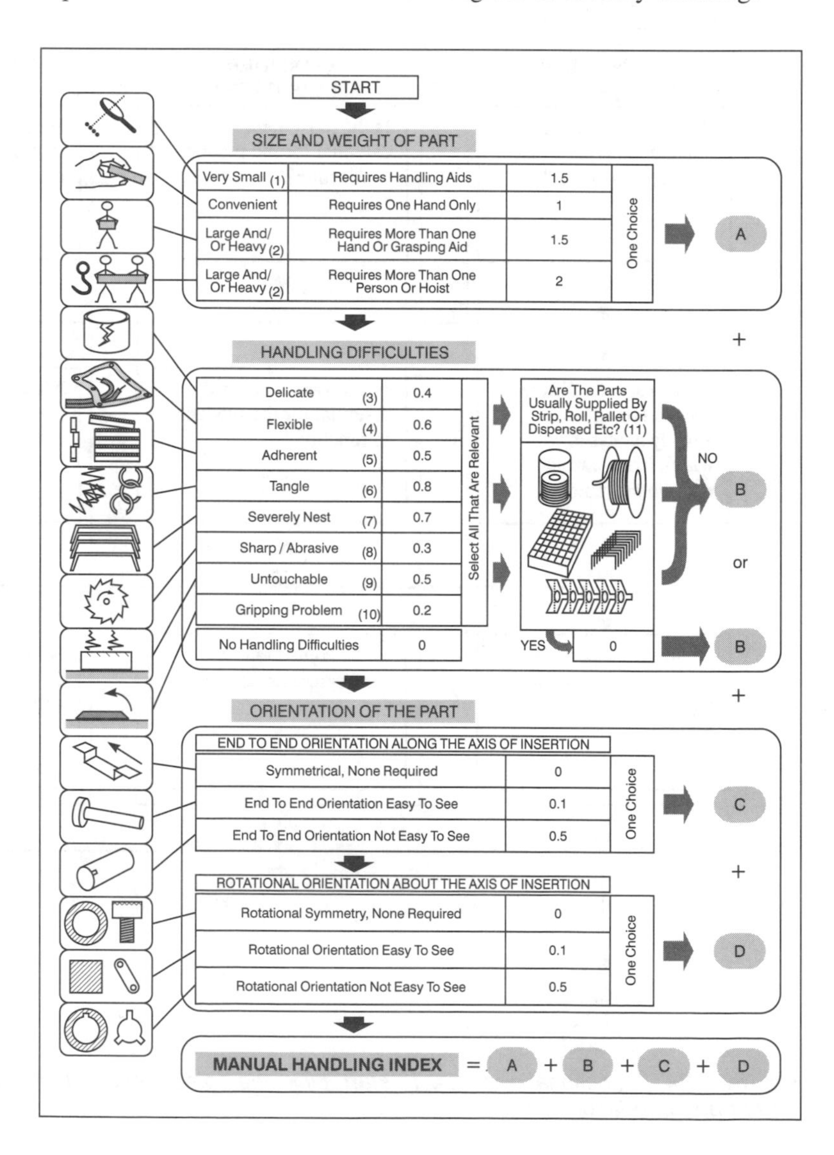

Figure 7.4.5 *(Reproduced courtesy of Lucas Industries Ltd)*

Manual handling (see figure 7.4.5)

In this analysis the part receives a total score based on the sum of its score under three separate headings:

- Size and weight – from 1.0 for a normal single-handed lift, up to 2.0 for size/weights above which a single person can reasonably handle.
- Handling difficulties – from zero for no difficulties, up to a sub-total of 4.0 for a combination of factors such as delicate, sharp, flexible, tangling, etc. If the part is supplied in a strip, roll or dispenser, then it attracts a zero score.
- Orientation (rotational and end to end scored separately) – from zero for exact symmetry (i.e. no orientation needed) up to 0.5 when the orientation needed is difficult to spot.

This means that each part can have a total score of between 1.0 and 6.9 for manual handling.

Automatic feeding

A similar procedure to manual handling is followed for parts using automatic feeding facilities with the following special case (see Figure 7.4.6).

Parts not suitable for individual mechanical orientation receive a score by analysing them under the manual handling path initially and this result is then multiplied by five to give a score for automatic feeding. Parts which have to be judged under this are either very large or heavy, very delicate or liable to tangle, or very flexible. All these present difficulties in automatic feeding.

Parts which are considered suitable for individual mechanical orientation are scored under the normal automatic feeding path as follows:

- General feeding considerations – from zero for no problems up to a sub-total of 11.5 for problems such as small, abrasive, overlapping, etc.
- Orientation (rotational and end to end scored separately) – from 1.0 for exact symmetry (i.e. no orientation needed) up to 3.0 when the orientation needed is difficult to achieve. In extreme cases, the manual score multiplied by five is used.

This means that each part can have a total score between zero and 14.5 (or the manual value multiplied by five in extreme cases) for automatic feeding. This is termed the feeding index.

The individual indices for manual handling or automatic feeding for all the parts are then added together to give an overall feeding score for the assembly. A ratio is then calculated:

$$\text{Feeding ratio} = \frac{\text{total of the feeding indices}}{\text{number of 'A' parts}}$$

The aim is to minimize this. A feeding ratio acceptable target is a maximum of 2.5, anything greater indicates that a redesign is required.

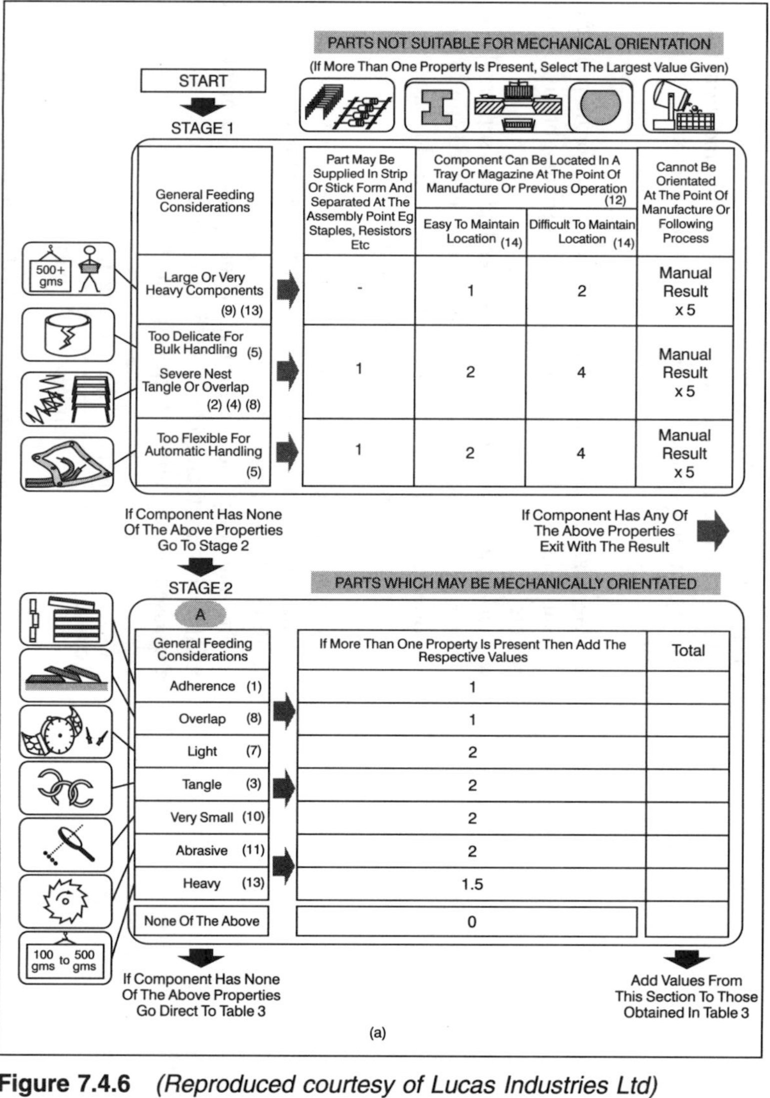

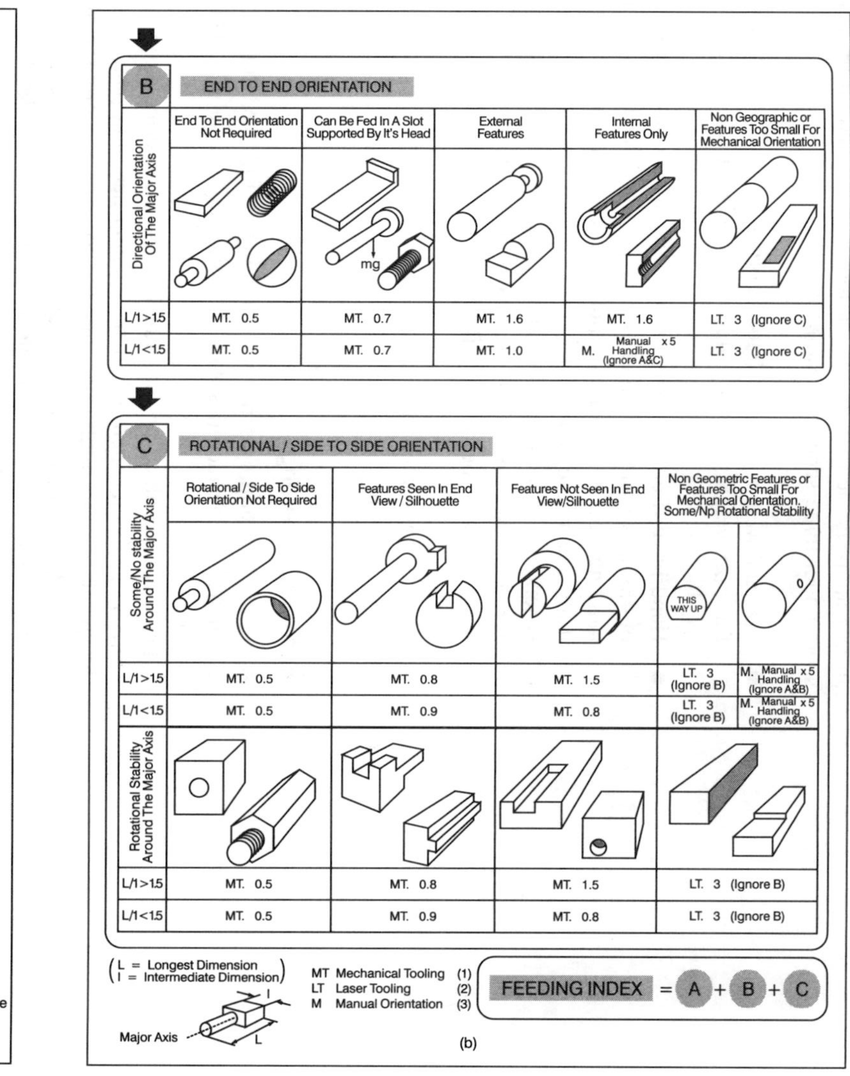

Figure 7.4.6 *(Reproduced courtesy of Lucas Industries Ltd)*

Mostly points can be eliminated or reduced through a change in handling and presentation to one that involves preloading the component into a dispenser or strip in a set orientation. The obvious place for this is at a previous operation where the part is normally already in a set orientation which is then captured and transferred onto the mode of transport. Doing this ensures that none of the problems impact on the assembly operation. The extra cost of doing so must be justified by the saving at the assembly.

Another answer is to change the profile of the part to make easy registration of the correct orientation.

After any redesign, the new design for the complete assembly must undergo this procedure again to prove that the assembly is now under the target feeding ratio limit.

Fitting analysis

Again each part is scored depending on whether the assembly is manual or automated. This time the analysis for manual and automated assembly are the same, but attract different scores. The scoring is done under three headings.

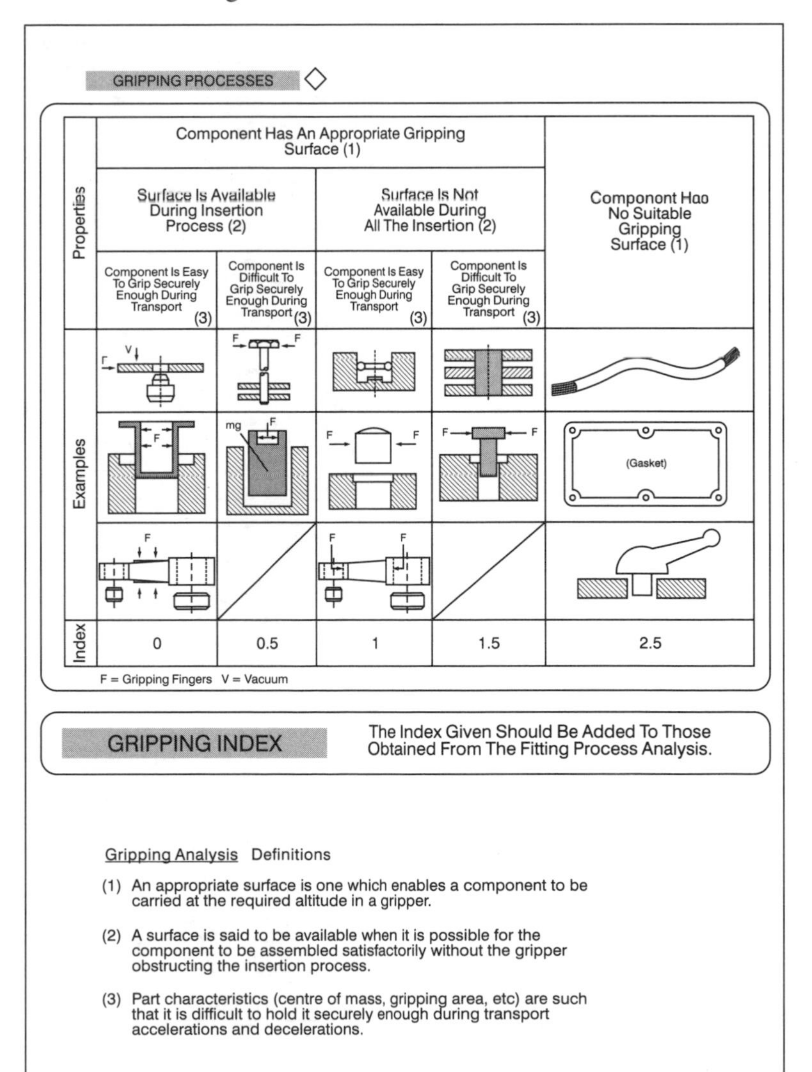

Figure 7.4.7 *(Reproduced courtesy of Lucas Industries Ltd)*

Gripping index (see Figure 7.4.7)

Scores from zero for an easy gripped and securely carried surface which is used at all times during assembly, up to 2.5 where no suitable surface is available. Note the score is increased if the gripping surface is not available for the complete assembly operation. This is termed the gripping index. Manual assembly need not be scored under this heading, but it is useful to do so.

Fitting index (see Figure 7.4.8)

This is scored under six sub-headings. The automated score is given first against the factor with the corresponding manual score in brackets:

- Part placing and fastening (i.e. snap fit, screwing, riveting and bending) – from 1.0 (1.0) for self-sustaining to 1.6 (4.0) for bending.
- Direction – from zero (zero) for straight from above up to 1.2 (1.6) for change of direction.
- Number of locations – from zero (zero) for a single location up to 1.2 (1.2) for multiple locations required.

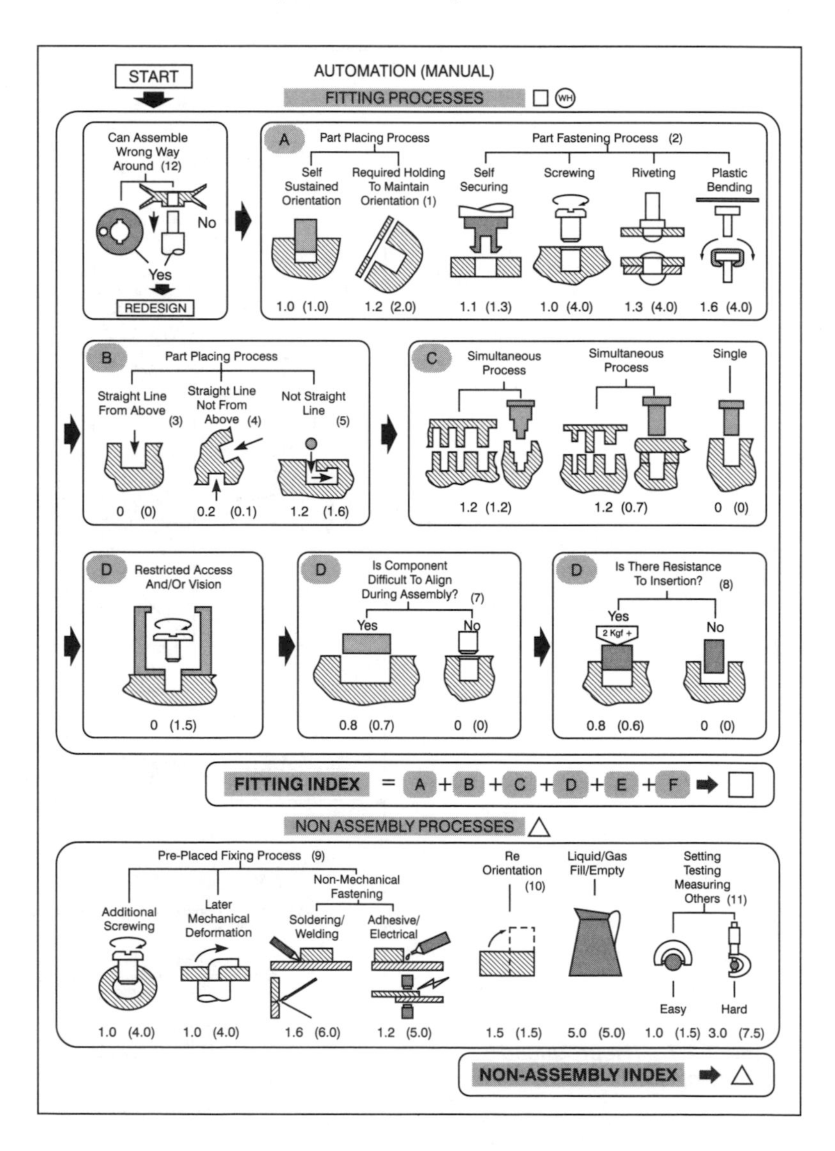

Figure 7.4.8 *(Reproduced courtesy of Lucas Industries Ltd)*

- Restricted access or vision – if effect then zero (1.5).
- Alignment difficulty – from zero (zero) for no difficulty up to 0.8 (0.7) if some difficulty.
- Resistance – from zero (zero) for no resistance up to 0.8 (0.6) when there is resistance.

The scores from all these sub-headings are then summated to give a fitting index for the part.

Non-assembly index

There are processes carried out during an assembly which are necessary but cannot readily be classified as an assembly operation. These also attract a score under a heading of a non-assembly index – again the manual score is given in brackets (see Figure 7.4.8):

- Fixing: Screwing per turn – 1.0 (4.0); mechanical deformation – 1.0 (4.0); soldering – 1.6 (6.0); adhesive application – 1.2 (5.0); spot welding – 1.2 (5.0).
- Reorientation – 1.5 (1.5).
- Fill/empty – 5.0 (5.0).
- Setting, measuring, etc. – from 1.0 (1.5) for easy up to 3.0 (7.5) for difficult.

Fitting assessment

The overall fitting ratio is achieved by adding together all the gripping indices, fitting indices and non-assembly indices and dividing the result by the number of 'A' parts:

$$\text{Fitting ratio} = \frac{\Sigma \text{ gripping indices} + \Sigma \text{ fitting indices} + \Sigma \text{ non-assembly indices}}{\text{number of 'A' parts}}$$

Again the aim is to minimize this. A fitting ratio acceptable target is again a maximum of 2.5, anything greater indicates that a redesign is required.

As before the high scores during the analysis indicate where gains can be made through a redesign. This may involve changing the shape of the component to ensure an assembly can only be done the correct way, making the gripping surface always available during assembly, assembling every part straight from above and ensuring that there are lead-ins, i.e. bevels to aid location. Often these changes are accomplished at a previous operation with no, or very little, extra cost. Any cost that does arise must be justified as previously by the gain at the assembly operation.

After the redesign, the new assembly operation must undergo this procedure again to prove that the assembly is now under the target fitting ratio limit.

Case study

DFA – headlamp trimscrew

The assembly of a headlamp trim-screw (see Figure 7.4.9 and Table 7.4.1) was under consideration for automation. As part of

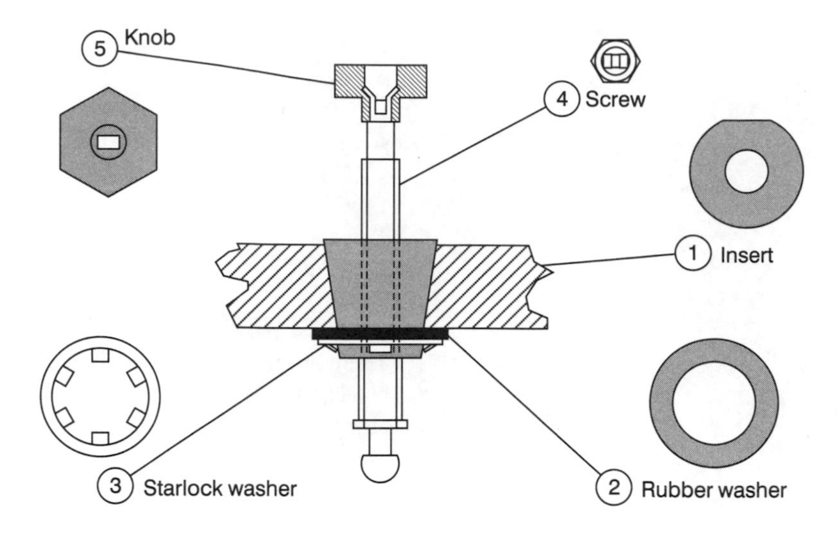

Parts Description

1. Insert Maintains grip to the screw.
2. Rubber washer Takes up assembly slack.
3. Starlock washer Retains insert.
4. Screw Provides adjustment.
5. Knob Provides adjustment leverage.

Figure 7.4.9 *(Reproduced courtesy of Lucas Industries Ltd)*

Table 7.4.1 Parts list for the headlamp trim screw

Part number	*Description*	*Function*
1	Insert	Anchorage to screw
2	Rubber washer	Takes up slack 1 to 3
3	Starlock washer	Retain insert
4	Screw	Provides adjustment
5	Finger knob	Leverage for adjustment

this process, the assembly was subjected to the Lucas DFA procedure to highlight any potential difficulties in the assembly process.

Functional analysis

The parts were first submitted to the functional analysis stage.

The starting point was the lamp body (i.e. material holding the insert) that was taken as being an 'A'. The lamp body was made from a different material (i.e. from the insert) that was extremely brittle and could not be directly screwed. The lamp body material had been examined, and it was decided to leave it as it was.

The functional analysis (see Table 7.4.2) resulted in:

$$\text{Design efficiency ratio} = \frac{3As \times 100}{3Bs + 2As} = 40\%$$

This is below the threshold target of 60%, and immediately indicated that a redesign is called for. The exercise could have

Table 7.4.2 Functional analysis of headlamp trim screw parts

Part number	*Description*	*Rating*	*Comment*
1	Insert	A	Different material
2	Rubber washer	B	
3	Starlock washer	B	
4	Screw	A	Movement to part 1
5	Finger knob	B	

BEFORE (SEE FIGURE 7.4.9)

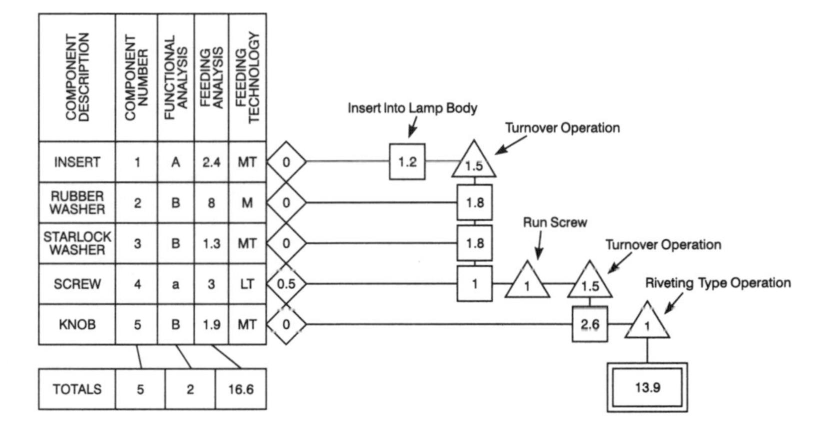

AFTER (SEE FIGURE 7.4.11)

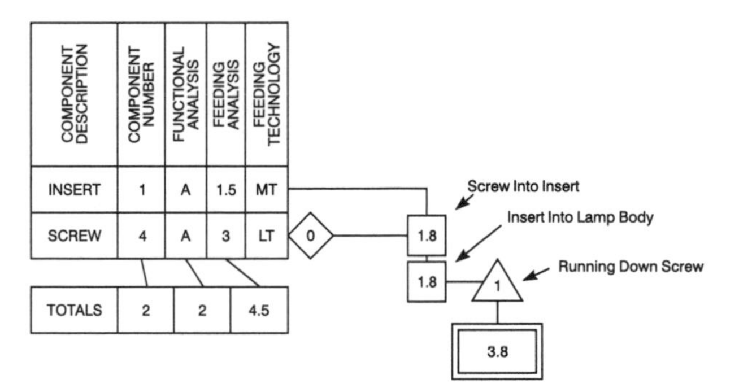

RESULTS SUMMARY	BEFORE	AFTER
Total Parts Count	5	2
Design Efficiency	40%	100%
Total Feeding Index	16.6	4.5
Feeding Ratio	8.3	2.3
Total Fitting Index	13.9	3.8
Fitting Ratio	7	1.9

Figure 7.4.10 *(Reproduced courtesy of Lucas Industries Ltd)*

finished here, but it was decided to press on and carry out the full analysis to gather information to make out the justification case for the redesign.

A full flow chart of the assembly procedure was done and that is shown in Figure 7.4.10.

Feeding Analysis

This was done using the automation route, giving the results shown in Table 7.4.3:

Table 7.4.3 Feeding analysis of headlamp trim screw parts

Part	*Description*	*Use figure*	*Score*	*Feeding index*
1	Insert	2	0	
		3	1.6 (B)	
			0.8 (C)	2.4
2	Rubber washer	2	Too flexible	
		1 (manual)	1.0 (A) × 5	
			0.6 (B) × 5	8.0
3	Starlock washer	2	0	
		3	0.5 (B)	
			0.8 (C)	1.3
4	Screw	2	0	
		3	3.0 (B)	3.0
5	Finger knob	2	0	
		3	0.7 (B)	
			1.2 (C)	1.9
			Total	16.6

$$\text{This gave an overall feeding ratio} = \frac{16.6}{2As} = 8.3$$

This is above the threshold of 2.5 – and would call for a redesign to reduce the total of the feeding indices to a maximum of 5, i.e. a reduction of 11.6. It may be possible to reduce the feeding ratio to acceptable limits through redesign:

- A replacement of the rubber washer with a less flexible material would contribute some of the present 8 against it (it would still have some score under the analysis).
- Additional, more prominent, features on parts 1 and 4 would reduce the high present score due to orientation problems.
- Combining parts will reduce the ratio.

Fitting ratio analysis

Table 7.4.4 Gripping analysis for headlamp trim screw

Part number	*Description*	*Gripping method*	*Rating*
1	Insert	Internal fingers	0
2	Rubber washer	Vacuum	0
3	Starlock washer	Vacuum	0
4	Screw	External fingers	0.5
5	Finger knob	External fingers	0
		Total	0.5

Table 7.4.5 Fitting and non-assembly analysis

Part number	*Description*	*Score*	*Rating*	*Non-assembly*
1	Insert	1.2 (A)	1.2	Turnover 1.5
2	Rubber washer	1.0 (A) 0.8 (E)	1.8	
3	Starlock washer	1.0 (A) 0.8 (F)	1.8	
4	Screw	1.0 (A)	1.0	Turnover 1.5 Screwing 1.0
5	Finger knob	1.0 (A) 2.0 0.8 (E) 3.0 0.8 (F)	2.6	Mech deform 1.0
		Total	8.4	5.0

$$\text{This gives an overall fitting ratio} = \frac{0.5 + 8.4 + 5.0}{2As} = 6.95$$

This is above the threshold of 2.5 and would call for a redesign to reduce the total of the fitting indices to a maximum of 5, i.e. a reduction of 8.9. It may be possible to reduce the fitting ratio to acceptable limits through redesign:

- Preventing the turnovers would achieve a reduction of 5.0.
- Changing the method of attaching the knob gives an opportunity of reducing by 3.6.
- Combining parts will reduce the ratio.

Therefore all stages show a need for a redesign – the redesign was carried out as shown in Figure 7.4.11. In the new design there are only two parts – both 'A' rated. The screw is partially sub-assembled into the insert before insertion of this sub-assembly into the lamp body. After insertion, the screw is run down forcing open the tangs on the insert which hold the insert tightly to the lamp body.

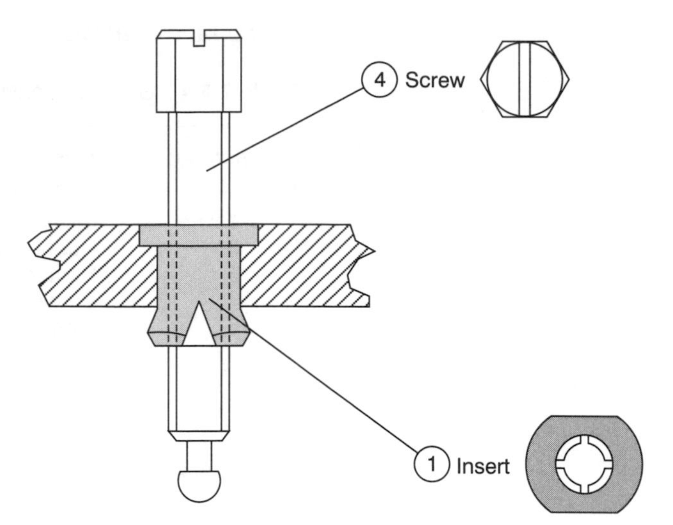

Figure 7.4.11 *(Reproduced courtesy of Lucas Industries Ltd)*

Table 7.4.6 Comparison of before and after DFA analysis

Results	*Before*	*After*
Number of parts	5	2
Design efficiency	40%	100%
Feeding ratio	8.3	2.3
Fitting ratio	6.95	1.9

The summary of the changes, given in Table 7.4.6, shows substantial benefits from this example.

Reductions in direct assembly costs have been achieved from 20% to 70% using this DFA procedure in the automobile, aerospace and other industries. There have also been substantial gains on the logistic and other costs.

To re-emphasize, these savings were achieved without any reduction in the assemblies' functions and come from:

- Reducing the number of parts.
- Simplifying the design for quick and easy assembly.
- Reducing the need for complex handling equipment.
- Reducing the possibility for errors.

Problems 7.4.1

(1) Take apart a common household item, such as a torch. If you were to manufacture this product, how many parts could you easily purchase from specialist manufacturers? Could any of these parts also be used in various designs of the product?

(2) Examine the complete range of cars available from any manufacturer. How many common components can you identify which are used in more than two models?
(3) Often a different material is used to fulfil a particular function, e.g. wear resistance. How can you impart different characteristics to parts which have the same chemical composition?
(4) Describe the changes in the headlamp adjusting screw design (Figure 7.4.11) and calculate the new indices and ratios?
(5) Examine an old design of a common household product against a modern design. Compare the number of 'A' parts and calculate the savings in feeding and fitting ratios achieved by the new design.

8 Managing design

Invention is 1% inspiration and 99% perspiration.
(Thomas Edison)

Summary

Being able to manage the process is just as important as understanding how it works and being able to do the individual parts in it. This final chapter provides an introduction to the management of the design process. It explains why the logical, sequential design process is ill-advised and how it can be improved. The use of skills, information technology and trends are also explored in detail.

Objectives

At the end of this chapter, you should:

- know how and why the design process should be managed;
- understand how both design practitioner and design manager can improve their effectiveness.

8.1 Planning and co-ordination

Like any process, the design process is prone to weaknesses and not always easy to do. As few as 1 in 10 SMEs may actually be capable of innovating. Reasons for this include:

- Components, consumer goods, industrial goods and saleable items might all be considered as the outputs of the design process. Design could also be defined by inputs too; the results of management discourses and strategies, group dynamics, design politics, social organisation, organizational context, the societal role of artefacts, actor networks, ambiguity and uncertainty. All of these inputs can have a negative effect on the way a design evolves.
- Business and design decisions must be made with inadequate levels of information. Usually there is less information than desired but often there may be too much, particularly as the design process becomes ever more data intensive.

- Different individuals and departments may have different goals.
- Quality, cost, time, risk and reward are in permanent conflict.
- Companies can have difficulties in supporting change, overcoming the status quo, paperwork, bureaucracy, fear of failing and of spoiling the company's image.
- Lack of skills and resources.
- Outside influences such as competition, customers and suppliers.

Worse still, it is not enough just to be capable of innovating. It is also necessary to be capable of innovating quickly. Speed is of the essence and it is therefore vital that the lead time (the time from idea to sale) is as short as possible.

A number of ideas have been developed which aim to improve the management, planning and co-ordination of the design process. Some of these are formal such as the British Standards guides and some are informal.

Key point

First movers into a market usually maintain the market lead and even after competitors respond retain the largest segment of the market at around 40%.

A six month delay in new product development can cost up to 50% of the product's value during its lifetime.

Example 8.1.1

Examples of formal management tools for managing the design process include Integrated Product Development (IPD), Total Value Management (TVM), Lifecycle Management, Concurrent Function Deployment (CFD), the European Design Innovation Tool (EDIT), Six Sigma Design, Productivity Growth Technology and Hoshin Kanri.

Example 8.1.2

The British Standards Institution (BSI) produces national standards for use in Britain and many other parts of the world, particularly Commonwealth countries. The mission statement of the British Standards Institution is 'to increase U.K. competitive advantage, and to protect U.K. consumer interests'. The BSI also represents UK interests at the European Committee for Standardization (CEN) and in the development of European (EN) and international (ISO) standards. The role of a standard, be it ISO/IEC, CEN/CENELEC or BSI, is defined as 'a document, established by consensus and approved by a recognized body, that provides, for common and repeated use, rules, guidelines or characteristics for activities or their results, aimed at the achievement of the optimum degree of order in a given context'. Standards have a life of five years, after which they are reviewed and a decision made as to whether they should be revised, rewritten or discontinued.

British Standard Guidelines for the management of design include BS EN ISO 9001: 1994 for Quality, BS 7000 for Design management systems Part 1, BS 7000 Part 2 for the Guide to managing the design of manufactured products and BS 7000 Part 10 for Design terminology. BS 7000 is one of the best selling standards in the BSI.

BRITISH STANDARD | BS 7000-1:1999

Design management systems —

Part 1: Guide to managing innovation

ICS 03.100.01

BSi

Figure 8.1.1 *BS 7000 Part 1 front page, BS 7000 Part 2 front page and BS 7000 Part 2 contents page reduced and tiled. (Extracts from BS 7000 Part 1: 1999 and BS 7000 Part 2: 1997, reproduced with permission of BSI under licence number 2000SK/0400. Complete British Standards can be obtained by post from BSI Customer Services, 389 Chiswick High Road, London W4 4AL)*

BRITISH STANDARD

BS 7000 : Part 2 : 1997

Design management systems

Part 2. Guide to managing the design of manufactured products

ICS 03.100.99

Figure 8.1.1 *(Continued)*

Contents

Figure 8.1.1 *(Continued)*

Concurrent engineering

The design process outlined so far and illustrated earlier in Figure 1.1.4 appears logical. It is followed in the layout of this book. In reality, to follow a design process in a sequential, step-by-step fashion is not acceptable. Particularly because it is too slow. If the process takes too long then the customer information gathered at the start of the process may be out of date by the end of the process. Any changes to the design would also have to go through the same time-consuming loop.

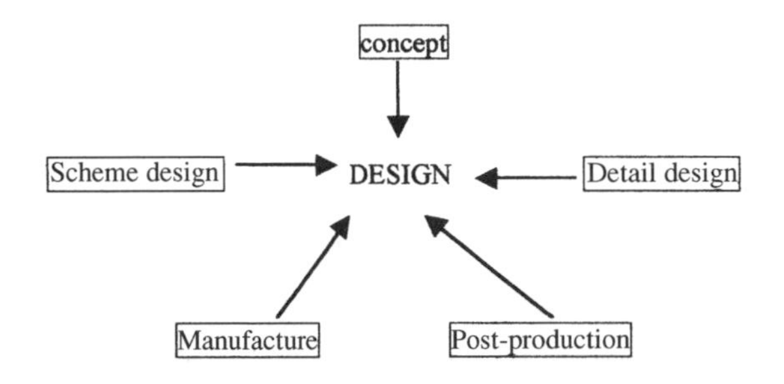

Figure 8.1.2 *Concurrent engineering approach to the design process*

Concurrent engineering is the term given to undertaking all of the design process activities at the same time rather than sequentially. All concurrent engineering initiatives have features to encourage this simultaneous approach to design:

- Focusing on the entire product life not a particular stage at a time.
- Using and supporting multidisciplined design teams.
- Recognizing that processes are just as important as the product. For example, if information is slow in forthcoming then not just finding the information but finding out why it was slow and putting the system right for next time.
- Planning ahead for information-centred tasks so that the information is available when needed.
- Developing product requirements carefully to avoid mid-course corrections.
- Using multiple concept generation and evaluation.
- Ensuring that quality ('customer requirements') is designed in.
- Communicating the right information at the right time.
- Keeping things simple (the Edisonian principle).
- Having stamina and conviction.
- Keeping in total control.

8.2 Communication and information

Concurrent engineering relies on good communication and information. The following section briefly introduces topics behind improving communication and information and uses exercises to demonstrate some of the basic principles.

Group dynamics

The design process often requires a number of specialists to be involved: market researchers, scientists, designers, engineers, project managers, production engineers, purchasing managers, sales and marketing staff. These groups may have different perspectives, aims and ways of working but for the sake of any design project must work together as a group or team. The way that they function when working together is termed 'group dynamics' and this covers:

- The way the group is formed.
- The way the group is organized.
- The way the group develops.
- Whether the group is controlled or self-managed.
- Whether the group is led by one person, by a few, or by everybody.
- If one person leads, is it through function, seniority or by rotation?
- Is it a real group that meets face to face regularly or a virtual group meeting across time zones?
- The mechanics of the relationships.
- The behaviour norms (e.g. lax, peer policed, multicultural, multigendered).
- The way decisions are reached.

Key point

In any group activity it is important to attend to the process (the way the group functions) as well as the task that the group has.

Understanding the dynamics within a group can help you to perform better in the group and for the group to perform better as a whole.

Activity 8.2.1

Form teams of people with a minimum of six in each team. You are to be the editorial board for a new daily newspaper meeting to discuss which current topics should be in the front pages. There is room for only six articles. The articles must be selected by debate – no voting or selection by majority rule.

Pay reduction for civil servants and government workers proposed.
EEC to debate expulsion of Britain over lack of support for policies.
Environmental ship vessel rams Navy warship.
Record balance of payments deficit forecast for next quarter.
Newsreader cited in divorce case.
Honda declared bankrupt.
Baby batterer gets suspended sentence.
Chancellor announces tax reduction of 5%.
An American scientist claims link between microwaves and cancer.
Pop star in secret romance with soap star.

The leader of the Conservative Party is rumoured to think of quitting.
Arctic ice cap is said to be one-third thinner than 50 years ago.

Afterwards describe the way your team functioned in the newspaper editorial exercise in terms of its group dynamics.

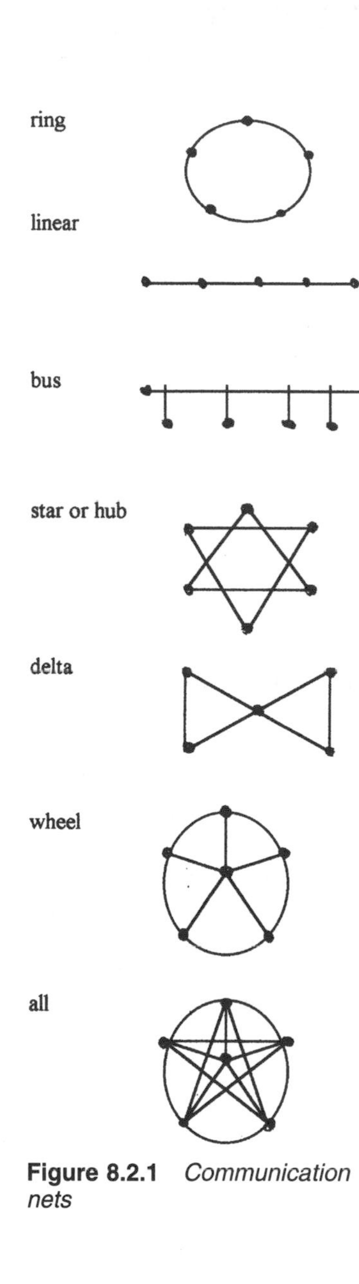

Figure 8.2.1 *Communication nets*

Information flow

The way that information flows in a group can be modelled. A communication net is a pattern describing the way that information travels between different members of a group (Bavelas and Leavitt). A number of net shapes are possible (see Figure 8.2.1).

The shape of the net has an influence on the team's leadership, its ability to learn, its ability to find solutions, the originality of the solutions, its flexibility, the number of messages and the team's morale. A democratic style communication network such as 'all' is susceptible to collapse under the pressure of the group when rapid action is needed. A wheel could give the authority desired and the group the efficiency and motivation it needs.

Activity 8.1.2

Form two design teams. Team 1 should adopt a linear net and team 2 should adopt a wheel. Team member should only communicate in with other members of the team according to their position in the net. Each team should now attempt to solve a problem in competition with each other. The exercise should be competitive with teams comparing and voting for the best solution.

Suggestions are to design an environmentally sound, man powered lawn mower. Alternatively, solve a current political debate such as should Britain adopt proportional representation?

Compare not only the solutions to the problem but explore how well information was circulated and the feelings of the team members.

Teams should then adopt different net shapes and new design problems.

Working in a group

Working in groups is not always (some would say never) easy. You must trust your colleagues and accept that other people may benefit from the results of your efforts. Some people are better at working in groups than others (Belbin). Working in groups is, however, a skill that everyone can improve.

Activity 8.2.3

Form groups of six and again design a new product or tackle a new political issue. The exercise should again be competitive with teams, or an independent adjudicator comparing solutions and voting for a winner. Suggested tasks include designing a new type of bicycle for stowing in the space usually reserved for a spare wheel in the boot of a car or deciding whether to abolish the monarchy.

At the end of the exercise, describe your role in group work. Include how you felt, how you contributed, what type of comments you made and whether you concentrated on the task or the group.

Find out from your colleagues what they think of your group work. It may be best to do this anonymously.

Assignment 8.2.1

Create a plan that will show how you will improve your group working skills. This might include:

- Noting your own least preferred co-worker, asking why it is that person and concentrate on overcoming your opinions.
- Attend as many group activities as possible and role play to gain greater confidence to do the same thing again in group meetings that matter.
- Attend a meeting as an observer and watch how others manage or lead in their group situations.
- Practise your skills in:
  - listening
  - assertiveness
  - negotiating
  - motivating
  - handling conflict

Leadership

A great deal has been written on leadership theories. Modern theories suggest that the style of a leader is more important than the person's personality traits. Hence it is not necessary to change personality to lead a group but instead to adopt a style, such as delegating, supporting, coaching or directing (Hersey and Blanchard). The 'best fit' theory (Fiedler) suggests that the style should be adaptive and changeable as circumstances dictate. For example, in a group with low leadership power and poor relationships a directing, or a task orientated style might be most effective. A democratic style might help to diffuse a hostile group but an autocratic style might achieve success earlier on in the group development or when speed is needed. A good leader should:

- Consider the circumstance around the group and the style needed to lead it.
- Teambuild first, teamwork second.
- Attend to the group relationships and process, and trust that the group will attend the task.
- Try not to do too much too soon.
- Note the changeability of contribution levels and changing communication nets.
- Look for breakthroughs:
  - when the group identifies its goals
  - when the group begins to perform better
  - when the goals appear to become achievable.
- Set difficult targets occasionally.
- Not interfere all the time.

Activity 8.2.4

In groups of between four and ten, attempt to tackle a series of tasks. Each member of the team should play the role of the leader and should concentrate on that role rather than joining in the task. Suggestions for tasks include:

Build the tallest possible freestanding structure from ten sheets of A4 paper.
Put on a performance of a recent number one song.
Put on an (abbreviated) performance of a Shakespeare play.
Write a song about design.
Design and build packaging for a single egg that can be dropped from 2 m.
Design and build packaging for a light bulb that can be dropped from 2 m.

Afterwards, assess each other's performance as leader.

Meetings

Meetings are a function that is easy to arrange but easy to do badly. Simple rules to follow include:

- *Preparation*
  Ensure that there is an agenda which is distributed well in advance of the meeting.
  Avoid inaccurate or outdated information.
  Avoid bringing new information to a meeting.
  Allocate sufficient time for discussions and summaries after discussion.

Avoid where possible new items which are raised at the meeting under Any Other Business (AOB) and which unfocus thoughts.

- *Record the proceedings in the form of minutes.*
  Concentrate minutes on conclusions.
  Agree the action points with names.
  Distribute minutes as soon as possible.
- *Control the meeting.*
  Call the meeting.
  Arrive for meetings first.
  Chair the meeting.
  Make sure everybody can contribute.
  Stamp out ulterior motives.
  Keep to time.
  Stand up.
  'Steer' the group rather than the 'drive' it.
  Use contributory comments liberally (building, clarifying, seeking information).
- *Encourage success.*
  Deal with easy matters first.
  Keep it short (in Finland, any meeting expected to last more than an hour is held in a sauna!)
  Ignore distractions and maintain concentration during long speeches.
  Avoid inconsistent action and talk.
  Avoid personalization.
  Avoid alienation.
  Avoid inequality and unfairness.
  Avoid mistrust.
  Avoid disempowering (a single leader workgroup, not a team).
  Seek support for ideas.
  Use pro formas to analyse performance as quickly as possible and devote energy to addressing problems rather than recording them.

Activity 8.2.5

Plan a meeting which would be aimed at reducing graffiti in your town.

Managing information

Knowledge is the retention of information. As the level of information grows, so the importance of 'knowledge management' grows. In design, this might include the simple recording of operational information such as log books, notes and thoughts, to the more complex organisation of drawings, documentation, product structures, and planning. Increasingly it is also extending into strategic areas that underlie the entire design process, such as management information, databanks, ideabanks, links to external information services and networking-based contacts underlying the entire design process and over the life cycle of the product.

Key point

The time and effort necessary for the management of knowledge is now one of the limiting factors to further improvements in business performance. Only 2% of companies consider knowledge management to be a passing fad.

Example 8.2.1

The operating manual alone for a power station will fill two containers whilst the paperwork necessary to create an oil rig stacks up higher than the oil rig itself.

Information technology (IT) has a wide variety of uses within the design process including word processing, drawing in two and three dimensions, computation, and communication but its role management of information is one of the most important. Product data management (PDM) systems is a term used to denote such systems.

A PDM system comprises a storage vault, incorporating product classifications, product structure, group technology, project management, process management, termed workflow; configuration management, access management, data transportation, translation and integration.

PDM helps to reduce the costs of photocopying and reprographics, drawing storage and engineering change but there are also more significant organizational benefits. These include reduced time to market, reduced product cost, improved product quality, conformity of product ranges to standards, reduced product life-cycle costs, security and back-up, reduced environmental impact and improved process flexibility of processes, shared, secured and managed access to data across the product development domain and along product life cycle, improved consistency and integrity of data.

Key point

Sales of PDM systems rose by 20% in the latter part of the 1990s.

Example 8.2.2

Claims for the success of PDM include a 95% reduction in the product change cycle, savings through re-engineering of the engineering change process, an 80% improvement in right-first-time design, new product introduction times cut from 36 to 22 months, design-to-manufacture lead times reduced from 20 to 22 days, stock and work in progress reduced by 5%, material costs reduced 1.5%.

All PDM systems work on a computing principle known as the client/server principle. This means that the user sits at a client's workstation and all of the work and data resides on a server's computer. Put simply, the user has a cheap PC and shares a powerful computer with others on an as-needs basis. The Internet also works on the client/server principle. The Internet is what connects the user to the server so PDM also has a natural bond with web technology. An important rule of the Web is that all applications must speak a common language, called HTML. This language describes the screen that the user sees. The program running on the client workstation – for example, a PC – is called the web browser and is usually either Netscape or Microsoft's Internet Explorer.

This browser interprets the HTML characters into the screen that you see when you attach to a particular website. Thus, if you connect to a PDM server, you see the PDM interface which the server has sent to your PC. Each screen sent using HTML is called a page. So any interface that you have, or have seen on a PDM system, can be described in HTML and used to access the PDM system over a local or wide areas network (LAN or WAN). Upgrades are expensive and usually all pervasive but web-based PDMs can overcome this problem if the software generation is stored on the server rather than natively client generated.

Example 8.2.3

Boeing's maintenance engineers can obtain the latest information applicable to a specific plane that has just arrived in a hangar needing maintenance. This includes any service bulletins for the plane, downloaded as the latest version from the Seattle design headquarters. When finished, an engineer indicates the work carried out, and the records in Seattle are updated so that the next engineer to work on that plane has an 'as built' set of information to work from, specific to that plane.

This communication technology also enables controlled and information with both customers and suppliers. Catalogues of components as seen in most design rooms are replaced as engineers ask for detailed specifications, local availability and the prices of the components that they wish to consider. Orders can also be placed where mail order services are available.

Alternative view

Many PDM implementations have been less successful than anticipated when their cost has been measured against system functionality. PDM implementation is a complex process which can easily run out of control and end in a poorly performing system. Common difficulties include:

- Developing a specification which is derived from the features and functionality available in commercial PDM solutions when business requirements should determine the system's specification.
- Knowing where to start and what to do first.
- PDM can be seen as a large project requiring significant resource input over long periods of time.
- Lack of confidence in sharing information.

Activity 8.2.6

Conduct a communications audit within your working environment. Looks at the flow and storage of information, group working and the use of IT.

Assignment 8.2.2

Use a spreadsheet to create a product data management system and enter the details of a product selected from your kitchen. Enter its details into your PDM. Swap your details with two colleagues and enter their product details into your PDM. See if you can agree on a common form of PDM to be used in your group.

Case study

Clyde Material Handling Ltd

Clyde Material Handling (CMH) employs around 140 staff and is a key player in the design and installation of pneumatic bulk handling equipment. Pneumatic conveying is a bulk handling process that uses air pressure to move materials. It requires a hopper at the pick-up point, a hopper at the collection point and intermediate pipework, filters and pumps (turbos and exhausters). The process finds applications in industries as diverse as plastics, chemicals, food and pharmaceuticals. The materials to be moved therefore vary not only in size and basic structure but often differ from day to day according to temperature and humidity. Applications also vary in terms of layout and customer requirements. The final design of any system is hence often a combination of science and experience. It is often referred to as the 'grey art'. The design process involves lengthy analysis so that even a scheme design in response to a sales concept can be a long, costly and inaccurate process.

CMH wanted to increase sales but its lead time from concept to scheme design was taking ten weeks after which they were often uncompetitively overpriced or uneconomically underpriced. The firm was also only returning one order for every seven enquiries. Increased advertising might benefit CMH's competitors who might reap the benefits of the extra demand. It was therefore decided to reduce the enquiry/order ratio so that more than one in seven orders were held. Cutting prices or employing more staff may have appeared as simple and easy options but were rejected in favour of a knowledge management system that made the design process more efficient.

Stage One began with a programme of product rationalization and modularization. In summary this was achieved by:

- Standardizing hopper ranges as far as possible.
- Standardizing accessories.
- Eliminating unnecessary design features.
- Designing a universal seal to allow more cross component mixing.
- Selecting cheaper production processes such as metal spinning.
- Shedding manufacturing processes to concentrate primarily on assembly.
- Creating a product family for accessories. (Allowing volume discounts to be negotiated.)
- Creating more accurate costing data.
- Creating detailed flow diagrams, and part numbers.

Simply having more accurate cost data alone was estimated to save the company £80 000. The next stage of the process was to create a database for the product data.

A number of proprietary software packages were examined but it was decided the company would build its own database. This was because the system would need to act not only as a data warehouse but also provide the logic for information processing. It is this aspect that would make the system more than a product data management system and the 'mode of arrangement' that would justify its nomenclature as a 'configurated database'. This would allow parameterization of the design data to make the drawing stage of the design quicker. Create better links for sales staff in communicating directly with the design team. Allow the establishment of benchmark and referential data so that future projects could be assessed. A number of platforms for achieving this were considered, but Microsoft Access was eventually chosen. It was recognized that scalability was a key factor in the failure of IT projects and that there was the chance that Access would not ultimately be of sufficient size to cope with the task but the package was considered easy to learn. The company had also decided to use a young graduate who was inexperienced with both the engineering and computing processes.

The methodology followed was distinctly soft and followed a bottom-up prototyping life cycle. This may not have been a totally efficient prescription, in that occasional amendments had to be implemented. The most significant constituted a redesign to include the goal of making the configurator available to a number of users on the road towards a company intranet. Other amendments included:

- Improving performance by splitting the growing database into 'static' tables on an application.mdb and a shared data.mdb.
- Improving performance by moving the shared data mdb file to a server and application.mdb files to the workstations.

- Enabling multi-user accessibility on non-Windows 95 work-stations by moving from Access to Visual Basic with jet engine 3.0.
- Fine tuning the operation, for example file locking, buffering and inserting a print table during data entry time event code saving repeated complex and time-consuming queries from running before each print.

However, the ultimate aim was an effective configurator, rather than one that had been put created in the most efficient way. The completed configurator utilized 1200 top level and 5000 sub level components, grouped into 160 tables by product families. Dynamic tables contain all the elements of a product and its own particular sub-components. This created some duplicated tables of near similar parts which is unsatisfactory from a simple database perspective but for a configurated database allows system components to be amended without having to regenerate entire systems.

The final database allowed single component or system selection, adjustable sales margins, accumulated costs and customer and operational details. It achieved actual cost reductions, increased orders through faster quoting and a reduction in design errors, computed as:

22% reduction in costing enquiries by sales staff.
80% reduction in outstanding customer quotations with the response period down to 11 days.
100% increase in the number of order per enquiry ratio.
40% decrease in design time.
34% decrease in project time by negating the need to re-input data.
30% increase in resource time with a redeployment of four staff from the engineering department.

Other, intangible results include a 'quality' quotation process including 100% repeatability, co-ordinated, accurate and adjustable pricing policies, reduced staff training time and rapid responses to customer alterations.

As a result of the database, the role division between designer and salesman was changed to an intermediate 'concept engineer' role. The company is hoping to narrow even further the division between the concept engineer and production engineer by including a knowledge-based module in the configurator. Connection with MRP modules and internet modules are also planned to allow the database to interface directly with the production process and with customers. Hence the configurator is turning from an operational tool into a strategic tool and is a driving force behind the re-engineering of the entire business process towards a virtual enterprise.

8.3 What next?

This book has looked at a vast range of topics that help in the production of a successful design. It is important to mention that there are others: strategies, organization, culture, structure, resources and systems of the company all play a major role. There are many business related books that deal with these issues.

It is also important to note that having got this far, and produced a successful design, it is essentail not to sit back and rest. KEEP INNOVATING. The shelf life of new designs is reducing so that new designs must always be forthcoming. Fortunately the more a company designs, the better it becomes at managing the process of design and the more it can design. It is a virtuous cycle.

Key point

Only 40% of new products hitting the market are still around five years later.

Finally having striven through the design process, it is vital that the work is protected to prevent anybody else from simply copying your designs without having gone through the hard work. Intellectual property rights are the legal tools for protecting your designs. They include patents, design rights, copyrights and trademarks.

New designers must also be capitalized on. The UK has a tremendous history of design and innovation but an equal reputation for failing to exploit it. What might British society look like today had the ideas generated in Britain for the jet engine, penicillin, the computer and the internet been properly protected and exploited?

Further reading might hence take the designer to look at the topics of intellectual property rights, business and entrepreneurship.

Example 8.3.1

The Internet was 'fathered' by Tim Berners-Lee, a British telecommunications engineer working in Cern. He wanted to make his research findings available to others working in the same field. During the mid-1990s, this grew into the Internet that we know today: a large number of computers connected to a common, and public, communications network that obey certain rules and can thereby live in harmony making exchanges.

Key point

For every 1000 research papers published by the UK, only about 87 obtain a patent whilst the equivalent figure in Japan, for example, is 488. It is estimated that 57% of current Japanese technology originated in the UK.

Review questions

(1) What sort of leader would be best suited to a research and development organization. How would this differ to a leader of a lifeboat crew?
(2) How might you create categories for different types of workers in a group?
(3) Discuss why moving the design process in-house can increase a company's profitability.
(4) Discuss the issues you would think about before creating a team to design a new washing machine.

Assignments

(5) Undertake an analysis of knowledge management software available. Your analysis should include the different types of systems available, costs, features and scope.

(6) Discuss whether people have natural leadership skills or whether these are developed.

Solutions to Problems

Chapter 2

1. Visual characteristics, tactile contact, expected response, recognition, positive/negative response.
2. As a study of events or changes over a period of time or of use.
3. Identifies what *is* and what *is not* known. Identifies objectives for client and for design process. Identifies the research that needs to be carried out. Provides a common reference for all parties to the process, as a basis for interdisciplinary working.
4. Because they link us with the experiences that underpin our knowledge, i.e. the way in which we came to understand things.
5. Because vision devoid of tactile confirmation has little meaning. People use and enjoy things by coming into contact with them and using them over time.
6. To gain from others' perceptions. To check if our perceptions coincide with those of other people. To gain clarity by expressing ideas in a form accessible to others.

Index

THE LIBRARY
GUILDFORD COLLEGE
of Further and Higher Education
Author CATHER H, MORRIS R et al
Title DESIGN ENGINEERING
Class 620.0042 CAT
139938